Digital Twin Technology and Applications

The Fourth Industrial Revolution is being accelerated by the digital twin technological revolution, which converges intelligent technologies and defines the connectivity between physical and digital items. The Internet of Things (IoT) connects the real and digital worlds, allowing connected items to deliver a vast array of services to internet users. IoT devices create large amounts of data that may be fed into AI systems for decision-making. In a decentralized architecture, digital twin technology may be utilized to protect platforms and create smart contracts. Digital twins decentralized ledger, immutability, self-sovereign identification, and consensus procedures hold a lot of promise for improving AI algorithms. Furthermore, leveraging smart contracts in a digital twin system to facilitate user interaction via IoT might have a big influence, and this integrated platform is expected to revolutionize many fields.

Digital Twin Technology and Applications examines the problems, issues, and solutions for using big data to enable streaming services using IoT and AI with digital twin technology. The IoT network concept is the key to success, and to establish a solid IoT platform on which large data transmission may take place, it must handle protocol, standards, and architecture. The book provides insight into the principles and techniques of IoT and AI. It explores the idea of using blockchain to provide security in a variety of sectors. The book also covers the application of integrated technologies to strengthen data models, improve insights and discoveries, innovate audit systems, as well as digital twin technology application to intelligent forecasting, smart finance, smart retail, global verification, and transparent governance.

Digital Twin Technology and Applications

Edited by

A. Daniel, Srinivasan Sriramulu, N. Partheeban, and Santhosh Jayagopalan

CRC Press
Taylor & Francis Group
Boca Raton London New York

CRC Press is an imprint of the
Taylor & Francis Group, an **informa** business

First edition published 2025
by CRC Press
2385 Executive Center Drive, Suite 320, Boca Raton, FL 33431

and by CRC Press
4 Park Square, Milton Park, Abingdon, Oxon, OX14 4RN

CRC Press is an imprint of Taylor & Francis Group, LLC

ISBN: 9781032430591 (hbk)
ISBN: 9781032745176 (pbk)
ISBN: 9781003469612 (ebk)

DOI: 10.1201/9781003469612

Typeset in Garamond
by Newgen Publishing UK

Contents

About the Editors

A. Daniel is an associate professor in ASET-CSE, Amity University, Madhya Pradesh, India. He has 13 years of experience in academics. He has expertise in AIML, deep learning, and quantum computing.

Srinivasan Sriramulu is a professor in the School of Computing Science and Engineering, Galgotias University, India. He has more than 22 years of experience of teaching. He has expertise in image processing, big data, cloud, IoT, and artificial intelligence.

N. Partheeban is an associate dean and professor at the School of Computer Science and Engineering, Galgotias University. He has more than 23 years of experience in academics. He has expertise in artificial intelligence and machine learning.

Santhosh Jayagopalan is with the Faculty in Computer Science, British Applied College, Umm Al Qwain, United Arab Emirates. He has 14 years of academic experience. He has expertise in IoT, networks, and machine learning.

Contributors

M. Arvindhan
Galgotias University
Greater Noida, Uttar Pradesh, India

Ajay Sudhir Bale
New Horizon College of Engineering
Bengaluru, India

Surendiran Balasubramanian
National Institute of Technology Puducherry, India

M.J. Carmel Mary Belinda
Institute of AI & ML, Saveetha School of Engineering
SIMATS University
Chennai, India

Sunil Kumar Boran
Galgotias University
Greater Noida, India

R. Baby Chithra
New Horizon College of Engineering
Bengaluru, India

A. Daniel
Amity School of Engineering and Technology
Amity University Madhya Pradesh, India

S.S. Darly
Anna University
Chennai, India

P.R. Joe Dhanith
Vellore Institute of Technology
Chennai, India

Mrinalika Durairaju
Shiv Nadar University
Chennai, India

Mrinalini Durairaju
Shiv Nadar University
Chennai, India

S. Geetha
Pondichery University, Kalapet
Puducherry, India

S. Geerthik
Agni College of Technology
Chennai, India

P.V. Gopirajan
SRM Institute of Science and Technology
Chennai, India

R. Indrakumari
Galgotias University
Greater Noida, India

Pallavi Jain
SCSE, Galgotias University
Greater Noida, India

B. Jaison
RMK Engineering College
Kavaraipettai, Tamil Nadu, India

Salna Joy
New Horizon College of Engineering
Bengaluru, India

D. Kadhiravan
University College of Engineering
Tindivanam
Anna University
Chennai, India

R. Sujithra Kanmani
Vellore Institute of Technology
Chennai Campus
Chennai, India

S. Dinesh Krishnan
B V Raju Institute of Technology
Narsapur, Telangana, India

Vallidevi Krishnamurthy
Vellore Institute of Technology
Chennai, India

A. Selva Kumar
SIMATS University
Chennai, India

N. Vel Murugesh Kumar
RMK Engineering College
Kavaraipettai, Tamil Nadu, India

V. Sheeja Kumari
SIMATS University
Chennai, India

J. Madhusudanan
Pondichery University, Kalapet
Puducherry, India

K.V. Mahalakshmi
B V Raju Institute of Technology
Narsapur, Telangana, India

T. Manikandan
SIMATS University
Chennai, India

Raghvendra Ajay Mishra
Galgotias University
Greater Noida, Uttar Pradesh,
India

S. Nivetha
RVS College of Engineering &
Technology
Coimbatore, India

K. Prabu
Galgotias University
Greater Noida, Uttar Pradesh, India

S. Premkumar
Galgotias University, Greater
Noida
Uttar Pradesh, India

V. Sathya Priya
B V Raju Institute of Technology
Narsapur, Telangana, India

S. Ponmaniraj
SIMATS University
Chennai, India

R. Radhika
S.A.Engineering College, Chennai
Tamil Nadu, India

K. Rajasathiya
P.S.R.R College of Engineering
Sivakasi, India

Rajkumar Rajavel
Department of CSE
Christ University
Bangalore, Karnataka, India

R. Rajesh
Vellore Institute of Technology
Tamil Nadu, India

N. Rekha
National Institute of Technology
Puducherry, Karaikal, India

V. Sakthivel
Vellore Institute of Technology
Chennai Campus
Chennai, India

A.M. Sermakani
S.A. Engineering College, Chennai
Tamil Nadu, India

S. Sivakumar
SRM Institute of Science and Technology
Tiruchirappalli, India

Bhairvee Singh
GLBITM
Greater Noida, India

Pankaj Singh
SCSE, Galgotias University
Greater Noida, India

Dyagala Naga Sudha
B V Raju Institute of Technology
Narsapur, Telangana, India

V. Sujatha
S.A. Engineering College
Chennai, Tamil Nadu, India

V. Prasanna Venkatesan
Pondichery University
Kalapet, India

Chapter 1

Digital Twin Past, Present, and Future

A.M. Sermakani, R. Radhika, and V. Sujatha

1.1 Introduction

A digital twin (DT) is a modern technology for representing actual assets and corporate operations. Real-time monitoring of data and systems is made possible by this fusion of the virtual and physical worlds. Sensors and various wireless technologies are utilized to gather structure of the data for digital twin models of physical items. They lower maintenance costs while also enhancing project design and quality. An exact clone of a physical functionality is known as a "digital twin," which simulates the performance, identifies inefficiencies, and creates improvements for the counterpart in the real world. In addition to a simulated environment, digital twins function in virtual worlds and are unavailable in the outside world, they are exact replicas of certain physical objects integrated with sensors that constantly update their virtual counterparts in real-time with accurate, and highly valuable information. Organizations utilize digital twins to plan, design, develop, operate, and explore products across the course of their lives. Since digital twins have access to the most recent information on actual objects, they can be utilized in conjunction with intelligent technology like machine learning to generate exact prediction models and anticipate outcomes that are more accurate than most simulations. Data scientists and IT specialists can run simulations on digital twins before actual devices are built and deployed, which are virtual reproductions of pre-built devices. Furthermore, digital twins can employ AI and data analytics to improve performance by utilizing real-time IoT data. A digital twin is an indication of pre-built actual devices that

DOI: 10.1201/9781003469612-1

work along with entities in the digital world. Buildings, factories, and even cities can now have digital twins thanks to technology. Some people have even suggested that processes handled by people might have digital twins, further broadening the idea. A digital twin is an accurate and constantly updated logical real product or procedure that is used to test different scenarios, forecast problems, and identify optimization opportunities. A DT handles distinctive, real-world implementation of a counterpart, receives real-time data from it, and adjusts as necessary to replicate the origin throughout its existence, in contrast to conventional computer-aided design and engineering (CAD/CAE) models [33]. However, the twinning is not a random occurrence. There are many parts to this process, but they all function as one cohesive one. A digital twin is essentially computer software that makes simulations using actual data regarding a physical entity or system as inputs or predictions of how those inputs will impact the actual data or system as outputs.

1.2 History of Digital Twin Technology

The concept of digital twin technology was originally introduced with the publication of David Gelernter's book [37] "Mirror Worlds" in 1991. However, Dr. Michael Grieves, a professor at the University of Michigan, is credited with introducing the concept of digital twin software and using it in production for the first time in 2002. Finally, in 2010, NASA's John Vickers created the phrase "digital twin." The fundamental procedure to use a digital doppelganger to analyze a physical object was incorporated earlier. In fact, NASA was the first to utilize digital twin technology for space exploration missions during the period of the 1960s. Because of this, NASA's traveling spacecraft was exactly invoked with replicas of its object in an earthbound version. The effect was hugely inherited by flight crews with NASA personnel observing for research and simulation purposes. The idea of a digital twin first came about at NASA, when full-scale replicas of early spacecraft were used to diagnose and mirror issues in orbit. Eventually, fully digital simulations took their place. After Gartner suggested "digital twins" as one of the major technologies on the highest utilized lists for 2017, the predictions were geared up with the specific nature of dynamic software models for physical models. Within 3 to 5 years, the predictions came true and billions of things happened through the use of digital twins. With an anticipated 21 billion sensor connections and important devices, "digital twins" were identified by Gartner as a top trend for 2020.

1.3 Digital Twin System Architecture

A digital twin majorly consists of hardware and software components connected with network devices [15]. This system contains data handling concepts with middleware

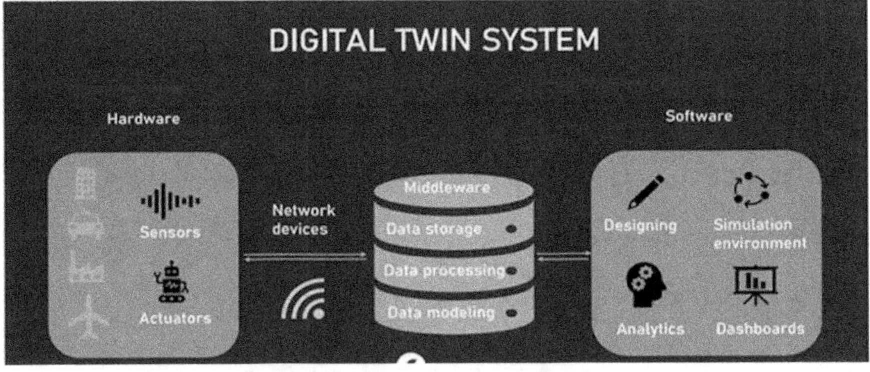

Figure 1.1 System architecture for a digital twin system.

for data management in between. Figure 1.1 shows the system architecture for a digital twin system.

Hardware components: the major components include exchange of information between assets and software integration through Internet of Things (IoT) sensors [11]. The basic technology behind DTs that use actuators to convert digital impulses into hardware-based powered motions. The hardware component also includes network devices, such as edge servers, routers, and IoT gateways, among others.

Data management middleware: its core component is a centralized repository that aggregates data from many sources. In an ideal scenario, the middleware platform would also manage connectivity, data integration, processing, quality assurance, data visualization, data modeling, and governance, as well as other related responsibilities. Such systems include common IoT platforms and industrial (IIoT) platforms, which usually feature pre-built tools for digital twinning.

Software components: because it translates simple observations into relevant business information, the analytics engine is a critical component of digital twinning. It frequently draws power from machine learning models. Dashboards for real-time monitoring, modeling design tools, and simulation software are all important pieces of the DT puzzle.

1.4 Digital Thread: A Bridge Between Physical and Virtual Worlds

In a closed loop known as a "digital thread," physical systems and their virtual counterparts can be linked as soon as you have all the required components [16]. Within it, the following iterative procedures are performed (see Figure 1.2).

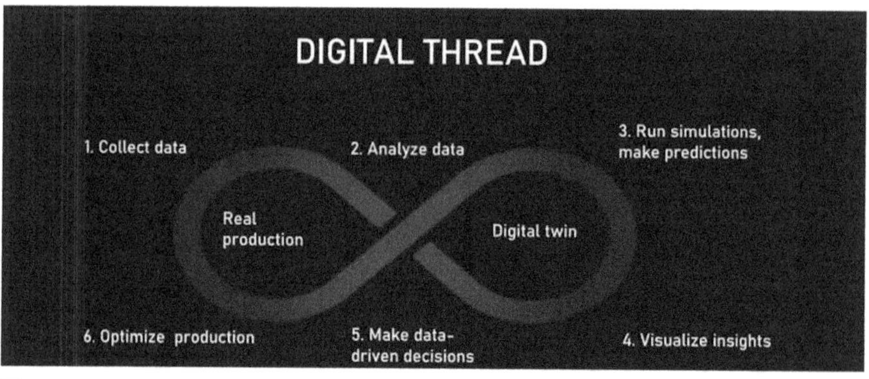

Figure 1.2 Steps within a digital thread.

1. A physical object's environment and environment at large are both sources of data that are transferred to the central warehouse.
2. The information within data is produced for updating the DT after evaluation.
3. The digital twin uses new data to assess and decide what would occur if the environment changed, bottlenecks were found, and the object's operation was replicated in real-time. At this point, algorithms can be used to alter product designs, spot dangerous trends, and prevent costly downtimes.
4. The control panel displays and visualizes analytic perceptions.
5. Participants make decisions based on actionable data.
6. Accordingly, the specifications, practices, or upkeep programs of a physical thing are altered.

The procedure is then iterated using the updated information. The complexity of the physical world is distilled down to its core components via digital twins. The technology is embraced by many industries as a result.

Digital twin examples

The exponential rise of digital twin technologies in the near future is projected to be significantly impacted by the development of the IoT, AI, extended reality (ER), and cloud computing [26]. Nevertheless, it has already gained popularity across a wide range of industries, where it is used for anything from production to maintenance and product creation. Here are a few instances of how the digital twins are put to use on a daily basis in the real world.

Manufacturing:

The complex manufacturing sector includes a significant amount of the physical production of commodities using both manual and automated labor, raw material procurement and transportation across global supply networks, and product research and development (R&D). Manufacturing is a suitable industry for the introduction of digital twins because to its complexity, which can assist companies in boosting efficiency and reducing costs at practically every stage of the production process [14]. In the early stages of designing new automobile parts, engineers at an automotive company might, for instance, model and test prototypes using digital twins. They may then simulate prototypes in different simulations, iterating the design until they reach the final result. Analysts may utilize, before actual production starts, digital twins to simulate and subsequently enhance the supply chain and manufacturing process required to generate the finished product [29].

Healthcare:

In just a few decades, data-driven technology has fundamentally changed the healthcare sector. Digital twinning can model treatment plans, enable the creation of cutting-edge medical equipment, and give individualized care for specific patients whether it is utilized in biomedical engineering, healthcare analytics, or healthcare management [34]. However, monitoring, managing, and improving physical healthcare facilities, such as hospitals, may be the most widespread application of digital twins in healthcare [38]. Healthcare providers can utilize digital twins to simulate the daily activities of their staff members and the availability of vital resources, such as hospital beds or ventilators, to improve patient care and their overall organizational effectiveness.

1.5 Types of Digital Twin

- *Stand-alone digital twins:* these are digital twins of specific goods or pieces of equipment.
- They can aid in keeping track of and improving the effectiveness of particular assets, people, and other physical resources.
- *Duplicated digital twins:* this aids in monitoring and optimizing the use of several related discrete digital twins, such as virtual representations of complex multipart systems like vehicles, machinery, or structures.
- *Enhanced digital twins:* these are virtual representations of large-scale, complicated entities (e.g., an entire organization or a city). They are constructed from digital twins of the parts that make them up. They aid in analyzing and improving higher-order performance.

1.5.1 Standalone Digital Twins

A standalone digital twin will have enough information to build a model of a system component.

A view of the asset's health, for instance, can be obtained by tracking sensors connected to a heating, ventilation, and air conditioning (HVAC) system's digital twin. The standalone digital twin typically contains a small number of data sources. Now, a toolkit that may be built can be used to enhance the twin utilizing accurate data that is now available from a variety of sources [27]. To match sensor output with data from CAD systems, three-dimensional (3D) models, or point clouds are produced by laser scanning, for instance. To monitor important measurements or states from a single asset, such as HVAC or environmental systems, limited digital twins can be helpful. They can also be used to confirm that important building site components comply with regulations and adhere to the original design.

1.5.2 Duplicated Twin

A digital twin's duplicate version includes all the significant and quantifiable data sources needed to replicate the entire asset. The limited digital twins' components could be combined to create the duplicate model. This is when the model's level of detail may prove to be too fine for your purposes.

To prevent collecting excessive amounts of data, your aims must be clear. By switching from gauges and dashboards to enhanced, virtual representations of actual assets, it is possible to increase complexity while maximizing the potential of duplicated digital twins. The value and utility of this type of digital twin can be multiplied while the dangers are reduced by modeling and executing the proper digital twin [28]. Duplicate models have an interesting foundation thanks to building information modeling (BIM) database-first models. BIM was created to promote better teamwork in the building sector. Additionally, it offers a platform with a single source of truth that enables flexible data viewing. These features would also aid digital twin models in providing the appropriate information to the appropriate individuals at the appropriate time. As changes are made to the physical twin, Artificial intelligence (AI) and software analytics would update the digital twin. Duplicated digital twin computer simulations can aid in averting actual issues. Sensor data can be collected by smart components that are connected to a cloud-based system, allowing for real-time analysis and performance comparisons.

1.5.3 Enhanced Twin

Digital twins are capable of going much beyond just one building or one piece of machinery. Scale will determine the digital twin's destiny. The connected asset's additional data sources are added to the enhanced digital twin. Correlated environmental data from outside sources as well as information from outside analytics and algorithms would fall under this category. The difficulty of managing and controlling every component of a system increases with its complexity. Because of this, digital twins represent a huge possibility for future, networked cities [35]. The information and analysis required maximizing personnel, planning, and resources will be provided by them.

Digital twins can be used to model anything, from lone components to complete systems. While every kind of digital twin accomplishes the same basic task—virtually imitating a physical system or object—their goals and application domains are significantly different. The following are the four main categories of digital twins.

1.5.4 Component Twins

Digital representations of a particular system or product component, like a gear or screw, are known as component twins. Component twins are typically used to represent integral parts, such as those under specific stress or heat, rather to simply modeling all the various components of a product [39]. Designers and engineers can discover how the parts might be modified to maintain their integrity in foreseen conditions by digitally modeling the parts and running dynamic simulations on them.

1.5.5 Asset Twins

Digital reproductions of an entire product rather than simply its parts are called asset twins, also referred to as product twins [41]. Though theoretically made up of many component twins, asset twins' main goal is to show how the many components of a real-world product interact with one another. For example, a wind turbine's asset twin may be used to track performance and spot component breakdown due to normal wear and tear.

1.5.6 System Twins

System twins, commonly referred to as unit twins, are computer simulations of actual product systems. System twins represent these several items as components of a bigger system, whereas asset twins simulate real-world products made up of numerous parts [40]. It is possible to enhance the relationships between assets by having a better understanding of how they interact, which will boost productivity and efficiency.

1.5.7 Process Twins

Digital representations of interconnected systems are called process twins [42]. For instance, a process twin could simulate the complete factory down to the workers working the machinery on the factory floor, whereas a system twin might simulate a production line.

1.6 Working Principles—Digital Twin

A digital twin is first invented by specialists, frequently data scientists or applied mathematicians.

These programmers investigate the physics underlying the physical system or object being imitated, and then use the information they learn to create a mathematical model that replicates the original in digital space. The twin is designed to be able to take information from sensors collecting data from a physical counterpart. Because of this, the twin can duplicate the physical object in real-time and offer information into its functionality for inculcating potential issues [13]. The physical counterpart's prototype may have served as the basis for the twin's design, in which case the twin can offer input as the product is developed or even act as a prototype before the physical counterpart is constructed. Eniram, a company that develops digital twins of the enormous container ships that transport a large portion of global trade, describes the procedure in some depth in this post. This kind of application for digital twins is quite complex. The amount of data you utilize to create and update a digital twin, though, will decide how well you are imitating a physical object [25]. A digital twin can be as complex or as straightforward as you wish. For instance, this lesson shows how to create a straightforward digital replica of an automobile that computes mileage using only a few input variables.

1.6.1 Digital Twin vs. Simulation

Although they are two distinct concepts, the words "simulation" and "digital twin" are frequently used interchangeably. A simulation can be created using a CAD system or other comparable platform and put through its paces, but it might not be a perfect duplicate of an actual object in the real world. However, data usage by utilizing digital twin through IoT sensors on actual equipment, which allows it to duplicate real-world systems and adapt to those systems' changing needs [36]. A digital twin gives all elements of the organization insight into how a product or system they are presently using is performing right now, unlike simulations, which are typically used during the planning phase of a product's lifecycle to predict how a future product would work.

1.6.2 Digital-Twin Use Cases

The car and the cargo ship, the digital twin examples we highlighted earlier, give an idea of possible use cases. Before being built physically, items like turbines, railways, offshore oil platforms, and aeroplane engines can be developed and tested digitally [43]. The utilization of these digital twins for maintenance tasks is another potential application. For instance, technicians could test a suggested patch for a piece of equipment using a digital twin before actually performing the fix.

1.6.3 Technology Involved in Digital Twin

Digital twins incorporate a range of technologies, including artificial intelligence and machine learning. Future advancements in 3D laser scanning and IoT sensors will significantly influence the development of digital twins. Processing of a cloud framework is the main focus of the construction of the digital twin model framework, scanning technologies, and LiDAR when there is a visual component, such as in the built environment [44].

1.6.4 IoT Sensors

IoT sensors are primarily responsible for making digital twins conceivable [3]. Directly integrating IoT functionality into modules or hardware is expected to significantly disrupt the market [9]. This will cut the cost of deploying IoT devices and simplify setup and deployment considerably [19].

1.6.5 3D Laser Scanning Software

One of the digital twin technology's unsung heroes is reality capture. Using vector-based, multi-stage point cloud processing, the visual data for digital twins is created (or stitched). Additional details are available in our point cloud processing guide.

1.6.6 AI and Machine Learning

The only techniques available are artificial intelligence and machine learning [23] to analyze the model of operations represented by the digital twin because of the massive amount of data generated [5–8]. If digital twins are to fulfil their potential, they must be able to perform analysis in real-time or more quickly, provide a high level of prediction accuracy, and incorporate data from numerous distinct and usually incompatible sources [10].

1.6.7 5G Connectivity

Every use case for a digital twin is supported by connection. 5G is increasingly important. Low latency, high bandwidth, large capacity, excellent reliability, enhanced mobility, and extended battery life are just a few of the features that 5G claims will revolutionize digital twin capabilities. For instance, a single cell on a 4G network may accommodate 5500–6000 IoT devices [19, 20]. Up to one million devices can be supported by a 5G network [21].

1.6.8 Cloud Computing

Digital twins can be processed and executed using cloud technologies. Microsoft already offers an expanding selection of Azure Digital Twin application to assist in line with cloud deployment [18]. The concept of "device shadow" offered by Amazon Web Services (AWS) is implemented as a division of AWS IoT [22].

1.6.9 Benefits of Digital Twin

Digital twins have as many uses as there are advantages to using them. But some of their more well-known advantages are as follows:

- By developing, testing, and fine-tuning systems or products in virtual environments before putting them into mass production or launching them, you can cut expenses overall [30].
- Improve operational efficiency by modeling systems with the most recent data and testing changes in dynamic simulations before implementing them in the real world [31].
- Continually monitor the performance of physical assets and functioning systems, such as structures or aircraft, and identify issues as soon as they arise.
- By simulating the customer's journey, you may enhance the experience they have while making a purchase or entering a store.

1.6.10 Advantages of Digital Twin Technology

- *Improved R&D*
  Utilizing digital twins generates a lot of information about anticipated performance outcomes, supporting more effective product research and development. Businesses can utilize this data to get insights that will help them make the necessary product adjustments before they start production [17].
- *Enhanced effectiveness*
  Even after a new version of the product has gone into production, digital twins can help in monitoring and mirroring application deployment with the aim of achieving and targeting peak efficiency for the entire manufacturing process [32].
- *Product implementation lifecycle*
  Digital twins can also assist producers in determining how to handle products that have reached the end of their useful lives and require final processing, such as recycling or important further actions [46]. By using digital twins, they can choose which product materials can be gathered.

1.6.11 Digital Twin Market and Industries

Despite the advantages that digital twins offer, not every business or every product manufactured must use them. Not all items are complex enough to require the constant and intense influx of highly processed sensor data that digital twins precisely demand [45]. Additionally, investing a significant amount of money in the creation of a digital twin is not always profitable (remember that a digital twin is an exact replica of a physical thing that has been pre-defined; as such, creating one might be costly). However, there are certain clear benefits to employing digital models for various applications:

Physically substantial projects: buildings, bridges, and other complex structures are governed by strict technical standards.
Mechanically difficult projects: vehicles, jet turbines, and aircraft. Digital twins can contribute to increased productivity in massive engines and intricate machinery.
Power devices: this includes both the transmission and generation of power.
Manufacturing initiatives: digital twins are effective at increasing process efficiency [24], much like industrial settings with cooperating machine systems.

As a result, industries that handle large-scale products or projects benefit the most from using digital twins, such as the following: system engineering; manufacturing; Power utilities; railcar design; automobile manufacturing; aircraft production.

1.6.12 Digital Twin Market: Poised for Growth

Although digital twins are already utilized in a variety of industries, the market for them is expanding swiftly, which indicates that demand will continue for a while [47]. In 2020, the digital twin market was expected to be valued USD 3.1 billion. Some industry analysts predict it will develop quickly through at least 2026, reaching an estimated USD 48.2 billion.

1.6.13 Applications

Digital twins are utilized frequently and invoked in the following applications:

1.6.14 Power-Generation Equipment

Large engines, including jet engines, locomotive engines, and power-generation turbines, benefit greatly from the use of digital twins, particularly when assessing whether routine maintenance is necessary [4, 12].

1.6.15 Systematic and Structures

Digital twins can improve massive physical structures, notably when designing large structures like offshore drilling rigs or tall towers [2]. Furthermore, they are useful for designing the HVAC systems that function inside those structures.

1.6.16 Manufacturing Operations

Given that they are designed to simulate a product's whole lifecycle, digital twins have spread throughout all stages of manufacturing, from product concept to finished product and every step in between [1].

1.6.17 Medical Services

Similar to how items can be profiled, patients seeking medical treatment can be as well.

The same type of sensor data that the system generates can be used to track various health markers and generate crucial insights [49].

1.6.18 Automobile Sector

In the automotive business, digital twins are frequently used to increase vehicle performance and production efficiency. Automobiles stand in for a number of complex, interrelated systems [48].

1.6.19 Urban Design

Civil engineers that work in the field of urban planning greatly benefit from the use of digital twins, which may show real-time applications that use 3D and four-dimensional (4D) spatial data and also incorporate augmented reality systems into constructed surroundings.

1.7 The Future of Digital Twin

There is no denying that the present operational paradigms are drastically changing. A disruptive digital revolution is currently taking place in asset-intensive industries, and this revolution asks for an integrated physical and digital perspective of assets, machinery, facilities, and processes. Digital twins are a key element of that readjustment. The potential of digital twins is essentially limitless because more and more cognitive resources are continuously being allocated to their use. Given that they are always learning new skills, digital twins can keep producing the insights needed to improve products and streamline processes [50–51]. In this context, on using digital twins to transform asset operations, learn how change will affect your industry. Similar to royal food tasters or stunt doubles for movie stars, digital twins serve the same purpose for complex technologies and processes: they guard priceless assets from harm that may otherwise befall them. Duplicates have entered the virtual world, saving many enterprises time, money, and effort while preserving the health and security of valuable resources.

1.8 Conclusion

With the establishment of IoT devices, there is a huge number of distributed levels in local and remote storage processes. There are powerful tools in generating data using data analytics and AI algorithms to process batch and real-time approaches. Digital twin technology and the latest recent machine learning and artificial intelligence tools, are assisting businesses across numerous industries in lowering operational costs, boosting productivity, and changing the way that predictive maintenance is carried out. For product producers, in particular, to attain more productive processes, digital twin technology is vital and has a shorter time to be established in the market.

On the other hand, there are digital twins being developed for IoT devices to fully comprehend and describe their structural characteristics and behavior in various settings. Therefore, the range of applications and advantages of digital twins include continually increasing IoT devices. However, digital twin applications can be remotely penetrated. And also, smart hackers cause irreversible damage to the device networks, and damage to IoT applications and systems. With blockchain's increased stability and maturity, IT professionals are looking into the prospect of connecting it to digital twins to fend off security concerns. These two technologies working together to offer a promising future for both the commercial and IT sectors. By combining cutting-edge technologies including machine and deep learning (ML/DL) algorithms, computer vision (CV), and natural language processing (NLP), digital twins are also regularly updated and improved. Thus, the cool synchronization with blockchain will enable the digital twin industry to grow over the upcoming years.

References

[1] A. A. Malik, T. Masood and A. Bilberg, "Virtual reality in manufacturing: Immersive and collaborative artificial-reality in design of human-robot workspace", *Int. J. Comput. Integr. Manuf.*, vol. 33, no. 1, pp. 22–37, Jan. 2020.

[2] A. Bilberg and A. A. Malik, "Digital twin driven human–robot collaborative assembly", *CIRP Ann.*, vol. 68, pp. 499–502, 2019.

[3] A. Čolaković and M. Hadžiali, "Internet of Things (IoT): A review of enabling technologies challenges and open research issues", *Comput. Netw.*, vol. 144, pp. 17–39, Oct. 2018.

[4] A. Coraddu, L. Oneto, F. Baldi, F. Cipollini, M. Atlar and S. Savio, “Data-driven ship digital twin for estimating the speed loss caused by the marine fouling”, *Ocean Eng.*, vol. 186, 1–14, Aug. 2019.

[5] A. El Saddik, “Digital twins: The convergence of multimedia technologies”, *IEEE MultimediaMag.*, vol. 25, no. 2, pp. 87–92, Apr. 2018.

[6] A. M. Karadeniz, I. Arif, A. Kanak and S. Ergun, “Digital twin of eGastronomic things: A case study for ice cream machines”, *Proc. IEEE Int. Symp. Circuits Syst. (ISCAS)*, pp. 1–4, May 2019. DOI: 10.1109/ISCAS.2019.8702679

[7] A. Madni, C. Madni and S. Lucero, “Leveraging digital twin technology in model-based systems engineering”, *Systems*, vol. 7, no. 1, 1–13, Jan. 2019.

[8] A. Moujahid, M. ElAraki Tantaoui, M. D. Hina, A. Soukane, A. Ortalda, A. ElKhadimi, et al., “Machine learning techniques in ADAS: A review”, *Proc. Int. Conf. Adv. Comput. Commun. Eng. (ICACCE)*, pp. 235–242, Jun. 2018.

[9] A. Reyna, C. Martín, J. Chen, E. Soler and M. Díaz, “On blockchain and its integration with IoT. Challenges and opportunities”, *Future Gener. Comput. Syst.*, vol. 88, pp. 173–190, Nov. 2018.

[10] A. S. Modi, “Review article on deep learning approaches”, *Proc. 2nd Int. Conf. Intell. Comput. Control Syst. (ICICCS)*, pp. 1635–1639, Jun. 2018.

[11] B. Schleich, N. Anwer, L. Mathieu and S. Wartzack, “Shaping the digital twin for design and production engineering”, *CIRP Ann.*, vol. 66, no. 1, pp. 141–144, 2017.

[12] C. Brosinsky, D. Westermann and R. Krebs, “Recent and prospective developments in power system control centers: Adapting the digital twin technology for application in power system control centers”, *Proc. IEEE Int. Energy Conf. (ENERGYCON)*, pp. 1–6, Jun. 2018.

[13] D. Burnett, J. Thorp, D. Richards, K. Gorkovenko and D. Murray-Rust, “Digital twins as a resource for design research”, *Proc. 8th ACM Int. Symp. Pervasive Displays*, pp. 1–2, Jun. 2019.

[14] D. Curry, *ARM: One trillion IoT devices by 2035 \$5 trillion in market value*, Jul. 2017.

[15] D. Howard, “The digital twin: Virtual validation in electronics development and design”, *Proc. Pan Pacific Microelectron. Symp. (Pan Pacific)*, pp. 1–9, Feb. 2019.

[16] D. Shangguan, L. Chen and J. Ding, “A hierarchical digital twin model framework for dynamic cyber-physical system design”, *Proc. 5th Int. Conf. Mechatronics Robot. Eng. ICMRE*, pp. 123–129, 2019.

[17] E. Negri, L. Fumagalli and M. Macchi, “A review of the roles of digital twin in CPS-based production systems”, *Procedia Manuf.*, vol. 11, pp. 939–948, Jan. 2017.

[18] F. Tao, H. Zhang, A. Liu and A. Y. C. Nee, “Digital twin in industry: State-of-the-art”, *IEEE Trans. Ind. Informat.*, vol. 15, no. 4, pp. 2405–2415, Apr. 2019.

[19] F. Tao, J. Cheng, Q. Qi, M. Zhang, H. Zhang and F. Sui, “Digital twin-driven product design manufacturing and service with big data”, *Int. J. Adv. Manuf. Technol.*, vol. 94, no. 9, pp. 3563–3576, Feb. 2018.

[20] H. Boyes, B. Hallaq, J. Cunningham and T. Watson, “The industrial Internet of Things (IIoT): An analysis framework”, *Comput. Ind.*, vol. 101, pp. 1–12, Oct. 2018.

[21] H. Brandtstaedter, C. Ludwig, L. Hubner, E. Tsouchnika, A. Jungiewicz and U. Wever, “Digital twins for large electric drive trains”, *Proc. Petroleum Chem. Ind. Conf. Eur. (PCIC Europe)*, pp. 1–5, Jun. 2018.

[22] H. Laaki, Y. Miche and K. Tammi, "Prototyping a digital twin for real time remote control over mobile networks: Application of remote surgery", *IEEE Access*, vol. 7, pp. 20325–20336, 2019.

[23] H. Pargmann, D. Euhausen and R. Faber, "Intelligent big data processing for wind farm monitoring and analysis based on cloud-technologies and digital twins: A quantitative approach", *Proc. IEEE 3rd Int. Conf. Cloud Comput. Big Data Anal. (ICCCBDA)*, pp. 233–237, Apr. 2018.

[24] I. U. Din, M. Guizani, J. J. P. C. Rodrigues, S. Hassan and V. V. Korotaev, "Machine learning in the Internet of Things: Designed techniques for smart cities", *Future Gener. Comput. Syst.*, vol. 100, pp. 826–843, Nov. 2019.

[25] J. David, A. Lobov and M. Lanz, "Leveraging digital twins for assisted learning of flexible manufacturing systems", *Proc. IEEE 16th Int. Conf. Ind. Informat. (INDIN)*, pp. 529–535, Jul. 2018.

[26] J. Liu, X. Du, H. Zhou, X. Liu, L. ei Li and F. Feng, "A digital twin-based approach for dynamic clamping and positioning of the flexible tooling system", *Procedia CIRP*, vol. 80, pp. 746–749, Jan. 2019.

[27] K. E. Barkwell, A. Cuzzocrea, C. K. Leung, A. A. Ocran, J. M. Sanderson, J. A. Stewart, et al., "Big data visualisation and visual analytics for music data mining", *Proc. 22nd Int. Conf. Inf. Visualisation (IV)*, pp. 235–240, Jul. 2018.

[28] M. Joordens and M. Jamshidi, "On the development of robot fish swarms in virtual reality with digital twins", *Proc. 13th Annu. Conf. Syst. Syst. Eng. (SoSE)*, pp. 411–416, Jun. 2018.

[29] M. Kritzler, J. Hodges, D. Yu, K. Garcia, H. Shukla and F. Michahelles, "Digital companion for industry", *Proc. Companion World Wide Web Conf.*, pp. 663–667, 2019.

[30] M. Lohtander, N. Ahonen, M. Lanz, J. Ratava and J. Kaakkunen, "Micro manufacturing unit and the corresponding 3D-model for the digital twin", *Procedia Manuf.*, vol. 25, pp. 55–61, Jan. 2018.

[31] O. Elijah, T. A. Rahman, I. Orikumhi, C. Y. Leow and M. H. D. N. Hindia, "An overview of Internet of Things (IoT) and data analytics in agriculture: Benefits and challenges", *IEEE Internet Things J.*, vol. 5, no. 5, pp. 3758–3773, Oct. 2018.

[32] O. Novo, "Blockchain meets IoT: An architecture for scalable access management in IoT", *IEEE Internet Things J.*, vol. 5, no. 2, pp. 1184–1195, Apr. 2018.

[33] P. Jain, J. Poon, J. P. Singh, C. Spanos, S. R. Sanders and S. K. Panda, "A digital twin approach for fault diagnosis in distributed photovoltaic systems", *IEEE Trans. Power Electron.*, vol. 35, no. 1, pp. 940–956, Jan. 2020.

[34] Q. Qi and F. Tao, "Digital twin and big data towards smart manufacturing and industry 4.0: 360 degree comparison", *IEEE Access*, vol. 6, pp. 3585–3593, 2018.

[35] R. He, G. Chen, C. Dong, S. Sun and X. Shen, "Data-driven digital twin technology for optimized control in process systems", *ISA Trans.*, vol. 95, pp. 221–234, Dec. 2019.

[36] R. R. B and I. S V Engineering CollegeTirupati, "A comprehensive literature review on data analytics in IIoT (Industrial Internet of Things)", *HELIX*, vol. 8, no. 1, pp. 2757–2764, Jan. 2018.

[37] R. Stark, C. Fresemann and K. Lindow, "Development and operation of digital twins for technical systems and services", *CIRP Ann.*, vol. 68, no. 1, pp. 129–132, 2019.

[38] R. Vishwakarma and A. K. Jain, "A survey of DDoS attacking techniques and defence mechanisms in the IoT network", *Telecommun. Syst.*, vol. 73, no. 1, pp. 3–25, Jul. 2019.
[39] S. Gahlot, S. R. N. Reddy and D. Kumar, "Review of smart health monitoring approaches with survey analysis and proposed framework", *IEEE Internet Things J.*, vol. 6, no. 2, pp. 2116–2127, Apr. 2019.
[40] V. Damjanovic-Behrendt, "A digital twin-based privacy enhancement mechanism for the automotive industry", *Proc. Int. Conf. Intell. Syst. (IS)*, pp. 272–279, Sep. 2018.
[41] V. J. Mawson and B. R. Hughes, "The development of modelling tools to improve energy efficiency in manufacturing processes and systems", *J. Manuf. Syst.*, vol. 51, pp. 95–105, Apr. 2019.
[42] W. Kritzinger, M. Karner, G. Traar, J. Henjes and W. Sihn, "Digital twin in manufacturing: A categorical literature review and classification", *IFAC-PapersOnLine*, vol. 51, no. 11, pp. 1016–1022, 2018.
[43] W. Kuehn, "Simulation in digital enterprises", *Proc. 11th Int. Conf. Comput. Modeling Simulation ICCMS*, pp. 55–59, 2019.
[44] X. Fei, N. Shah, N. Verba, K.-M. Chao, V. Sanchez-Anguix, J. Lewandowski, et al., "CPS data streams analytics based on machine learning for cloud and fog computing: A survey", *Future Gener. Comput. Syst.*, vol. 90, pp. 435–450, Jan. 2019.
[45] Y. He, J. Guo and X. Zheng, "From surveillance to digital twin: Challenges and recent advances of signal processing for industrial Internet of Things", *IEEE Signal Process. Mag.*, vol. 35, no. 5, pp. 120–129, Sep. 2018.
[46] Y. Liu, L. Zhang, Y. Yang, L. Zhou, L. Ren, F. Wang, et al., "A novel cloud-based framework for the elderly healthcare services using digital twin", *IEEE Access*, vol. 7, pp. 49088–49101, 2019.
[47] Y. Lu, T. Peng and X. Xu, "Energy-efficient cyber-physical production network: Architecture and technologies", *Comput. Ind. Eng.*, vol. 129, pp. 56–66, Mar. 2019.
[48] Y. Umeda, J. Ota, F. Kojima, M. Saito, H. Matsuzawa, T. Sukekawa, et al., "Development of an education program for digital manufacturing system engineers based on 'digital triplet' concept", *Procedia Manuf.*, vol. 31, pp. 363–369, Jan. 2019.
[49] Y. Xu, Y. Sun, X. Liu and Y. Zheng, "A digital-twin-assisted fault diagnosis using deep transfer learning", *IEEE Access*, vol. 7, pp. 19990–19999, 2019.
[50] Y. Zheng, S. Yang and H. Cheng, "An application framework of digital twin and its case study", *J. Ambient Intell. Humanized Comput.*, vol. 10, no. 3, pp. 1141–1153, Jun. 2018.
[51] Z. Liu, N. Meyendorf and N. Mrad, "The role of data fusion in predictive maintenance using digital twin", *Proc. Annu. Rev. Prog. Quant. Nondestruct. Eval.*, vol. 1949, no. 1, April 2018.

Chapter 2

Digital Twin Types and Design

Pallavi Jain, Sunil Kumar Boran, Bhairvee Singh, and Pankaj Singh

2.1 Introduction

Many scholars are now interested in the function of Industry 4.0 and intelligent developing. The Industry 4.0 enabler is the digital twin (DT), which combines the real physical system with a virtual counterpart utilizing prototypical, sensor devices, facts, and software to track and examine data. To predict the upcoming corresponding abstract twin, DT is a model of the arrangement or physical strength that can continuously adjust to operative changes grounded on the information and facts gathered online. It is nondestructive testing, which was introduced by Industry 4.0.

NASA's Apollo program is where the "twin/twins" manufacturing idea first appeared. NASA was tasked with creating two identical spacecraft for the project. The term "twin" is used to describe the reputation or state of the spacecraft while it is still in service and refers to the spacecraft that is left behind. Before a journey, the twin spacecraft are frequently used for training. To assist the astronauts in space in making the best choice in an emergency, the twins were hired to recreate the modeling approach on the surface during the voyage. This model can as closely as possible reflect and take into account the operating environment of the space automobile.

It is clear from this vantage point that the twins are obviously a sample or version that faithfully replicates the operational environment in reality. It has two main benefits: (1) the twins of the objects to be taken into account are almost the same in

DOI: 10.1201/9781003469612-2

terms of presentation (the Bezier curves and size of the object), subject matter (the entity's framework and its macro- and components of the system characteristics), and residence (the object's character trait and ultimately results), and (2) it enables you to make realistic operations and nations and reflect them, among other things. It is important to stress that the twin is still in good physical condition at this time.

This creative prototype significantly improves the "twins" in the Apollo software in three ways: (1) it turns the twin model into a digital one and uses virtual expression to create a simulated item with the same subject matter and essence as the item object in style; (2) it presents the immersive area and establishes a connection in comparison to the actual area so that information and records can be shared between the two; and (3) it demonstrates the idea of fusing the corporal and simulated creations. The idea and extension might be strengthened to incorporate a wider range of goods, manufacturing facilities, workstations, production lines, and production resources (painting positions, systems, the bodies of workers, substances, etc.). A comparable digital twin model can then be developed in the online world.

2.2 Problem and Motivation

Industry 4.0 has become one of the most popular topics in creation engineering in recent years. Industry 4.0 techniques are now underutilized in industrial processes nonetheless. On the one hand, this is predicated due to the inconsistent meaning of Industry 4.0, a problem which recent announcements combat. Nevertheless, recurring issues like the lack of standards and uncertainties surrounding the economic benefits while addressing the requirement for frequently significant expenditures point to a clear need for alternative approaches to the implementation of a cyber-physical production system (CPPS). Its primary goals are to provide real-time production control and to improve transparency in the production system.

Production engineering is dominated by the theme of "Industry 4.0," particularly the applied sciences that are linked to each other in recent automation. The value chain's digitalization demonstrates its necessity for the industrial future, supporting, for instance, the power of adapting something from various systems that are automated. Vertical integration, or integration inside firms, in businesses, and hence in the manufacturing system as it is created by its center value, must be secured in order to achieve Industry 4.0's horizontal integration, or integration between companies. Though most future business initiatives will center on issues like Industry 4.0, online real systems, systems that automatically make, etc., the lack of specialized expertise is severely impeding the adoption of digitalization technologies at all levels of experience (Figure 2.1).

Thus, professional and academic education is crucial for the adoption and success of Industry 4.0. In order to inform industry users about the advantages and disadvantages of using inter innovations and, consequently, Industry 4.0,

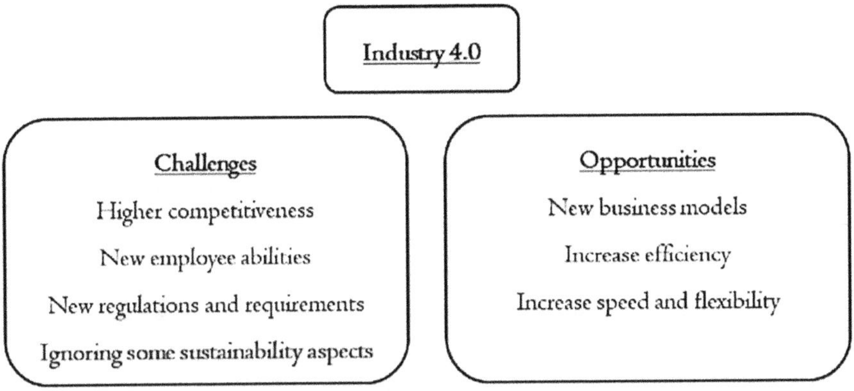

Industry 4.0

Challenges
Higher competitiveness
New employee abilities
New regulations and requirements
Ignoring some sustainability aspects

Opportunities
New business models
Increase efficiency
Increase speed and flexibility

Figure 2.1 Industry 4.0's challenges and opportunities.

demonstrators are crucial. The chapter puts forth the idea of a "studying place" to contrast the benefits of a digital twin of the contribute series to the traditional method of mapping the value streams for enhancing transparency and discovering opportunities for improvement actions in workflows.

2.3 Definition and Characteristics

Digital twin technology is fast emerging and has gained fame in the current scenario. It digitally signifies an actual resource, procedure, or setup that may be used for simulation, analysis, optimization, and monitoring, among other things. It is possible to construct a digital twin for anything, from a single machine to a whole city, and use it to simulate a variety of physical systems, from power grids to industrial processes.

2.4 Definition of Digital Twin Technology

A digital twin is a signifier of an actual item, process, or setup in technology. It is a computer-generated presentation of the factual thing, which may be applied to model, evaluate, improve, and track its work. Although the idea of digital twins has been around for a while, new developments in sensors, analytics, and machine learning have made it possible to create more advanced and precise digital twins.

2.4.1 Features of Digital Twin Technology

Features of a digital twin are demonstrated in Figure 2.2.

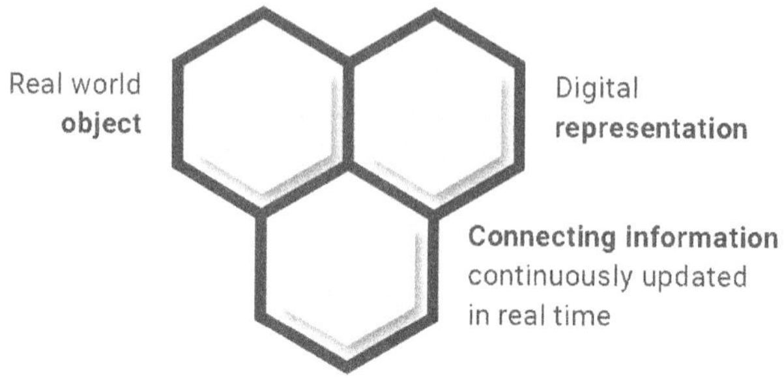

Figure 2.2 Digital twin.

2.4.2 Virtual Representation

A system's or asset's virtual representation is made using digital twin technology. The information gathered from instruments and other information, such as computer aided design (CAD) replica or blueprints, is applied to generate this virtual representation. Engineers and operators may view and test different situations using the digital twin, that may be utilized to replicate and model the corporal system.

2.4.3 Real-Time Monitoring

Real-time monitoring of physical arrangements is the major advantage of digital twin technology. The digital twin can be connected to sensors or other monitoring equipment to track the physical system's performance in real-time. Operators may take corrective action before any damage is done by using this real-time monitoring to spot problems before they become critical.

2.4.4 Statistical Analysis

The behavior of physical systems can also be predicted using digital twin technology. The digital twin can find patterns and trends that might hint at potential future issues or failures by analyzing data collected from instruments and other sources. Predictive analytics can be used to physical systems to improve performance, decrease downtime, and avert expensive repairs.

2.4.5 Modeling and Simulation

The capability of the digital twin technology to mimic and model physical systems is another important characteristic. The digital twin may be used by engineers and operators to test various situations and spot possible problems before they arise.

This modeling and simulation may be used to optimize the functionality of physical systems, boost productivity, and save costs.

2.4.6 Communication and Collaboration

Collaboration and communication between engineers, operators, and other stakeholders are also made possible by digital twin technology. To improve the performance of physical systems multiple users can share and access the digital twin, enabling them to collaborate. By working together and communicating openly, issues may be identified and resolved more quickly, resulting in less downtime and more productivity.

The intriguing and quickly developing notion of digital twins has the potential to completely alter how we create, manage, and maintain physical systems. It is an effective tool for enhancing the performance of physical systems because it can enable real-time monitoring, predictive analytics, simulation and modeling, collaboration, and communication. We may anticipate seeing increasingly more advanced and precise digital twins as technology develops, allowing us to build physical systems that are more effective, dependable, and sustainable.

2.5 Relationship between Digital Twin and Digital Thread

Two technologies that have attracted a lot of interest recently are digital twin and digital thread. These two ideas are connected, and their combination can offer a previously unheard-of level of understanding into the processes involved in product design and manufacture. This chapter will examine the connection among digital twin and digital thread and also how these technologies might be applied to boost output and efficiency in various sectors of the economy.

A digital twin signifies a real-world thing or organization. It is a computer representation of the real item that contains all the pertinent information, including its geometry, composition, and environmental settings. Manufacturing, aerospace, and the automotive sectors all employ digital twin technology to generate, test, and improve products before they are constructed. The quality of digital twin technology to model how an actual system might behave under various operating scenarios is one of its key advantages. This cuts down on the time and expense of physical testing by enabling engineers and designers to test and optimize their products in a virtual environment. Predictive maintenance is also made possible by digital twin technology since it can track a physical system's performance and spot prospective problems before they become critical.

A structure called "digital thread" links data from many phases of a product's lifespan, including design, manufacture, and maintenance. It offers a comprehensive digital record of a product, including details on its design, production methods, and maintenance history. Several sectors employ digital thread technology to enhance product quality, save costs, and boost productivity. Data from several systems, including CAD, PLM, MES, and ERP, are connected by the digital thread to give a comprehensive perspective of the product's lifespan. This makes it possible for engineers and designers to monitor a product's development from conception to manufacture to maintenance and to spot potential problems before they become serious. Real-time manufacturing process monitoring is another benefit of digital thread technology, which may enhance quality assurance and lower failure rates.

Digital twin and digital thread are related technologies that when used together, can offer considerable advantages. Whereas the digital thread offers a complete digital record of the product's lifespan, the digital twin offers an online copy of a physical entity. Engineers and designers may get a full picture of the product's design, production, and maintenance processes by merging these two technologies.

By simulating a physical system's behavior under various operating conditions, the digital twin can offer important insights into the performance and design of the final product. The product's development, from design to manufacture to maintenance, may be followed via the digital thread, giving important details about the creation's lifespan. Engineers and designers may find possible problems and improve the creation design and manufacture procedures by merging data from the digital twin and the digital thread.

2.6 Applications of Digital Twin and Digital Thread

Many sectors in the world are using digital twin and digital thread. Although digital thread technology is used to follow the development of the aircraft from design to maintenance, digital twin technology is employed in the aerospace sector to build and test aircraft engines. Although digital thread technology is used to monitor the progression of the vehicle from design to manufacture to maintenance, digital twin techniques are applied in the automobile area to replicate the behavior of cars under various operating situations. While digital thread technology is used to track the development of the product from design to production to maintenance, digital twin technology is used in the manufacturing sector to optimize manufacturing processes. Although digital thread technology is used to monitor the development of the device from design to manufacture to maintenance, digital twin technology is utilized in the healthcare sector to replicate the behavior of medical equipment.

Powerful tools like digital twin and digital thread technologies can offer important insights into the processes used in product design and manufacture. Engineers and designers may get a comprehensive picture of the product's lifespan, from design to maintenance, by merging these two technologies.

2.6.1 Classification Schema

The digital twin enables continuous adaptation to modify the operations or atmosphere with the use of supporting techniques like multi-physics virtualization, cloud services, and machine learning. The ability to simulate different situations, improve machine visibility, and link with corporate processes to assist supply chain operations, financial decisions, and other activities are all significant benefits. In order to get the best business results, this technology is essential for product optimization, preservation, throughout manufacturing and after-sale services.

Additionally, the digitization of production processes offers businesses a viable option to respond to changing consumer demands, greater uncertainty, and resource costs. Additionally, the application of DT models in after-sale services, another crucial stage of product lifecycle management, has received less attention than other applications. The study focuses on the review of DT application fields described above while assessing the review of literature use of this model. This paper also highlights difficulties with DT model implementation in real-world situations (Figure 2.3). When viewed from the perspective of the digital twin's beginnings and present growth, its applications primarily focus on the steps of product design, exercise, and preservation. But the digital twin has exceeded the traditional constraints of product design and operation due to the swift absorption of contemporary technical solutions like the IoT, mobile services, and computing infrastructure. The digital dual generation can be made easier to recognize; this phase offers a definition.

The procedures and techniques used to imitate and explain the creation, behavior, and building of physical items using virtual technology are known as the "virtual dual period." A "virtual twin model" is a virtual representation that accurately matches and is regular with the items in real-life in modern society and that accurately mimics their working and performance in a workflow. While digital twin models are things like samples, records, and procedures, virtual twins may be described as strategies, tactics, and approaches. Digital dual generation may be utilized to find and guess the nonreal place, search new routes, and continually promote human original inquiry. It does this by employing human principles and information to create digital fashion. Therefore, the quest for optimization and advancement in digital twin technology offers fresh perspectives and new resources for the innovation and advancement of contemporary production (Figure 2.4).

The following facts about the invention digital twin prototypical can be inferred from its definition: (1) the brand digital twin edition is a modeling edition in which creation corporal objects are combined with the statistics area, a virtual record of the full life expectancy of the tool neurological units, and they offer additional facts about all of the product offerings facts and full cost wire facts; (2) the item digital twin edition is refined through constant communication of facts and figures with the business; (3) the item digital twin version is improved by constant data and

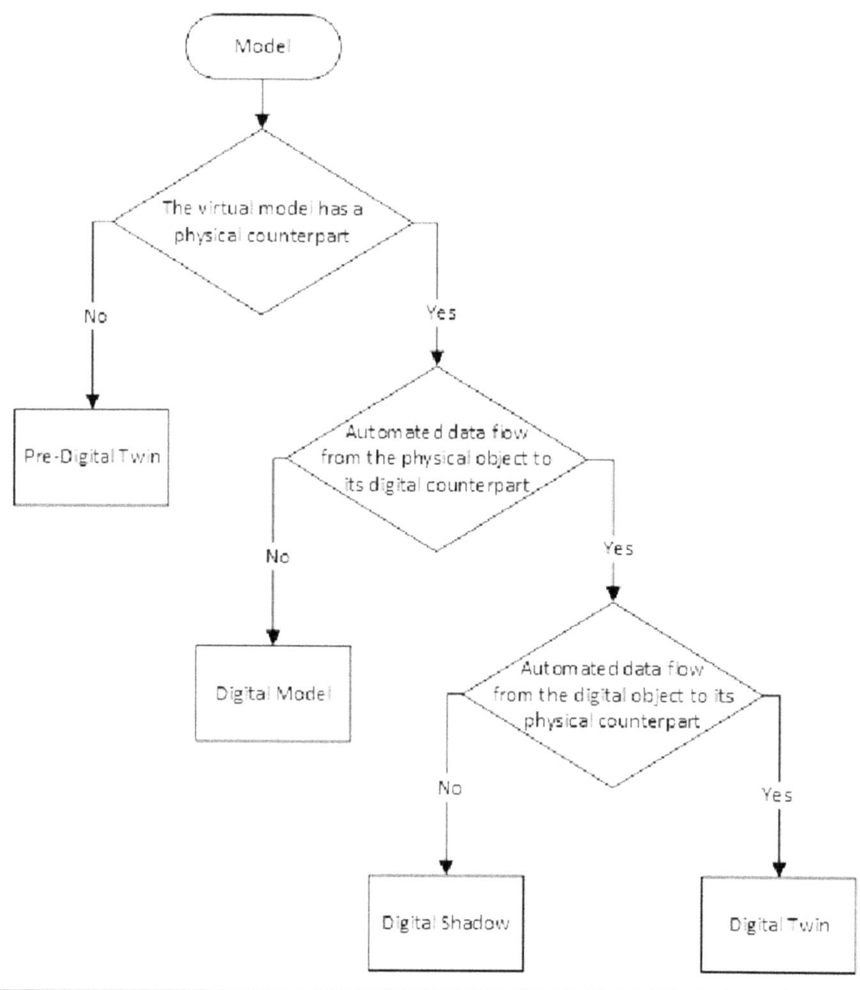

Figure 2.3 Classification of the digital twin model.

data collaboration with the creation; and (4) the product digital twin model can be applied to pretend, display, identify, expect, and manage the development steps. Future digital twin models of physical organisms will be exact replicas of one another in cyberspace. For instance, a real factory has a virtual complement in the form of a digital twin, and a physical workshop has a virtual counterpart in the form of a digital twin. The equivalent digital twin paradigm for physical manufacturing lines exists in the simulated world, and so on. Digital twins serve as the framework for intelligent manufacturing systems. The digital representation of cyberspace's ability to receive input from the physical system is the digital twin's most important educational benefit.

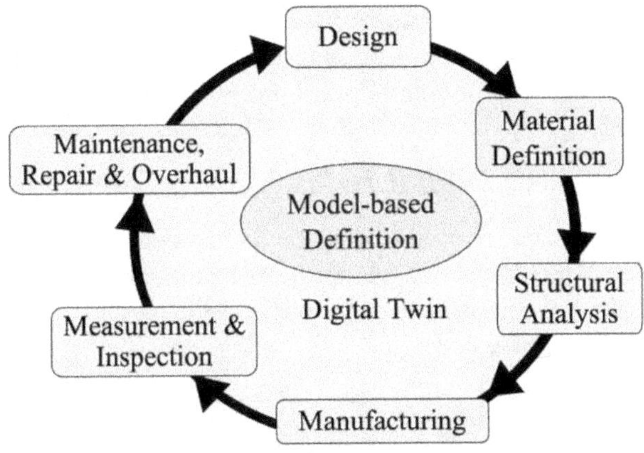

Figure 2.4 Digital twin model's definition.

2.7 Digital Twin Types

Product, production, and performance are the core categories of digital twins, and each is described here. The three digital twins working together as a whole to progress is referred to as the "digital thread."

2.7.1 Product Digital Twins

You can use digital twin to visually verify product performance and demonstrate how your items are now behaving in the actual world. You can use "product digital twin," to connect virtual and physical worlds, you can examine how a product functions in various situations and make changes in the virtual world to guarantee that the following actual product will function exactly as intended in the field. No matter how complicated your systems and materials are, product digital twins can help you traverse them and reach the best judgments. All of this removes the requirement for several prototypes, cuts down on overall development time, raises the quality of the finished product, and enables quicker iterations in response to client input.

2.7.2 Production Digital Twins

Before anything really goes into production, a production digital twin can assist in assessing how effectively a manufacturing process will perform on the shop floor. Companies may develop a production approach that is effective under a range of circumstances by recreating the process using a digital twin and studying why things are occurring using the digital thread. The production may be improved even further by building digital twins of every piece of manufacturing machinery. Businesses may avoid expensive equipment downtime and even forecast when preventive maintenance will be required by using the data from the product and production digital twins. Faster, more effective, and more dependable manufacturing processes are made possible by the continuous flow of precise information.

2.7.3 Performance Digital Twins

Massive volumes of data are produced on the use and efficacy of smart goods and smart plants. In order to give actionable knowledge for wise decision-making, the performance digital twin collects these data from active items and plants. Performance digital twins enable businesses to

- make new business connections;
- gain knowledge to enhance virtual models;
- collect, assemble, and evaluate operational data;
- boost the effectiveness of the product and manufacturing systems.

2.8 Advancements in Digital Twin Technology

The field of "digital twin" technology is fast evolving and is altering how systems and products are created, used, and maintained. The idea behind digital twin technique is to generate a digital copy of a physical asset or scheme. A number of uses, including design optimization, proactive maintenance, and real-time monitoring, are possible for this digital copy.

2.8.1 Design and Development of Digital Twin Models

Designing and creating digital twin models is one part of digital twin technology research advancement. For the purpose of building precise and realistic digital twin simulations of physical assets and systems, researchers are investigating novel methodologies and technologies. To do this, a range of data sources will be used, including simulation data, CAD models, and sensor data. To increase the precision and dependability of digital twin models, researchers are also investigating the application of artificial intelligence and machine learning methods.

2.8.2 Time Control and Monitoring

Real-time control and monitoring is another area of digital twin technology study advancement. Real-time performance monitoring of physical assets and systems can be done using digital twin models. This makes it possible to identify potential issues early and take remedial action before they worsen. New methods for real-time monitoring and control, including the use of cutting-edge sensors and data analytics, are being investigated by researchers.

2.8.3 Predictive Maintenance

Another area of digital twin technology study advancement is predictive maintenance. For determining whether physical assets and systems need maintenance, digital twin models can be employed. As a result, preventive maintenance is possible, which can increase dependability and save downtime. Predictive analytics and machine learning algorithms are two novel approaches being investigated by researchers for predictive maintenance.

2.8.4 Virtual Optimization and Testing

Virtual testing and optimization are other uses of digital twin technology. To replicate the performance of actual assets and systems under various circumstances, digital twin models are used. This makes it possible to optimize the design and operating characteristics, which can boost efficiency and cut expenses. New methods for virtual testing and optimization are being investigated by researchers, including the use of sophisticated simulation tools and machine learning algorithms.

Although digital twin technology has advanced significantly, there are still numerous issues that need to be resolved. Data management and integration is among the most difficult tasks. Digital twin models need a lot of data from many sources, and organizing and integrating this data may be challenging. The absence of standards and criteria for digital twin technology presents another difficulty since it might be challenging to compare and assess various methods. Future study in the field of digital twin technology offers many fascinating potentials. Digital twin technology is being used in novel ways by researchers, particularly in the fields of smart cities, healthcare, and agriculture. Also, they are investigating fresh methods for data management and integration as well as for real-time monitoring, proactive maintenance, and virtual testing and optimization. The design, use, and upkeep of physical resources and schemes are anticipated to become more and more dependent on digital twin technology as it develops.

2.8.5 Uses of Digital Twin

The idea of a "digital twin" intended for a creation goes much beyond the definitions of a virtual sample (or electronic concept) and a model online. The product's form, characteristics, and general performance are described in the product's virtual dual version, together with the circumstances of the item's whole lifecycle, including the manufacturing and upkeep procedures. A virtual prototype, also known as electronic design, is a computer representation of a technical item or a subsystem with independent operations. In addition to its geometrical features, the item's productivity and characteristics in at least one area are also considered. In the product design segment, a virtual prototype is developed and may be utilized throughout the whole product lifecycle, including structural engineering, production, integration,

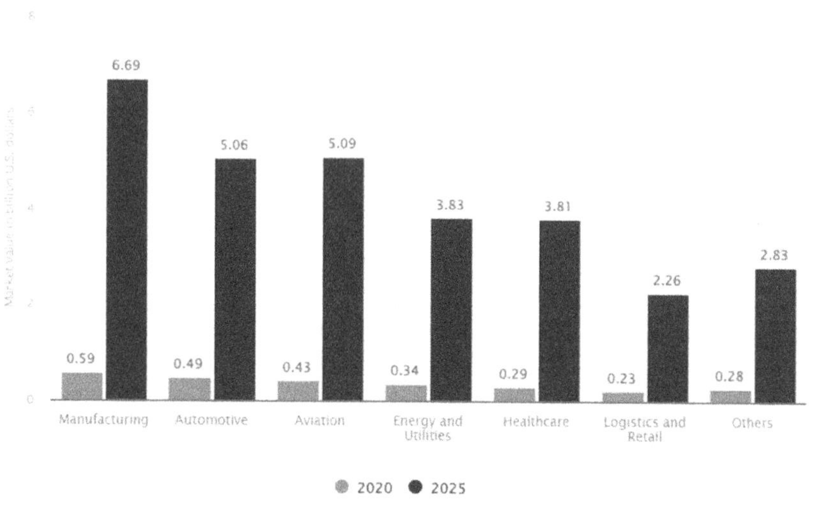

Figure 2.5 Global digital twin market by industry, 2025.

examination, sales, consumption, after-sale, rehabilitation, and various connections. The global digital twin market by industry is shown in Figure 2.5 for 2025.

The concept of an "online item" means to digitalize the activities' usefulness, efficiency, and real qualities of various goods. When it comes to the definitions of a virtual design (or virtual blueprint) and a virtual product, those terms typically refer to the characterization of a product's configuration, characteristics, and performance at the product design level rather than the creation method and states of different lifespan levels, like item production or keeping it maintained. The hyperreality, robustness, metaphysical, multi-dimensional, multi-level, connected, dynamic, outstanding, computability, chance, and interdisciplinary characteristics of the item's digital twin model are only a few of them.

2.8.6 Designing Products with a Digital Twin as a Foundation

The phrase "design work based on artificial intelligence twins" describes the merging of already-present real and digital products, with the appearance being influenced by the virtual information that the item generates. This method is employed to continually find fresh, valuable, and distinctive item views and transform the item into challenging goods. The technique of making the item gradually makes the disparity between the product's real behavior and the projected performance of the layout less appealing. The importance of product design associated with digital twins is on the complete enhancement of layout quality and performance via the merger of digital and the creation of exceptionally realistic digital simulation models.

2.8.7 Digital Twin-Based Virtual Prototype

Before building a synthetic physical prototype, the device's functioning may be evaluated utilizing multi-area thorough modeling and system effectiveness reduction analysis. Additionally, by addressing design flaws earlier, the time required for design development may be reduced. A virtual model is a representation that simulates the accuracy of a real-world object. The machine parts of a wiring machine and a fluid complex molecular machine of the system are precisely specified depending on the virtual twin. Predictive preservation may be used to map the lifecycle of a physical object, which can subsequently be used to plan the object and offer powerful analytical decision support.

2.8.8 Using a Digital Twin to Accurately Distribute Production Logistics

For high output performance and lower product prices, production logistics—which includes the organization's internal and company transportation and external transportation between companies—are essential. Figure 2.6 shows humans and AI with digital twins.

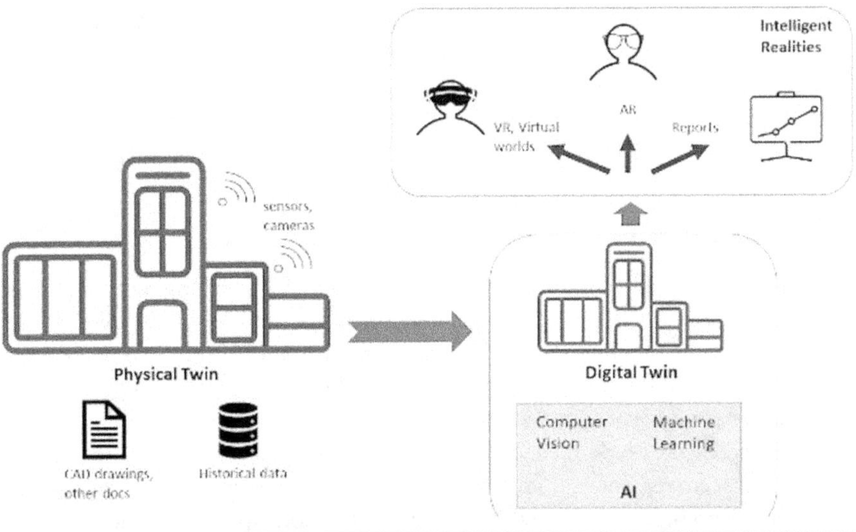

Figure 2.6 Humans and AI with digital twin.

They also ensure consistent production. Transportation between businesses—both internal and external—is crucial. They also make sure that production is constant. Thanks to meaningful engagement, confined regulation, and the mapping of actual bodily entities and virtual models, digital twin production logistics is able to perform tasks like deciding to produce structure-based enhancement, shipping path planning, and mobility kind of way regulation in both the online and offline platforms, as well as achieve production supply chain seam.

2.8.9 Digital Twin-Based Man-Machine Interaction

Interaction between people and computers may increase a device's flexibility and lighten the burden of manual labor. When a digital twin workshop is being made that is perfectly compatible with genuine physical industries is known as the "workshop person-system interaction." The robot can also instantly change the work schedule by sensing commands from the workers by touch, gesture, or sound, utilizing high-speed and reliable communication technology, enabling collaboration with the workers. The industry receives modifications in real-time to the production system of the cardinal workshop.

2.8.10 Using Digital Twins to Manage Manufacturing Energy

Monitoring, analysis, control, and optimization of the energy use of water, energy, gasoline, heat, and raw ingredients within the production technique are all part of the process of generating electricity consumption. This is done to ensure the effectiveness of the manufacturing system and the organization's financial advantage. Due to the enterprise's competitiveness, this provides precise control over energy consumption, results in electricity savings, lowers the cost of the producing firm, maintains the commercial enterprise, and supports the corporation.

2.9 Conclusion

Academics have mostly conducted a significant study on modeling, data mechanics unification, communication and cooperation, and the service usage of digital twins as part of the concept's ongoing evolution and refinement. Although there has been studying the structure and building method of digital twin modeling, no conclusive findings have been reached. In connection with modeling concepts, research has advanced in the areas of physical behavior, nondestructive material evaluation, numerical inaccuracy, and confidence assessment. These supporting technologies will aid in setting up behavior restrictions, determining model parameters, and assessing model correctness. The digital twin technology can not only apply social theories and acquaintance to create simulated models, but it can also use the replication technology of computer-generated models to discover and expect the unknown world and seek better solutions, always inspire human creativity, and keep pursuing optimization and growth, which are the innovations of the present production sector. The goal, implications, and applications of digital twin technology are outlined in this chapter.

Bibliography

Anderson, S., Barvik, S., Rabitoy, C. *Innovative digital inspection methods. In Proceedings of the Offshore Technology Conference*, Houston, TX, USA, 6–9 May 2019. [Google Scholar]

Boje, C., Guerriero, A., Kubicki, S., Rezgui, Y. Towards a semantic construction digital twin: Directions for future research. *Autom. Constr.* 2020, 114, 103179. [Google Scholar] [CrossRef]

Dröder, K., Bobka, P., Germann, T., Gabriel, F., Dietrich, F. A machine learning-enhanced digital twin approach for human-robot-collaboration. *Procedia CIRP* 2018, 76, 187–192. [Google Scholar] [CrossRef]

Fera, M., Greco, A., Caterino, M., Gerbino, S., Caputo, F., Macchiaroli, R., D'Amato, E. Towards digital twin implementation for assessing production line performance and balancing. *Sensors* 2020, 20, 97. [Google Scholar] [CrossRef][Green Version]

Jacoby, M., Usländer, T. Digital twin and Internet of Things—Current standards landscape. *Appl. Sci.* 2020, 10, 6519. [Google Scholar] [CrossRef]

Lee, J., Lapira, E., Bagheri, B., Kao, H.A. Recent advances and trends in predictive manufacturing systems in big data environment. *Manuf. Lett.* 2013, 1, 38–41. [Google Scholar] [CrossRef]

Li, Y., Lu, Y., Li, Z., Li, Y. Digital twin-based intelligent maintenance: A review. *J. Intell. Manuf.* 2021, 32(3), 647–669.

Lu, Y., Wang, L., Wang, Y. Digital twin technology for intelligent manufacturing: A review. *J. Intell. Manuf.* 2021, 32(2), 339–357.

Madni, A.M., Madni, C.C., Lucero, S.D. Leveraging digital twin technology in model-based systems engineering. *Systems* 2019, 7, 7. [Google Scholar] [CrossRef][Green Version]

Malik, A.A., Masood, T., Bilberg, A. Virtual reality in manufacturing: Immersive and collaborative artificial- reality in design of human-robot workspace. *Int. J. Comput. Integr. Manuf.* 2020, 33, 22.

Nåfors, D., Berglund, J., Gong, L., Johansson, B., Sandberg, T., Birberg, J. Application of a hybrid digital twin concept for factory layout planning. *Smart Sustain. Manuf. Syst.* 2020, 4, 231–244. [Google Scholar] [CrossRef]

Negri, E., Fumagalli, L., Macchi, M. A review of the roles of digital twin in CPS-based production systems. *Procedia Manuf.* 2017, 11, 939–948. [Google Scholar] [CrossRef]

Pallavi, J., Nipun, P.S., Roushan, K. "Surveillance Based Hostel Security Measurement using Data Analytics and Machine Learning Technique RFID" in *International Conference on Innovative Computing and Comunication (ICICC)*, 2022.

Pang, T.Y., Restrepo, J.D.P., Cheng, C.T., Yasin, A., Lim, H., Miletic, M. Developing a digital twin and digital thread framework for an 'Industry 4.0' Shipyard. *Appl. Sci.* 2021, 11, 1097. [Google Scholar] [CrossRef]

Perez, G.C., Korth, B. "Digital Twin for Legal Requirements in Production and Logistics Based on the Example of the Storage of Hazardous Substances" in *Proceedings of the 2020 IEEE International Conference on Industrial Engineering and Engineering Management (IEEM)*, Singapore, 14–17 December 2020; pp. 1093–1097. [Google Scholar]

Pramod, K.S, Pallavi, J. "Single Camera based Real Time Framework for Automated Fall Detection" in *Second International Conference on Computer Science, Engineering and Applications (ICCSEA)*, 2022.

Siew, C.Y., Ong, S.K., Nee, A.Y.C. Improving maintenance efficiency and safety through a human-centric approach. *Adv. Manuf.* 2021, 9, 104–114. [Google Scholar] [CrossRef]

Tao, F., Liu, A., Hu, T., Nee, A.Y.C. *Digital Twin Driven Smart Design*; Academic Press: Cambridge, MA, USA, 2020. [Google Scholar]

Tao, F., Qi, Q., Zhao, D. Digital twin-driven product design, manufacturing and service with big data. *J. Intell. Manuf.* 2020, 31(6), 1371–1391.

Tao, F., Zhang, M., Cheng, Y., Hu, X., Liu, Y. A survey of digital twin: concepts, characteristics, and applications. *IEEE Access* 2021, 9, 125157–125179.

Vimala, K., Pallavi, J. "Automated Testing Tools for Mobile Applications: A Study of Different Types of Tools" in *International Conference on Communication, Security and Artificial Intelligence (ICCSAI)*, 2022.

Zohdi, T.I. A digital twin framework for machine learning optimization of aerial firefighting and pilot safety. *Comput. Methods Appl. Mech. Eng.* 2021, 373, 113446. [Google Scholar] [CrossRef]

Chapter 3

Real Issues, Opportunities, and Open Investigations in Digital Twins

V. Sheeja Kumari, T. Manikandan, M.J. Carmel Mary Belinda, A. Selva Kumar, and K. Prabu

3.1 Introduction

The virtual dual is at the vanguard of the revolution ushered in by Industry 4.0, which is made viable by using advanced records' analytics and the connection provided by using the Net of Things. This revolution became made feasible by using state-of-the-art fact analytics and the connection supplied with the aid of the Internet of Things (IoT) as a result of the Net of Things, there is now a greater extent of records that can be utilized in contexts along with industrial manufacturing, hospital therapy, and the development of smart towns. While blended with information analytics, the rich surroundings of the Internet of Things offers a useful resource that is crucial for predictive protection and fault detection, to name examples, as well as the future fitness of manufacturing techniques and the improvement of clever cities [1]. This useful resource additionally provides a resource that is critical for the future fitness of manufacturing tactics and the development of clever towns. The Internet of Things has the potential to be helpful in a variety of areas, including the detection of faults, the management of visitors in smart cities, and the identification of anomalies in patient care. The digital dual is ready to triumph over the undertaking of achieving

DOI: 10.1201/9781003469612-3

seamless integration between the Net of Things and information analytics through constructing a connected bodily and digital dual. This lets in the virtual dual to gain its complete capacity (digital dual) [2].

An environment that permits the usage of virtual twins makes it feasible to perform accurate analyses in a short amount of time and to make choices in actual time that are primarily based on the effects of those analyses.

In this chapter, a comprehensive evaluation of the utilization of digital twins for programs in the fields of healthcare, manufacturing, and clever metropolis environments is supplied, together with a discussion of the technology that support digital twins, the demanding situations that they face, and open research inside the subject. This evaluation makes an effort to cowl latest guides, which are applicable to all three of the following regions: production, clinical remedy, and technologically advanced cities. Considering that the producing software is the number one cognizance of most of the studies carried out, this overview makes an attempt to cowl recent guides, which might be applicable to the manufacturing application. Even though the thing does employ a huge variety of instructional resources that were found by using keywords related to the Net of Things (IoT) and information analytics, the number one objective of the item is to locate publications, which are related to virtual twins.

This evaluation is being carried out as a part of our efforts to find out responses to the study questions that might include the following: What is precisely meant by the time period "virtual dual," and which commonplace misunderstandings of this concept can be traced further back to advance definitions of the time period? What sorts of applications can be evolved through utilizing data analytics, virtual twins, and the Net of Factors/industrial Net of Factors (IIoT) [2]? Additionally, what types of challenges and technological advances are vital to triumph over these barriers, and the way these can be executed? Is there a connection between the Net of Things, the economic Net of Factors, and fact analytics carried out with the help of the generation called a virtual dual? When it comes to the idea of digital twins, what kinds of clinical questions and demanding situations are there that have yet to be resolved? The cutting-edge kingdom of virtual twins is the primary recognition of this text, and numerous allowing technologies, such as statistics analytics, the net of factors/business Internet of Things, and the Net of Things in preferred, are discussed.

3.2 Digital Twin Definitions

"Virtual twin," as well as updated an evaluation of related thoughts and applications and brought up to date a discussion of the commonplace misunderstandings that surround definitions of this type. For ease of analysis, the most recent findings from the studies have been divided into three updated subsections. After that, there is a condensed analysis of a ramification of articles on digital twins that span a multitude of disciplines, and the very last segment gives an impression from the most up-to-date perspective of the industry. In this segment, we can talk about the origins of the concept behind the digital twin. At some point in the course of the review, precise definitions are mentioned, and the evaluator additionally debunks a number of myths that are connected to updated digital twins that might be misidentified. Formal ideas regarding virtual twins were kicking around in the early years of the brand-new millennium [3]. Having stated that, given the fluid nature of terminology, it is far from impossible that the definition of virtual twins could have been possible faster.

3.2.1 Definitions

In a presentation in 2003, Grieves hooked up the inspiration for the vocabulary used in the updated virtual twins. This nomenclature was subsequently posted in a white paper that paved the way for the introduction of digital twins.

NASA published a file in 2012 with the title "The digital twin paradigm for future NASA and U.S. Air Force automobiles." This document represented a major breakthrough in the manner of defining digital twins and became "The digital dual paradigm for future: NASA and U.S. Air Pressure up to date." NASA 2012 [4]: in line with the definition provided by Wikipedia, a "digital twin" is "an integrated mannequin of a digital image" and can be described as follows by this definition: "A digital dual is an included Multiphysics, multiscale, probabilistic simulation of an as-built up to date or machine." This type of simulation makes use of the maximum number of bodily fashions, sensor data, and other information. Chen 2017 updated Wikipedia as follows: "A virtual dual is a computerised representation of a bodily iteration."

The information that is kept within the digital representation originates from a wide variety of sources that measure the entirety of the product's lifecycle. This information is continuously streamlined and is visualized in a variety of different ways in order to read present and unborn situations in both design and functional surroundings in order to grease further effective decision timber. "A digital twin is a virtual case of a physical system (binary) that's continually updated."

3.3 Digital Twin Misconceptions

3.3.1 Digital Model

NASA published a report in 2012 titled "The Virtual Twin Paradigm for Future NASA and U.S. Air Pressure Cars," which was a significant advancement in the definition of digital twins. A digital twin is a computerized representation of a bodily item or system that displays all practical capabilities and links with the operating system. The information that is kept within the digital representation originates from a wide variety of sources that measure the entirety of the product's lifecycle. This information is continuously streamlined and is visualized in a variety of different ways in order to read present and unborn situations in both design and functional surroundings in order to grease further effective decision timber. A digital twin is a virtual case of a physical system (binary) that is continually updated.

3.3.2 Digital Shadow

A digital shadow is the digital representation of an object that only flows in one direction between the physical and digital objects. Only when there is a one-way flow of information can a digital shadow be produced. Casting a digital shadow can only be done in one direction. This particular mode of representation is referred to as a "digital shadow," which is an all-encompassing term. Instead of a change in the state of the digital object being the cause of the change in the state of the physical object, the cause of the change is the state of the physical object itself. Figure 3.1 illustrates a digital shadow to help explain this idea.

3.3.3 A Copy in the Form of a Digital Twin

The phenomenon that is being described here is referred to as a "digital twin," which is a term that was developed specifically for the purpose of describing it. It takes place when data flows between an existing physical object and a digital entity, as well as when the physical object and the digital entity are fully integrated in both directions at the same time. When you make a change to the physical object, that change will immediately reflect itself in the digital representation of the object, and the same thing will happen when you make a change to the digital representation of the object: the physical object will immediately reflect itself in the digital representation of the object. These three definitions are helpful in identifying the widespread misunderstandings that are found in the study because they make it more clear what is being referred to in the research. Despite this, there are a number of common misconceptions, and these misconceptions are not limited to the examples that have been provided above alone. Rather, these misconceptions are widespread and include other examples as well. One of the most common misconceptions is the idea that digital twins have to be a three-dimensional, point-for-point reproduction of the physical object that they are meant to represent [6]. This is one of the most widespread misunderstandings that exists today. On the other hand, there are some individuals who are of the opinion that a digital twin is nothing more than a three-dimensional model of the thing that it is meant to be replicating. These individuals hold this opinion.

Figure 3.1 is a diagram that outlines the various stages of integration that can take place for a digital twin, along with a description of what each stage entails. This diagram can also be found in the accompanying text. Figure 3.1 contains this

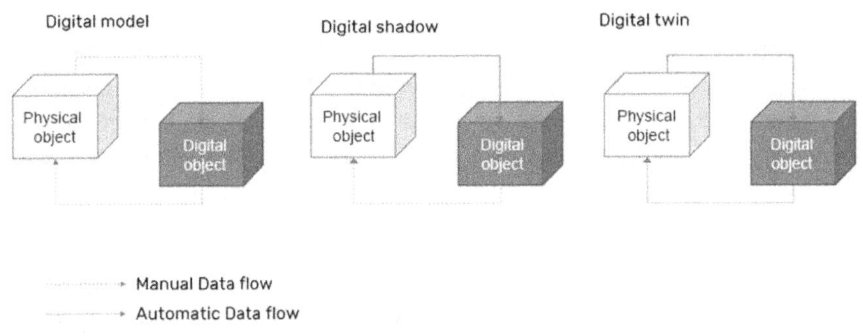

Figure 3.1 Diagram for digital model.

example to illustrate the point. It is hoped that the definitions and numbers will be useful in the creation of future digital twins as well as in the identification of those future digital twins. In the following part of this investigation, we are going to zero in on the countless diverse applications that can be found for digital twins. In order to get started with this procedure, the very first thing that we are going to do is investigate the numerous applications that can be found for digital twins.

After that, we will discuss the various applications of digital twin technology across a variety of industries, as well as the specific challenges that come along with using this technology. The concept of a digital twin, in addition to the expression "digital twin," is currently enjoying a meteoric rise in popularity among academics all over the world. This rise in popularity comes after a period in which the idea of a digital twin enjoyed a meteoric rise in popularity. This expansion is now able to proceed at a faster rate than it was previously capable of doing so, as a direct result of the progress that has been made in fields such as artificial intelligence (AI) and theInternet of Things (IoT).

Despite the fact that some applications of digital twin technology have been found to relate to the healthcare industry, the primary areas of concentration at the moment are manufacturing and smart cities. (1) Intelligent cities: the number of smart cities that make use of digital twins and the likelihood that these digital twins will achieve remarkable levels of success within those cities are steadily growing from one year to the next. In addition, the likelihood that these digital twins will achieve remarkable levels of success within those cities is also steadily growing. This phenomenon is primarily due to the IoTs' rapid advancements in connectivity. The proliferation of smart cities will lead to an increase in the number of connected communities, which will in turn lead to an increase in the utilization of digital twins as a result of the increase in the number of connected communities that will have occurred as a direct result of the proliferation of smart cities.

Not only this, but also the more data that we collect from IoTs' sensors that are incorporated into our core services within a city, the more it will pave the way for research that aims to create advanced AI algorithms [7]. This is due to the fact that the more information we gather, the more sophisticated AI algorithms we will be able to develop. It should be noted that the phrase that came before is insufficient. It is of tremendous value for the services and infrastructures of a smart city to be able to have sensors and be monitored by devices that are connected to the IoT. This helps to future-proof a wide variety of different types of things in a beneficial way. It can be used to assist in the design and development of smart cities that are still in the process of being developed, as well as additional smart cities that are still in the process of being developed. It can also be used to assist in the design and development of smart cities that are still in the process of being developed. Smart cities that already exist could benefit from its implementation.

Planning allows for the acquisition of benefits related to the conservation of energy in addition to the benefits that are associated with planning. One is able to obtain these benefits if they plan and prepare for them in advance. These numbers offer a wealth of information regarding the distribution and usage of our many different kinds of utilities, which can be found in their respective categories. It is important to take into consideration the potential applications of technology that is capable of producing digital twins in order to make headway towards the creation of smart cities. This will allow for the creation of smart cities more quickly. It is able to create a living testbed within a virtual twin, which can accomplish two things: first, it can test different scenarios, and second, it can enable digital twins to learn from their surroundings by analyzing changes in the data that they collect. This is accomplished through its ability to create a living testbed within a virtual twin. This is made possible by its capacity to generate a living testbed within a virtual twin, which allows for the completion of the task at hand.

This capability can be beneficial to growth because it simplifies the process and makes it easier to finish. In turn, this makes the process easier to complete. The information that was obtained has a number of potential applications, one of which is data analysis. Another potential application is monitoring, and both of these are examples of applications. There will be an increase in the number of opportunities for connectivity as well as an increase in the amount of data that can be put to use as more and more smart cities are built [20–23]. As a result of this development, the idea of digital twins will move closer to becoming a reality, which is a very exciting prospect.

3.4 Manufacturing

Because the manufacturing industry is the next potential application area for the digital twin technology, testing and evaluation of the technology is currently taking place for the purpose of utilizing it in the manufacturing industry. The fact that manufacturers are constantly looking for new ways to monitor and keep track of their products in an effort to cut down on the amount of money and time spent doing so is the most important factor in this phenomenon. This is the most significant factor that drives and motivates any producer. As a consequence of this, it would appear that digital twins are having the most significant impact within the context of this conversation. In a similar vein, the construction of a smart city makes connectivity one of the most important motivating factors for the manufacturing industry to make use of digital twins. This is in line with the previous point. This is consistent with the sentence that came before it. This is in line with the sentence that came before it. The concept of Industry 4.0, also known as the "fourth industrial revolution," which has been referenced in a variety of contexts, is consistent with the expansion that is currently taking place. The term "fourth industrial revolution" has been given to this concept, which makes use of the interconnectivity of a variety of devices in order to make the concept of a "digital twin" a reality for the processes that are involved in manufacturing. This concept was developed in order to take advantage of the fact that manufacturing processes are becoming increasingly complex. [8]. The digital twin has the potential to provide feedback in real-time on the performance of the equipment as well as on the operation of the production line. This feedback might originate from the equipment itself or from the production line itself. It is possible that the equipment or the production line itself is the source of this feedback. Either one is a possibility. It enables the developer to anticipate potential problems at an earlier stage in the process of product development, which is a significant advantage.

The use of digital twins promotes connectivity and feedback between devices, which, in the end, leads to improvements in both performance and dependability. In the long run, these benefits may become apparent. AI algorithms that are connected with digital twins have the potential to achieve a higher level of accuracy. This is because the machine has the ability to store massive volumes of data, which is required for performance and prediction analysis. This is due to the fact that the device has the ability to store the data. Because of its capacity to make use of real-time data, the digital twin has the potential to be an extremely valuable asset within a manufacturing environment. This is because it can create an environment to test products as well as a system that acts on real-time data. This is because the digital twin can act on the data right away. The automotive industry, which is where digital twins have been most prominently demonstrated, in the case of Tesla, is another sector that could benefit from their use. The use of digital twins is also a possibility in this sector of the economy. It is anticipated that the application of digital twins will be beneficial to this sector of the economy. The ability to create a digital twin of an internal combustion engine or a component of a vehicle can be useful in a number of contexts. These contexts include the simulation of different scenarios as well as the examination of data that has been collected [9]. Having the ability to create a digital twin of an internal combustion engine or a component of a vehicle can prove to be useful in a number of different contexts. These contexts include a variety of different scenarios. Artificial intelligence is able to achieve a higher level of testing accuracy because it is able to perform data analytics on live vehicle data in order to predict the current and future performance of components. The business of building and construction is yet another industry that stands to benefit from the utilization of digital twins in a wide variety of different applications. This is because digital twins can simulate the behavior of physical objects more accurately. One scenario in which a digital twin might prove useful is as an additional set of eyes on a project, such as during the construction phase of a building or structure. This is one example of a scenario in which a building or structure could benefit from having a digital twin. This technology can not only be used in the construction of smart city buildings or structures, but it can also be used as an instrument for continuous real-time prediction and monitoring. This is one of the many potential applications for this technology. This is just one of many possible uses for the technology that can be developed. Because of the capabilities of this technology, either one of these applications could be implemented. The utilization of data analytics and the digital twin has the potential to deliver more precision when it comes to predicting and maintaining the condition of buildings and structures, with any adjustments being made electronically before being implemented physically. This could be beneficial in the event that more accuracy is required.

This may prove to be helpful in the event that increased precision is required. Because the algorithms can be implemented in real time within the digital twin, the construction crews are able to simulate the building process with a higher level of accuracy. This is due to the fact that the digital replica of the building, also referred to as the "digital twin," is crafted before any work on the actual structure is even started. Real-time simulation, in contrast to low-detail static blueprint models, has emerged as a common goal that many researchers working on the topic of digital twins have in common. This is because real-time simulation allows for changes to be made to a model as it is being run. The application of these models serves a purpose; however, their learnability and predictability are severely limited because they do not make use of parameters that are based on real-time data. The application of these models serves a purpose; however, the models themselves do not use parameters that are based on data that is collected in real-time. In addition to making use of machine learning and deep learning algorithms [10], the digital twin also has the capacity to learn while simultaneously monitoring its own progression.

3.4.1 Healthcare

The healthcare sector is another sector that could potentially profit from the implementation of digital twin technology. Things that were previously unimaginable are now becoming a reality as a result of the expansion and improvements brought about by enabling technology in the healthcare industry. These shifts have been made possible by the advancement of technology. This has produced an effect that is unlike anything that has ever been observed before. It is possible to attribute the rise in the total number of connections to the fact that the price of devices that are connected to the Internet of Things has decreased at the same time that the ease with which they can be set up has increased. The potential applications of using digital twins in the healthcare industry are only growing to become more extensive as a result of improvements in connectivity. This is one of the industries that could benefit the most from this technology. This is a direct consequence of the enhancements made to the connectivity. The creation of a person's digital twin is one of the applications that has the potential to be used in the future. This digital twin would make it possible to conduct an examination of the body in real-time, which would be a significant advantage. The use of a digital twin, which is a more realistic application that is currently being employed, is currently being employed in order to simulate the effects of various medicines. This is being done in order to further research and development in this area. This action is being taken in order to advance research and development in this particular field. The medical industry, and more specifically the planning and execution of surgical procedures, is another sector that could benefit from the use of digital twins [11].

This is something that can be accomplished, provided that one has access to a digital twin for assistance. Researchers, doctors, hospitals, and other healthcare providers now have the ability to simulate environments that are tailored to meet their individual requirements thanks to the use of a digital twin. This is a significant step forward in the development of the pitch. This can be done in real-time or with an eye towards potential future developments and applications. In addition to this, the AI algorithms are able to work together with the digital twin in order to generate more accurate projections and options for the future. Because there are a great number of applications within the healthcare industry that do not involve the patient directly but are beneficial for the patient's continuous care and treatment, the role that such systems play in patient care is one that is quite significant [10]. On the other hand, the potential applications for this technology are quite extensive, ranging from the management of individual beds to the management of large-scale wards and hospitals. Although the application of the idea of a digital twin in the field of medicine is still in its infancy, the potential applications for this technology are quite extensive and could be used in a wide variety of contexts. When it comes to the field of medicine, the ability to simulate various scenarios and react to them in real-time is of the utmost importance. This is due to the fact that it can quite literally mean the difference between living and dying. This is due to the fact that having this ability enables one to potentially save the life of a patient. Additionally, the digital twin may be helpful in terms of preventative maintenance as well as ongoing repairs to medical equipment. When applied to the field of medicine, both artificial intelligence and the concept of the digital twin have the potential to make decisions that could end up saving patients' lives by utilizing data obtained both in the here and now and from the patients' medical histories [12]. Uses of a digital twin are listed below, displaying some of the cross overs in the intended application to show how flexible predictive maintenance is from the perspective of manufacturing plant machinery to patient care. This demonstration aims to illustrate the various applications of a digital twin. It also demonstrates some of the applications in which they do not overlap, demonstrating that the employment of digital twins is limited to the purposes for which they were designed [11–13]. Recent developments in AI, the IoT, and Industry 4.0 have significantly aided the development of new applications for digital twins.

3.5 The Partner in Business That is Digital

In 2016, General Electric (GE) included a description of its use of a digital twin in an application for a patent that the company submitted. This application was for a power generation system. This was the first time that the general public had access to this description. They developed a programme that is now being utilized on a platform that is being referred to as "Predix" [42]. This programme serves as a tool for the creation of digital twins. This action was taken in order for them to be able to put into practise the concept that was outlined in the patent, so this action was taken. The software programme known as Predix is utilized whenever the tasks of data analytics and monitoring are carried out. Over the course of the past few years, General Electric has implemented a number of strategic changes that have resulted in a significant reduction in their plans to develop a digital twin. They have made the decision to pivot away from their legacy as a software company and towards that of an industrial global corporation in order to better honor their history. This will enable them to better preserve their past. On the other hand, Siemens is responsible for the creation of a system that is referred to as "MindSphere" [13].

This platform has embraced the concept of "Industry 4.0" by providing a system that is hosted in the cloud and connects machines and physical infrastructure to a "digital twin" [43]. This is accomplished through the utilization of all of the connected devices as well as billions of data streams in the pursuit of modifying organizations and providing solutions that are based on digital twins. The "ThingWorx" platform gives developers of artificial intelligence and digital twin technology an additional option for how to construct their products, expanding the range of possibilities available to them. The fact that the system is known as "ThingWorx" opens up the door for this particular option. This platform, which was developed by PTC and is known as an Industrial Innovation Platform, is primarily geared towards the harvesting of IIoT/IoT data and the presentation of that data through an easy-to-use, role-based user interface that provides users with valuable insights. This system was developed with a wide range of applications in mind, including those in the manufacturing, transportation, and energy sectors. This platform was created specifically to collect data from the IIoT and the IoT. The platform makes it simpler and more effective to create data analytics while simultaneously establishing an environment for a solution that is known as a digital twin. IBM developed a platform that it calls the "Watson IoT Platform," and it advertised it as an all-around IoT data tool that can be used to manage large-scale systems in real-time using data collected from millions of IoT devices. The Watson IoT Platform was marketed by IBM as an all-around Internet of Things data tool that can be used to manage large-scale systems. The Watson IoT Platform was created by IBM. This particular location has it available for purchase. IBM has marketed this platform as an all-encompassing data resource for the IoTs.

The platform comes pre-loaded with a wide variety of additional features and capabilities, some examples of which are cloud-based data analytics services, blockchain services, edge capabilities and capabilities, and so on and so forth. When put together, these components form a foundation that, when it comes to the creation of a digital twin system [14], is capable of being utilized as an option for a starting point that is both practicable and accessible. When talking about open-source software, there are two major projects that should be brought to the forefront of the conversation. Both of these projects are currently in the development stage, so it is appropriate to discuss them together. The first one is the "Ditto" project by Eclipse, which is a platform that is already ready to be used and can manage the various states of a digital twin. Additionally, it simultaneously gives the physical twin and the digital twin access and control. The platform is responsible for performing a back-end function in the form of facilitating the process of connecting digital twins and offering assistance to devices that are already connected. In addition to this, the platform offers support for devices that already have an established connection. This particular open-source project is a platform that can be used to generate, access, and construct digital twins. Bentley Systems is responsible for the development of yet another open-source project that goes by the name "imodel.js."

3.6 Challenges

It is becoming more and more apparent that the digital twin operates concurrently with technologies related to artificial intelligence and the Internet of Things, which results in shared challenges. Recognizing the problems at hand is the first step towards solving those problems, which is why this should be a top priority: it is the first step towards finding a solution. Finding issues that are common to digital twins is the ultimate goal. Both the field of data analytics and the Internet of Things are plagued with their fair share of issues, many of which are intertwined. Finding problems that are common to both digital twins is the ultimate goal.

3.6.1 Difficulties in Data Analysis

The following is a synopsis of some of the challenges that are currently being encountered in the field of deep learning and machine learning.

3.6.2 Information Technology Infrastructure

The first significant obstacle involves the entire information technology sector as well as the infrastructure that supports it. It will be necessary to have an infrastructure that can assist in the execution of the algorithms in order to keep up with the rapid development of artificial intelligence. In addition, this infrastructure will need to have a high level of performance (AI). Cutting-edge hardware and software will be used in this infrastructure. At the moment, the most significant obstacle that needs to be conquered by the infrastructure is going to be the extremely high cost of installing and maintaining these systems [15]. For instance, the cost of high-performance graphics processing units (GPUs) that are able to execute machine learning and deep learning algorithms can run into the thousands of dollars and can range anywhere from $1,000 to $10,000 in price. Due to the significant amount of power that they consume during operation, these GPUs can be quite pricey. In order for the algorithms to be put into action, these graphics processing units are an absolute necessity. In addition to this, the infrastructure itself must have up-to-date software and hardware in order for these kinds of systems to be successfully run. The utilization of GPUs "as a service," which provides on-demand GPUs at a cost through the utilization of cloud computing, is one strategy that can be utilized in order to circumvent this barrier. In this context, "as a service" refers to the utilization of graphics processing units. In this context, "a service" refers to "as a service" rather than "as a service." On-demand services are currently being made available by a number of companies, some of which include Amazon, Google, Microsoft, and NVIDIA, to name just a few. These services are completely unique and are not like any other cloud-based applications that have been developed in the past. As a consequence of this, the barrier that was preventing the demand from increasing has been removed; however, there is still a lack of adequate infrastructure, and the cost of data analytics is still too high. The use of cloud computing to carry out data analytics and the production of digital twins continues to raise some concerns with regard to the level of safety that is provided by cloud infrastructure. This is the case because cloud infrastructure is still relatively new.

3.6.3 Data

When it comes to the dependability of the data, it is an absolute requirement to check it thoroughly and make certain that it does not contain any errors in any way, shape, or form. This is because any error in the data could have a significant impact on the outcome of the analysis. Before feeding the data into the AI algorithms, it is necessary to organize and clean the data; this will ensure that the data is of the highest possible quality. Prior to the data being fed into the AI algorithms, this must be completed.

3.6.4 Protecting an Individual's Privacy and Ensuring that the Individual's Data is Kept Secure

When conducting data analytics, just like when conducting any other aspect of the computing business, essential concerns that need to be addressed include safeguarding the privacy of an individual and ensuring that an individual's data is kept secure. Both of these concerns are crucial and must be addressed. Because the field of AI is still in its early stages of development, the laws and regulations that pertain to AI have not yet been fully developed. This is due to the fact that AI is still in its infancy. It is unavoidable that, concurrently with the advancement of technology, there will be an increased demand for increased oversight and regulation, as well as for increased safety precautions. This is a situation that cannot be avoided. This presents a difficult obstacle to clear. In the not-too-distant future, regulations will require the implementation of safeguards in the form of algorithms designed to protect user information. These algorithms will be designed to protect user information. The safety of the user's information will be prioritized in the development of these algorithms. The General Data Protection Regulation (GDPR) is a new rule that was recently put into effect to protect individuals' rights to privacy and the security of their personal data across the entirety of Europe, including in the United Kingdom. This new rule was recently put into effect to protect individuals' rights to privacy and the security of their personal data. This regulation, which was just recently put into effect, is intended to protect people's rights to privacy and the safety of the personal data they provide. This sheds light on the difficulties that are associated with the handling of data when developing AI algorithms, despite the fact that it is a comprehensive piece of legislation that covers data and security [15].

Legislation is one step that can be taken to secure the protection of personal data, and another option that can be considered is federated learning, which is a decentralized framework for a variety of different training models. Both of these actions are possible. When data analytics are incorporated into a digital twin, concerns regarding the users' right to privacy and security can be alleviated. To achieve this goal, it is necessary to make certain that the data contributed by users to a learning model can be stored locally without the need for any data sharing to take place. This is done in order to ensure that the honesty of any data that may be involved is not compromised in any way.

3.6.5 Trust

The issue of trust is another barrier that must be traversed before significant progress can be made in certain subfields of the field of artificial intelligence. The implementation of artificial intelligence can be intimidating for a number of different reasons: first, it is still in its infancy; second, the complexity can be difficult to navigate without any prior experience. Together, these two aspects help to explain the phenomenon as a whole. There is a barrier to trust that is caused by the fear that artificial intelligence and robots will one day overtake humans as the dominant force on earth and take control of the essential infrastructure of the planet. This fear stems from the fact that artificial intelligence and robots are already here. This feeling of dread stems from the fact that at the present time, humans are the dominant force across the entire planet. There is a possibility that this will create a barrier in terms of trust. This is due to the fact that the portrayal of the AI focuses primarily on the negative effects that might occur in the future. This is due to how the AI is portrayed. Despite this, the problem is still evident, and there is a pressing need for increased exposure to AI and the positive uses of the technology, which would help overcome problems with trust. As time goes on, it is becoming an increasingly common occurrence for the media to cover positive developments in the field of artificial intelligence. Concerns regarding trust are made worse by issues relating to privacy and security; however, these concerns can be alleviated by instituting more stringent privacy and security regulations in the field of artificial intelligence [16].

3.6.6 Expectations

The assumption that data analytics can be used to tackle the challenge of finding answers to all of our issues is the final barrier that needs to be cleared by this field of study. This field has a long way to go before it can accomplish this goal. When working with artificial intelligence, careful consideration is absolutely necessary, and devoting a sufficient amount of time to this process enables one to determine which application would be the most appropriate. Because of this, it is extremely likely that traditional models will be unable to produce the same results as the newer models. They, along with other emerging technologies, have the potential to work hand in hand with the improvement of things like manufacturing and the development of infrastructure for smart cities. This is because of the potential synergy that can occur between these technologies.

This could be advantageous. The prospective customers are only aware of the benefits, and they believe and assume that as a result, they will immediately experience time and cost savings as a consequence; as a result, their expectations are extremely high. When putting data analytics into practise, it is essential to keep in mind that the field as a whole is still in its infancy, and that this represents a challenge that needs to be overcome before moving forward. This is because the infancy of the field represents a challenge that needs to be overcome before moving forward. It is made abundantly clear by the number of circumstances that utilize "AI" for operations that do not require it, in contrast to other circumstances where AI should be utilized. This is due to the numerous situations that make use of "AI" for tasks that do not call for it. People need to be exposed to and understand artificial intelligence more so that they can acquire the correct baseline knowledge of the field and, as a result, learn how it could be used. Increased awareness and understanding of artificial intelligence are necessary. Because of this, people will be able to acquire a deeper comprehension of the various applications that could be pursued (AI).

3.6.7 The Internet of Things and the Industrial Internet of Things: Challenges to Conquer

In the context of the Internet of Things and the industrial Internet of Things, the following is a list of potential challenges that may be encountered:

(1) *Data, Privacy, Security, and Trust.* It is difficult to collect significant quantities of data due to the enormous proliferation of Internet of Things devices in both domestic and commercial settings. This issue is caused by the proliferation of things connected to the Internet of Things, which is currently at an all-time high. The challenge lies in attempting to keep control over the flow of data while simultaneously ensuring that it can be successfully organized and put to use in productive ways. This presents a significant challenge. This is a significant obstacle to overcome. The advent of big data presents a challenge that is significantly more difficult to overcome than it was in the past when similarly difficult challenges were encountered. This is in contrast to the situation in the past, when similar challenges were encountered. The Internet of Things, which is an abbreviation that stands for the Internet of Things, is one of the factors that contributes to the expansion of the already enormous amounts of data that is unstructured. If the data is not first categorized and organized, the Internet of Things will not be able to effectively manage the volume of data that it generates. This is because the Internet of Things generates data in a continuous stream. Because of this, there will be an increase in the quantity of data that is not only valuable but also helpful. This will be a consequence of the situation.

In the event that this does not occur, the information that has been gathered through the Internet of Things will either be lost, or it will be unable to economically extract any value from the massive volumes that have been accumulated. Both of these outcomes are undesirable. Both of these outcomes are undesirable. Because the information has the potential to be regarded as sensitive, there is a greater chance that it could be of use to a criminal, which would make the threat significantly more severe than it would have been otherwise. When businesses are in a situation where they might have to deal with sensitive information about their customers, they put themselves in a significantly more precarious position than they would be in otherwise [17]. One of the many problems that can arise as a result of cyberattacks is the possibility that criminals will target computer systems and bring them offline in an effort to bring an organization's infrastructure to its knees. This is just one of the many problems that can arise as a result of cyberattacks. The occurrence of cyberattacks can lead to a wide variety of issues, including this being just one of them. Some businesses have thousands of Internets of Things devices connected, which puts them at risk of being targeted by cybercriminals who want to seize control of the devices and exploit them for their own purposes. Because of this, particular businesses run the risk of becoming the targets of cybercriminals. Because of this vulnerability, these companies are now at risk of being attacked by cybercriminals who are looking to steal their data. The controversy surrounding the Mirai botnet is a good illustration of this point and serves as an example of a controversy. In this particular instance, nearly 15 million Internet of Things devices that were located all over the world were compromised and used to launch a distributed denial-of-service attack. The attack was carried out by taking advantage of the vulnerabilities in the devices by means of hacking. A distributed denial of service attack is more likely to occur due to the growing number of connected devices. The acronym "distributed denial of service" (also abbreviated as "DDoS") is used to describe this type of attack. This is in addition to the danger that was discussed earlier. The most recent security features and protection are required to be installed on the devices during the installation process; failure to do so creates a vulnerability that provides criminals with a back door into a larger linked IoT ecosystem.

(2) *Infrastructure.* The exponential growth that has been observed in IoT technology in comparison to the older systems that are currently in place indicates that the current IT infrastructure is falling behind and needs to be updated because it is lagging behind because it is falling behind. The growth of the Internet of Things is helped along in part by the renovation of aging physical infrastructure and the implementation of recently developed technological capabilities. Both of these factors are helping to propel the growth of the Internet of Things. The development of the Internet of Things depends on the presence of both of these elements. Businesses now have the opportunity to take advantage of the most recent technological advancements and to use the applications and services that can be found in the cloud, all without having to spend money on expensive upgrades to their already existing systems and technology thanks to a modernized infrastructure for the Internet of Things. These companies stand to realize significant savings in operational expenses as a direct result of this development. Before they can be successfully implemented, solutions based on the Internet of Things must first overcome a number of obstacles. One of these difficulties is the difficulty of connecting legacy hardware to an environment that utilizes the Internet of Things. The modernization of older devices so that they are compatible with the Internet of Things is one of the potential solutions that could be implemented to solve this issue. Because of this, there will be no loss of data, and even machines that are quite a bit older will be able to carry out some sort of data analysis.

(3) *Location of Links and Connections.* Despite the Internet of Things' growing popularity, there are still issues with connectivity in many different locations. They are especially common during the phase of the project in which real-time monitoring is being worked towards as a goal. Due to the large number of sensors that are

components of a single production process, you will face a significant challenge when attempting to connect all of the sensors during the same production process at the same time. This overarching goal of connectedness is having a negative impact as a result of difficulties with aspects such as power outages, software issues, and continuous rollout errors. Even if just one of the sensors that are associated with a particular process is not properly linked, this could have a significant impact on the overall effort that is being made. AI algorithms, for instance, can consume data gathered from devices connected to the Internet of Things. The fact that the system needs all of the data in order to function properly means that this could become a significant problem. Furthermore, the fact that the system needs all of the data means that the absence of IoT data could have a negative impact on the system's ability to function normally. This can be difficult because the system needs all of the data in order to operate correctly. It is possible to ensure that each and every piece of data will be collected by first retrofitting machines and then harvesting the data that has already been provided by the machine. Imputation methods are a type of process that involve finding replacement values for data that was collected by Internet of Things sensors but was missing. Specifically, the data in question was missing some of its values. More specifically, some of the values that should have been present in the data in question were absent. This is a concept that is used to ensure that all connections are made, and it also helps to make it possible for AI models to be run with a high level of accuracy and very little or no missing data at all. Other uses for this concept include ensuring that all connections are made. This idea can also be used to make sure that all connections are made and to make it possible for AI models to operate with these qualities. Both the terms "imputation methods" and "imputing methods" refer to the same concept.

(4) *Expectations*. The expectations associated with the Internet of Things present a challenge for artificial intelligence when considered in the same context as the previous point. This is due to the fact that businesses and end-users do not have a solid understanding of what to anticipate from IoT solutions or how to make the most of them in their day-to-day lives. As a direct consequence of this, there is an inadequate uptake of these technological advancements. It is heartening to see that end-users and organizations are beginning to recognize the potential in the Internet of Things and the ways in which all of us may benefit from a world that is smarter and more connected. The fact that the Internet of Things is expanding at an exponential rate lends credence to this assertion. The assumption that the Internet of Things can be used indefinitely without any prior knowledge can be harmful, and as a knock-on effect, it can place more weight on issues regarding privacy and security, which further emphasizes the challenges associated with trusting other people. As is the case with AI, having a foundational understanding of IoT is required in order to guarantee that it is utilized to the fullest extent possible. This understanding is necessary in order to guarantee full utilization. To ensure maximum utilization, this

comprehension is essential. The majority of the time spent on discussion in this section of the chapter is dedicated to the challenges that are associated with digital twins. These challenges will need to be overcome within the digital twin. However, as the research continues, it is becoming more and more obvious that the challenges that are encountered in data analytics, the Internet of Things, and the industrial Internet of Things are very similar to those that are encountered in the challenges that are associated with digital twins. This is because the challenges that are associated with digital twins are very similar to those that are encountered in the challenges that are associated with digital twins. This is due to the fact that the difficulties that are associated with digital twins are extremely comparable to those that are encountered in the difficulties that are associated with data analytics, the Internet of Things, and the industrial Internet of Things. The following is a list of some of the challenges that have been identified as being present in the world today:

(1) *The Current Information Technology Infrastructure.* The current information technology infrastructure is the root cause of the problem; this holds true for analytics as well as the Internet of Things. This problem has an impact on both of these technological platforms. The presence of infrastructure that enables the Internet of Things and data analytics to function in the appropriate manner is necessary for the successful operation of a digital twin. This infrastructure is referred to as the "digital twin infrastructure." Because of the existence of this infrastructure, the administration of a digital twin will become noticeably less difficult. If the digital twin does not have an IT infrastructure that is connected and well thought out, it will not be able to effectively achieve the goals that it has set for itself.

(2) *Valuable Data.* The next challenge that needs to be conquered is the data that is necessary for a digital twin to operate in the correct manner. It is an absolute requirement that the data be of a high quality, that there be no background noise, and that they be transmitted in a continuous stream that is free of interruptions. There is a possibility that the digital twin will not perform as intended if the data are unreliable and inconsistent. This is due to the fact that the digital twin will base its decisions on data that is either incomplete or inaccurate. The digital twin may not function as effectively as it was intended to because of this. When it comes to determining the data that can be obtained from the digital twin, the number and quality of signals that are obtained from the Internet of Things are two highly influential factors (IoT). Planning and analysis of device use are required to determine which data should be collected and utilized for the most efficient operation of a digital twin. This is essential in order to select the data that should be gathered and utilized in the analysis that will follow. This is essential in order to determine which data ought to be gathered and utilized in the process of analysis, as it will help determine which data ought to be gathered.

(3) *Privacy and Security.* It is abundantly clear that a problem will emerge in the future regarding the protection of the privacy and safety of users when it comes to the application of digital twins in a commercial setting. This problem will arise in regard to the application of digital twins in a commercial setting. In the not-too-distant future, you will encounter this issue. First, because of the massive amounts of data that they use, and second, because of the risk that this poses to the sensitive data that is stored on the system. Both of these reasons are interrelated. These two facets are intertwined and dependent upon one another. The primary technologies that make digital twins possible, namely data analytics and the Internet of Things, will need to adhere to the most recent practises and revisions in the legislation that governs security and privacy in order to have any chance of successfully overcoming this obstacle [18]. When it comes to the protection of data and the security of digital twins, giving these concerns top priority can help alleviate some of the trust issues that are associated with the use of digital twins.

(4) *Trust.* There are a number of challenges that are associated with trust when viewed from the perspective of both the organization and the user. These challenges can be broken down into several categories. These difficulties can be divided into two categories: internal and external. The technology of a digital twin needs to be addressed further and explained on a foundational level in order to ensure that end-users and organizations are aware of the benefits of a digital twin, which will strive to overcome the barrier of trust. This is essential in order to ensure that end-users as well as organizations are aware of the advantages that a digital twin can provide. This will ensure that both end-users and organizations are aware of the advantages of utilizing a digital twin. The validation of models is an additional strategy that can be used in order to address the issues that are associated with trust. This can be done in order to ensure that the models are accurate. This is something that can be done to ensure that the models are faithful representations of the world as it actually exists. Checking to see if the performance of digital twins is on a par with what was expected of them is one of the most important steps in the process of winning the confidence of end-users. In tandem with the depth of their understanding, people tend to develop a higher level of confidence in digital twins. Because of this, it will be possible to overcome the difficulties associated with trust because the enabling technology will provide additional insight into the processes that ensure privacy and security procedures are adhered to throughout the development process. This will make it possible to overcome the difficulties associated with trust [19]. Because of this, it will be possible to get through all of these challenges.

(5) *Expectations.* Despite the fact that industry giants Siemens and GE are accelerating the deployment of digital twins, it is essential to exercise caution and highlight the problems that currently exist for the expectations of digital

twins, as well as the necessity for further understanding. Because it is necessary to lay rock-solid foundations for Internet of Things infrastructure and develop a deeper comprehension of the data required to carry out analytics, businesses will be compelled to adopt the technology behind digital twins. Before companies can make use of digital twins, both of these requirements must be satisfied. Because of this, the implementation of the technology known as the digital twin will be an absolute given. An additional barrier that needs to be conquered is the misconception that the digital twin should be used solely because the developments that are taking place at the moment are the driving forces behind its use. Because this is an incorrect assumption, there is a challenge that needs to be overcome in order to proceed. It is necessary to investigate both the positives and the negatives associated with the anticipation of digital twins in order to ensure that the appropriate steps are taken when establishing digital twin systems. This will ensure that the appropriate steps are taken. This will guarantee that the proper actions are taken. Making sure that these preparations are made before putting in place digital twin systems will guarantee that the necessary actions are taken. The challenges that are associated with the Industrial Internet of Things/Internet of Things and data analytics are also the challenges that are associated with the implementation of a digital twin. This is not a particularly difficult concept to grasp. The implementation of a digital twin does not present these challenges in a manner that is unique. In spite of the fact that digital twin, the Internet of Things, and data analytics all present challenges from the user's perspective, in addition to the challenges that digital twin brings to the table in terms of privacy and infrastructure, there are also specific challenges associated with the modeling and construction of the digital twin. These difficulties include the following:

(6) *Standardized Modeling*. The modeling of such systems poses the following set of difficulties for the development of any kind of digital twin. There is no one approach to modeling that can be used because there is no one method of modeling that is universally accepted. It is essential to have a standardized strategy from the early stages of design all the way through the simulation of a digital twin, regardless of whether the approach is based on physics or designed. These difficulties can be divided into the following three categories: This is the case regardless of the methodology that is being utilized. This is true regardless of the tactic that is utilized in the game.

(7) *Modeling of the Domain*. An additional challenge that arises as a result of the requirement for standardized application is the task of ensuring that information concerning the application of the domain is conveyed to each of the development and functional stages of the modeling of a digital twin [20]. This is a task that must be completed in order to satisfy the requirement for standardized application. To meet the requirements of the requirement for

standardized application, this is a challenge that must be overcome. This issue, which can be viewed as a challenge, is brought on by the requirement to implement a standardized application. This ensures compatibility with fields such as data analytics and the Internet of Things, which paves the way for the digital twin to be effectively utilized in the future within the context of their respective industries. Moreover, this paves the way for a more efficient use of the digital twin in the present as well. They are necessary for progress because doing so ensures that they will be taken into consideration in the future creation of digital twins as well as when employing IIoT/IoT and data analytics. They are therefore necessary for moving forward. As a consequence of this, having them is necessary in order to advance. As we move forward, it is essential that we keep in mind the significance of these things because doing so ensures that they will be taken into consideration. As we move forward, it is essential that you keep this particular point in mind.

3.7 Technologies That Make It Possible

In the following paragraphs of this article, we are going to discuss the technological advancements that make it possible to create digital twins of real-world objects.

3.7.1 A Summary Explanation of How the Internet of Things has Evolved

Over the past few years, this phenomenon is referred to as the "Internet of Things," which is a term that was made up specifically to describe the phenomenon. The various things that are connected to the internet are referred to collectively as the "Internet of Things." The purpose of using artificial intelligence is to give "things" the ability to think for themselves and gain knowledge about the environments in which they find themselves. This will be accomplished by endowing "things" with these capabilities. When Kevin Ashton was in the process of outlining his vision for the future of the network in the late 1990s, the term "Internet of Things" made its first appearance in print. Kevin Ashton was in the process of outlining his vision for the future of the network. Internet of Things is shortened to "IoT." The idea that all devices that are networked provide the programmer with the ability to follow and monitor everything that we do, which ultimately results in an environment that is more intelligent is referred to as the concept of ubiquitous computing. In this scenario, the programmer has the ability to follow and monitor everything that we do. The Carnegie Mellon University, which is located in Pittsburgh, is one establishment that can provide evidence of this. The period of time in the history of the university that is relevant to our discussion took place quite a few years before the present day. In this made-up scenario, a programme would connect a Coca-Cola vending

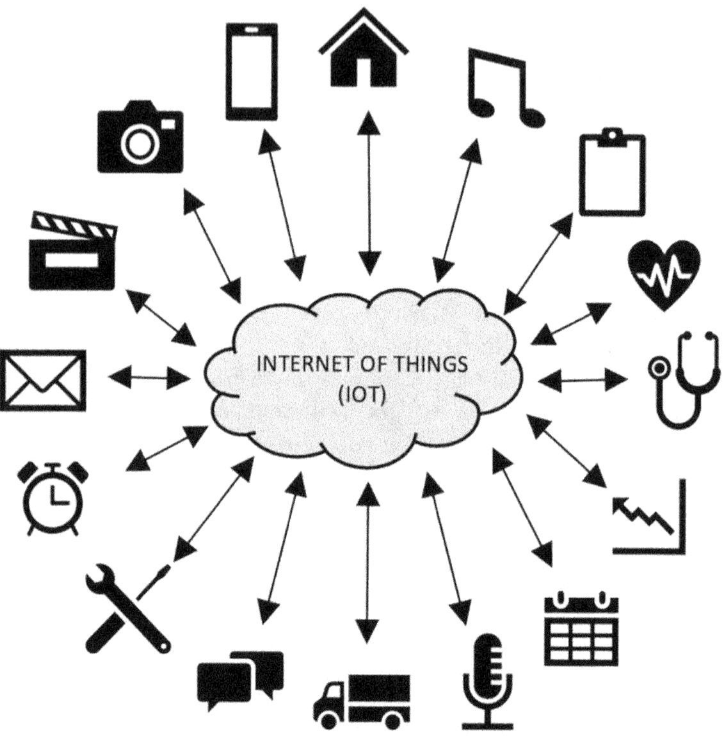

Figure 3.2 Internet of Things diagram.

machine to the internet in order to determine whether or not the beverage had sufficiently cooled down and was in a condition where it could be purchased by a customer and then enjoyed by that customer [21]. This is an illustration of a use case for Ashton's vision, which, despite the fact that it appears to be quite straightforward, actually adds a lot of value to the organization.

The exponential growth that has occurred in terms of this technology is amply demonstrated by the rising number of Internet of Things devices that are being registered on an annual basis. In 2018, the figure was greater than 17 billion. Estimates indicate that by the year 2025, there will be more than 75 billion electronic devices in use, and the value of the sector will have increased to more than $5 trillion by that time. The increase in the total number of devices that were connected to the Internet of Things in 2016 is depicted graphically in Figure 3.2. These numbers are evidence of the significant influence that these devices are having, which contributes further to Ashton's vision, which was presented earlier in this paragraph. Figure 3.2 shows how the Internet of Things can be used to realize the concept of connected services. Every time there is a sizeable increase in the number of devices that are capable of

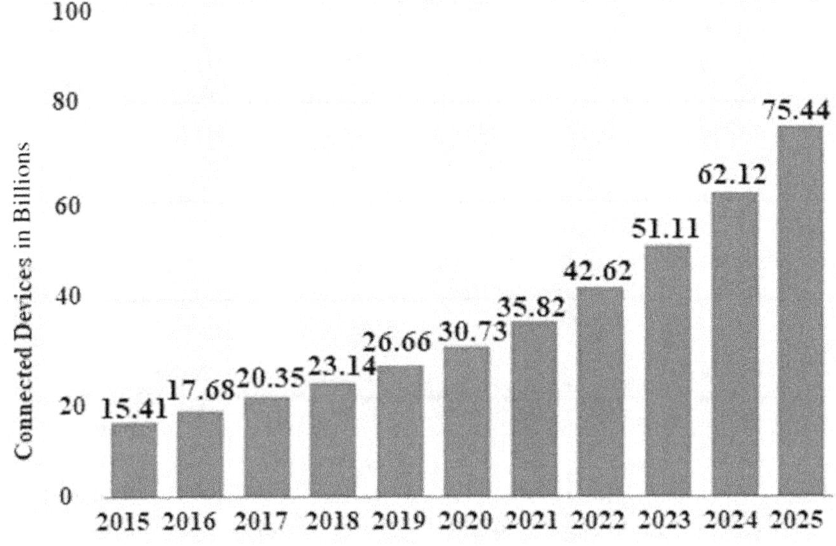

Figure 3.3 IoT device growth.

communicating with one another, the dream of a fully connected world moves one step closer to becoming a reality. The proliferation of Internet of Things devices is advantageous across the board and has an effect on the fundamental aspects of day-to-day life, including the communication industry, healthcare, the building and transport sectors, smart cities, and manufacturing.

3.7.2 An Introduction to the Industrial Internet of Things and its History in Brief

It was Ashton [22] who first proposed the term "Internet of Things," and this is also the location where the concept of the "Industrial Internet of Things" was initially conceived. The phrase "Internet of Things" has been interpreted in a number of different ways by a variety of academics; similarly, the phrase "Industrial Internet of Things" can mean a variety of different things depending on who you ask (IIoT). Figure 3.3 shows how the Internet of Things can be used. This phrase and the Internet of Things share many of the same characteristics, but the primary focus of this phrase is on the operations of industrial systems. Boyes *et al.* define the Industrial Internet of Things in a number of different ways, but they emphasize that the primary purpose of the IIoT is to increase industry productivity. The very first generation of control systems is referred to as "Industrial Control Systems" in the context of manufacturing and other types of industrial settings, where the term

"Industrial Control Systems" is used. This is despite the fact that there is a great deal of variation in the definitions that are offered for the IIoT (ICS). The implementation of IIoT might make it possible to observe the benefits that would result from these systems becoming self-sufficient and intelligent. This would be a potentially exciting development. These strategies have been the subject of extensive research and analysis, and a sizeable number of individuals currently employ them. Both the Internet of Things and the Industrial Internet of Things are deeply intertwined with a category of technology known as cyber-physical systems (CPS), which stands for the combination of digital and physical components as shown in Figure 3.4. Despite the fact that ICS and CPS are somewhat analogous to IIoT, there is absolutely no reason why the two distinct concepts should ever be confused with one another in any way, shape, or form.

The most significant difference is that in order for devices to be connected to the Industrial Internet of Things, they must first be connected to the internet [23]. This is an essential prerequisite for IIoT connectivity. Both the Internet of Things and the Industrial Internet of Things, which is very similar to the IoT, have the potential to have a significant impact on the overall improvement of manufacturing processes. This potential exists for both of these technologies because of their similarities. This is due to the fact that it makes it possible for tasks to be evaluated with increased knowledge and responses in real-time by making use of connected devices. This is a significant improvement over the conventional approaches. This, in turn, leads to improvements in performance, production rate, costs, waste, and a variety of other critical deliverables within an industrial setting. The Industrial Internet of Things not only has an effect on the manufacturing industry, but it also has a bearing on other large-scale processes, such as agriculture, the extraction of oil and gas, and other fields. This is because the Industrial Internet of Things enables more data to be collected and analyzed. In a manner parallel to this, the Internet of Things, and more specifically the Industrial Internet of Things, is having a sizeable influence on the world of business in the present day. This becomes abundantly clear when one takes into consideration the projection that was made by Morgan Stanley, which states that the total value of the market will reach $110 billion by the year 2021. This makes it abundantly clear that the market will reach this value. This emphasizes the point even more strongly. According to the data that can be found in the references, it has been estimated that the IIoT will have the potential to add an additional $14.5 trillion to the economy of the whole world by the year 2030. This figure was derived from an analysis of the potential economic impact of the IIoT.

3.7.3 Enabling Technologies and Building Blocks for the Internet of Things

Both the Internet of Things and the Industrial Internet of Things require different core components to operate connected systems. These parts are needed for connected

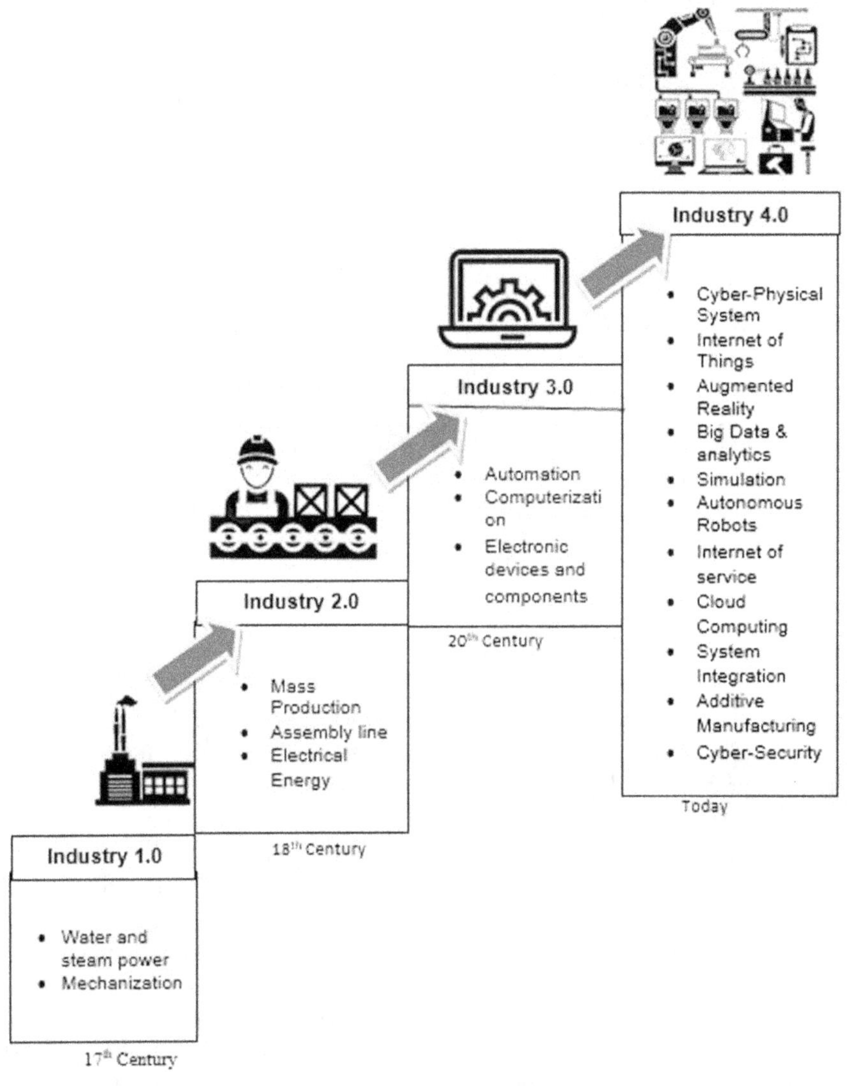

Figure 3.4 Development of Industry 4.0, which is also known as the incorporation of IIoT within the context of the industrial revolution.

systems to work well. Accordingly, these enabling technologies' core functional domains are divided into four divisions. These areas include network connectivity, hardware, and software; data processing, electricity, and energy storage. They individually have particular purposes, but the overall goal is to enable the full creation of an Internet of Things system that supports Industry 4.0 architecture. This purpose

Table 3.1 Enabling Technologies and Functional Blocks: I/IoT

Domain	*Enabling Technology*
D1 Application Domain	Application of I/O technology
	Architecture
	APIs and software
D2 Middleware Domain	The cloud platforms
	Mechanism for processing data
	Data retention
D3 Networking Domain	Protocol for communication
	Internet protocol
	Adoption bridge
D4 Object Domain	Equipment platform
	Objects embedded
	Electrical and mechanical components

drives all of their goals. Table 3.1 shows that the Internet of Things is supported by four types of enabling technology domains [24]: the application, middleware, network, and object domains are represented by D1, D2, D3, and D4 accordingly.

The D1 has three layers. Internet of Things applications make up the application layer, which is the topmost layer (IoT). There are intelligent farms, residences, even cities. The architecture layer may incorporate SOA or REST software architectures (representational state transfer). These two illustrate the functioning of the architectural layer. The third layer, which comprises of software and APIs, connects the application domain to the middleware domain (APIs). It maintains software and operating systems. Android and other specialized operating systems manage Internet of Things systems and devices. This may include application programming interfaces (APIs) designed expressly for constructing an Internet of Things system; both of these technologies are necessary to close the gap between D1 and D2. The middleware domain has added three levels. The first option is the cloud platform, which is made up of numerous services that work together to form the cloud and deliver computer resources when needed. Microsoft, Amazon, and Google hold the three greatest market shares for cloud service providers.

Data mining and BigQuery, Apache, and Storm illustration services enable the second tier, data processing. The data storage layer is the third and final enabling layer in the D2 architecture. MongoDB is an excellent example of a database that offers powerful storage engines, which is a crucial component of any Internet of Things design. The networking domain, sometimes known as D3 in some areas, is the third component of the Internet of Things architecture. There are three layers of facilitation. The uppermost layer, the communication protocol layer, enables flawless communication between all system components, including applications, transport, and networks. It consists of the network protocol, the transport protocol, and the application protocol. The network interface is the second layer in the stack that enables functionality. Again, the goal is to make IoT integration seamless. This facility houses the most essential Internet of Things standards, such as RFID. Adoption mechanisms make up the networking domain, the third and final tier of D3 in the domain hierarchy, consisting of the gateway layer, connection interface, and adoption layer, all of which are essential enabling technological standards for the development of an Internet of Things system. Both 6TiSCH and IEEE 1095 offer more reliable wireless connection. The D4 block concludes the Internet of Things. The object domain consists of three enabling layers. The first group includes all hardware systems (like Raspberry Pi or Arduino, for instance).

Sensors, radio tags, displays, and firmware are a few examples of the layers that come before this one in this domain's hierarchy. To connect the system, you will need all of these components. The last layer consists of the device's mechanical and electrical components, such as its batteries and any essential processing units. These elements make up the layer. This Internet of Things system has 12 distinct enabling sections, but by separating the domain into four functional blocks, each portion is easier to understand, creating an integrated and interconnected structure.

3.7.4 Summary of Data Analytics' History

The phrase "data analytics" acts as a catch-all for a variety of analytical principles utilized in academic writing and throughout the paper. These ideas are covered throughout the text. As a result, prior publications must be understood and examined. Data analytics comes from "data science," which collects and analyzes data to better understand our world. This field covers many topics. More information on a study of the data analytics market is provided below. Analyzing previous articles will assist evaluate how this research fits in by identifying and expanding on these issue areas [26]. This will assist in identifying how this research fits in with other studies.

(1) DATA. Access to raw data is required for data analysis. There are a few procedures to do before these data can be employed in algorithms and statistical analysis. These steps must be accomplished. These are the prerequisites for data collection, processing, and cleansing. The requirements explain the data's essential needs and how it is utilized to guarantee that particular criteria are stated while taking into account the intended use of the data. In the second stage, the place and method for data collecting are determined. This is done to fulfil the need for data acquisition. After collecting the data, they will be processed to meet specific requirements. The last and most crucial stage is data cleaning. Even if the data have been obtained and organized,

there may be substantial inaccuracies or missing data. Imputation processes, which have been problematic in data analytics, are applied here. These strategies ensure data completeness.

(2) STATICS. Data collection, classification, analysis, and conclusion-drawing are statistics. In a nutshell, the topic is vital to data analysis since statistical models are the cornerstone for machine learning techniques. Data analytics uses descriptive statistics and statistical inference to characterise data-based observations. AI and the issues outlined below [44, 45] highlight the growth of advanced data analytics.

(3) ARTIFICIAL INTELLIGENCE. AI is the most popular data analytics topic. The discipline of cognitive robotics got its start in the late 1950s, when the term "artificial intelligence" was first used. These are arranged into categories that may be crucial to our success [27].

(4) INTELLIGENT MACHINES THAT CAN LEARN. Machine learning methods enable computers to learn and perform tasks for users without being explicitly programmed. Machine learning allows computer programs to independently collect and analyze data. For a closer look, machine learning can be divided into two types:

A: ASSISTANT-DIRECTED LEARNING. For self-analysis and improvement, algorithms rely on tagged data. To precisely identify an activity, the programme analyzes labeled data. One example is photo classification [28]. The algorithms are trained with training data, then tested with test data to see how well they can predict image content. A percentage of accuracy is used to indicate test results. After analyzing these responses, the user will correct and re-learn any faults. This improves model training and algorithm accuracy.

B: UNSUPERVISED LEARNING. Machine learning also includes unsupervised learning. Unsupervised learning does not need expensively marked-up data because the predicted result for each input pattern has been set. Unlike supervised learning, which needs data.

Unsupervised learning systems recognize and emphasize data patterns without human involvement. Unsupervised algorithms learn faster than supervised ones. Clustering can classify data. Learning algorithms may categorize unlabeled data sets, revealing previously unseen patterns [29]. Data analytics is a multidisciplinary field that includes deep learning and machine learning. Complex neural networks that can autonomously extract input feature information are utilized by deep learning algorithms to learn from unstructured and unlabeled data. These networks use machine learning to construct deep learning models. These models take longer to train since their neural networks are larger than those employed in other models, but they are more accurate. Semi-supervised learning includes labeled and unlabeled data.

Semi-supervised learning uses labeled and unlabeled data to discover how algorithms can improve. Despite the fact that there are many more algorithms accessible, these are the ones that are utilised the most commonly in data science. DATA VISUALIZATION. Data visualization is the last topic under data analytics. Data type affects data visualization. Graphs and charts can display multidimensional data, the most frequent type. Bar charts and pie charts are examples. Geospatial data can be graphically represented in distribution maps, cluster maps, and contour maps. This data is collected straight from the ground using position data.

3.7.5 Data Analytics Enablement Technologies and Functional Blocks

Next, we will talk about data analytics as it relates to AI, ML, and DL. The framework of analytical data is laid forth, together with the tools that support it [30]. While the underlying technology share many similarities with IoT, the visualization and analytics techniques add an extra layer of complexity. Shown below is a synopsis, broken down into sections marked D5, D6, D7, and D8.

The object domain is represented by the symbol "D5," the middleware layer by "D6," the network layer by "D7," and the application layer by "D8" (D8). Each chapter lists analytical tools. Domain D5 is an item storage space with many tiers (at least three). Data collection is the first enabler since it prepares data for analytical solutions. Data sensing instruments enable data collecting. Digital signal processors ensure data collection reliability. Data repositories, which simplify database access and administration, make up the second level of the structure. The last layer of D5, which is connected to D6, has storage facilities. These facilities allow for huge data storage employing server storage on demand. This layer interfaces with the middleware domain, facilitating storage data processing. The middleware layer, or D6, has three layers that work together to make software work. The data representation code D5 is associated to archive processing. The second layer of the D6 model is dedicated to data processing, which supports data analytics, cloud services, and key middleware designs (such as software and database systems).

According to Table 3.2, the third layer is made up of D6 analysis and algorithms. This layer simplifies data mining, machine learning, statistics, and querying. Both supervised and unsupervised learning are useful in data analytics. Domain 7 (D7) shows how wireless and communication protocols can efficiently collect and handle higher-level data. Domain 7 is concerned about privacy and security standards. Domain 7 (D7) focuses on networking. In the eighth and final section, "Application," D8 uses a two-tiered technology stack to achieve its goals. The screen and hardware are the main layers. We can now record operational technology data and analyze it with machine learning, deep learning, or statistics. Because the layer is visual, it is easy to display data relevant to the user's current activities. On the application layer,

Table 3.2 Enabling Technologies and Functional Blocks: I/IoT

Domain	*Enabling Technology*
D5 Application Domain	Collecting information and undergoing preprocessing
	Storage facilities for information
D6 Middleware Domain	Facilities for storage
	Computing with data
	Methods of analysis and algorithms
D7 Networking Domain	Technology for wireless networks
D8 Application Domain	Data visualization and hardware
	Software for analyzing data

you will discover data analytics applications like autonomous cars, picture recognition, and virtual assistants like Amazon's Alexa [30].

Technologies that enable the creation of digital duplicates Domain 9 refers to applications, Domain 10 to middleware, Domain 11 to networking, Domain 12 to objects, and Domain 13 to objects in this additional synthesis of concepts for the functional domains and supporting technologies of a digital twin. Domain 14 discusses objects. The application domain, indicated by D9, is divided into three interrelated layers. Model architecture and visualization are first. This layer is crucial to developing realistic representations of the physical entity. This layer allows us to study digital twins and model their underlying structures. Because of this, digital twins will include not just physical entity behaviors but also other data in their models. Twin Builder and Simulink enable it [31]. The second layer of the digital twin architecture, commonly referred to as the software layer, consists of software and application programming interfaces (APIs). This makes room for the third layer, which includes data preparation and collecting. This final layer of the application domain is crucial for data collecting programmes like Predix, Mindsphere, and Storm. This layer bridges domains D9 and D10 to collect accurate data for the digital twin's IoT and analytics implementation. The middleware domain, also called D10, has two support levels. The first change stems from data storage developments. Data for digital twins can be stored in Mongo DB, MySQL services, and on-demand databases. Without the data processing layer, it is impossible to transport the data from D10 and D11. D11, or Domain 11, is the network domain and the two enabling levels. The first layer is called communication technology, and its job is to make data transfer as easy as possible. A layer that handles wireless connection makes up the second functional block that makes up the network domain of a digital twin. It communicates with the D12 domain and ensures wireless data transmission using the right protocol. There are two supporting levels in the object domain, often known as D12. The first two system components are sensor technology and a hardware platform. Both are required to ensure that the proper equipment is in place for conducting an analysis using a digital twin, which needs the use of sensors to collect data, and to facilitate the collection of data using sensors, respectively [31].

In this section, we will examine a range of publications to identify business opportunities in digital twins, the Internet of Things/Industrial Internet of Things, and data analytics. Our mission is to find these opportunities. Other researchers will benefit from this academic perspective because it will highlight areas that need further exploration and describe a digital twin more fully.

3.7.6 Analyzing the Data Using a Hierarchical Structure: Type A Categorical Review

In the first part of this section, a categorical evaluation of the relevant literature is undertaken to create a categorical review table of the selected articles. This review follows Kritzinger *et al.* [9] protocols. The three stages of digital twin integration described in Part II, subsection B of this study and Kritzinger *et al.* [9] provide the framework for the key concerns mentioned in the review. These descriptions can be found in both sources. This analysis uses Kritzinger *et al.*'s [9] classification of 43 digital twin articles published between 2001 and 2017. Using academic article finders like Google Scholar, the aforementioned publications were collected. The authors completed their research for this study using Google Scholar, and they only included results from ACM, IEEE, and Science Direct repositories. There are 177 papers to read between 2015 and now (December 31, 2019), but only 42 before 2017. Prominent search phrases like "digital twin" and "twin" were competitive (digital-twin, digital twins). The search included "digital twin" and other terms related to the scientific fields (industrial digital twin, healthcare digital twin, smart cities digital twin). It was compiled from 26 industry, healthcare, and smart city design sources.

A. RESTRUCTURING AND ANALYZING THE CATEGORY STRUCTURE. The following two rows define the user-accessible integration levels. The first step is to define terminology like "digital model," "digital shadow," and "digital twin" in the original work.

(1) The presence of a true identical twin (fourth etymology). In the next section, you will be asked to identify whether the digital duplicate, twin, or shadow is an accurate representation of the original.

(2) BROAD AREA. Manufacturing, healthcare, and smart cities are the current research priorities. The fifth column of this table lists these core research domains.

(3) SPECIFIC AREA. The sixth column illustrates how the conversation moved from broad to specific. The topic has become pretty specific. In the manufacturing industry, this would focus on smart factories or problem identification. In smart cities, the answer is either traffic or infrastructure.

(4) TECHNOLOGY. The fourth and final column will discuss the technologies used. These technologies include simulation, data analytics, and the Internet of Things [32].

B. ANALYSIS OF DOCUMENTS REVIEWED AFTER RECEIVED

A more in-depth analysis of category review papers will follow, bringing principles and cases to light. These parts' information came from category review articles. Although this is a constraint of the sections themselves, these sections will cover the most essential problems raised in the papers, which relate to healthcare, smart cities, and manufacturing. The order of the topics is a reflection of the scope of the most recent study, which was determined by the number of articles identified, and this order was chosen to best portray the information. The order of topics reflects the most recent study. We found few healthcare-related papers, however there is a lot of evidence that digital twins could change the healthcare industry. Following closely behind is the smart city region, which found a few items. Working labs and industries conduct most research. The percentage of papers categorized by subfield. Healthcare, manufacturing, and smart cities are subfields. Here are samples of relevant papers for each of these subfields of inquiry.

(1) HEALTHCARE. The idea that He [32] described digital twins as "digital replications" of a real entity is one that may be gleaned from the many definitions of what they are. Digital twin definitions suggest this. Consider this definition. Continue reading since this is a crucial point to remember. El Saddik reconsidered this idea by include digital twin replications of non-living elements in his presentation. He included digital twin replications to his presentation. The digital twin's potential for usage in the medical sector makes it clear that its uses are not limited to the business sector. One example is that it could improve patient care. Ross presents his work with Hewlett-Packard to create digital twin avatars of persons using AI and IoT. These methods are used throughout the lecture. When integrated with AI algorithms, digital twin technology may show the health implications of lifestyle changes and recommend improvements. This device can also show the effects of lifestyle modifications on a person without leaving home. This technology can also be used to determine a person's lifestyle changes' consequences. This application integrates data from the digital twin and the physical twin, also known as the human twin (The Replica) enabling people to see how their behaviors affect their physical twin and how lifestyle changes can affect them [41]. Laaki *et al.* [33] recently demonstrated a working autonomous surgical prototype in a different setting. This demonstration occurred elsewhere. This prototype creates a patient's digital twin using IoT and Industry 4.0 connectivity. This surgery took place in a different setting. The authors offer a mobile app for remote surgery.

The prototype can perform accurate surgery because they used a robotic arm, virtual reality (VR), and a 4G environment. The article discusses diverse research issues and why simulation was utilized instead of a physical prototype. Laaki *et al.* examine the challenges of combining the physical prototype with its digital analogue [33]. The authors examine contemporary AI and Industry 4.0 advancements. They emphasize how these innovations make connectedness, integration, and transdisciplinary research easier. Liu *et al.* [38] propose a new concept for the future of healthcare by merging cloud technology and digital twins to monitor, diagnose, and forecast patient health. This system was built to monitor, diagnose, and forecast patient health. This system was created to monitor, diagnose, and forecast patient health. This framework was created as an aid. We may implement this method soon. Liu *et al.* designed and applied the Internet of Things wearable technology and in-home sensors for senior adults. Hence, they attain their purpose. The authors propose a framework and demonstrate several applications, discussing each one's usefulness. Liu and his colleagues' [38] approach improved patient prediction. This is made possible by combining Internet of Things, cloud, and digital twins.

(2) INTELLIGENT CITIES. The most recent findings in digital twin and smart city research are the key topics of discussion in this field. Urbanization, the Internet of Things, and data analytics have all grown in recent years, according to study [3]. Mohammadi and Taylor [3] list the stages of spatio-temporal flux as one of their study's goals, emphasizing the need to understand them to maintain growing. Mohammadi and Taylor [3] also include this as a study goal. This is one of their goals for their work. This proposal uses a digital twin and virtual reality headgear to observe oscillations and forecast [3]. Ruohomaki *et al.* [13] present a conceptual framework called "mySMARTlife" to help build smart cities (IoT). The digital twin case study provides urban planning and built environment information. The digital twin may be used to monitor and compare energy usage based on the environment and human impact, hence it has specific applications in the field of energy consumption. Because the digital twin may be used to track and compare energy use, these applications are possible. This feature helps manage energy consumption. Both are used in planning for the future and current actions [13]. Mohammadi and Taylor [3] and Ruohomaki *et al.* state that Industry 4.0 concepts must be adopted to ensure that the twin receives adequate data to accurately perform its functions. This ensures that the twin receives enough information to perform its functions. This ensures that the volume of data flow is sufficient for the twin to accurately perform its functions. Supplying energy grid fuel while integrating alternative and renewable energy sources is difficult. As a result, smart cities will demand accurate delivery. Transport, monitoring, and assessment must be done with

renewable energy sources like wind power before they can be used. In their study [19], Pargmann and colleagues describe a digital twin monitoring method for wind farm development and monitoring. This system develops and monitors wind farms. This system is in the cloud. The authors propose a prototype that can operate and uses technology and business data streams and settings. This makes it possible to build a wind farm twin [19]. Sivalingam and his colleagues research and create a case study on how wind farms are used and how much power is utilized by a smart grid. This article discusses energy consumption dependability and wind turbine maintenance. These problems can occur for a variety of causes. The authors offer a system that can accurately perform wind turbine maintenance and predict its cost [18]. This strategy uses Internet of Things sensors and data analytics. Digital twins host this. Jo *et al.* [16] and Chen *et al.* [23] use digital twins to study smart cities. These articles explore the research's findings in respect to smart cities. A written report on digital twin applications for smart farms is the initial step. This will be the first item. Jo and her coworkers presented it [16], and the document is a presentation of a presentation. [Note:] Industry 4.0 is important for this project since it challenges digital twin implementation in tough contexts. Industry 4.0 challenges digital twin implementation in tough circumstances. This is why Industry 4.0 is so important. IBM's Watson, Eclipse's Ditto, and GE's Predix are three possible digital twin applications. GE develops Predix, and Eclipse develops Ditto and Watson. They also explain how a digital twin might improve cattle monitoring [16]. Chen *et al.* [23] explore digital twins in cars and traffic management. These uses are considered in relation to digital twins. The essay discusses driving challenges and the need for higher data flow within the car and connectivity to nearby cars. Chen *et al.* [23] use a digital twin and learning algorithms to analyze user behavior-based input. This strategy was created. They present this process to the reader as part of their research [23]. The algorithms produce a real-time digital twin of a driver's behavior, alerting them to potential threats and advising them on how to drive more safely. The algorithms construct a digital twin of a driver's behavior in real-time. In other words, the algorithms simulate a driver's behaviors to generate a digital twin of their behavior in real-time. Jo *et al.* and Chen *et al.* believe data exchange is necessary despite its challenges. Both groups of academics emphasize the importance of strengthening connectivity as a prerequisite for reaching the highest level of precision and precision-related results for each digital twin. This is because enhancing connection is necessary for the maximum precision.

(3) MANUFACTURING. This section will focus on digital twin research. This field will be covered in greater detail in the future section. It also includes a wide range of subfields, from massive intelligent factories to specialised equipment and instruments. This makes it a very broad pitch. This variability

is reflected in the industry's size. This section of the article has been divided into the following subsections in light of the information supplied above: (a) system design and development; (b) simulation and artificial intelligence; (c) energy efficient manufacturing; and (d) smart manufacturing reviews.

3.7.7 Evaluations of Intelligent Factoring Technology Reviews

a: EVALUATIONS OF INTELLIGENT FACTORING TECHNOLOGY. It provides a "360" view of digital twins for big data applications in industrial and manufacturing contexts. It compares digital twin technologies and emphasizes the importance of developing technology for smart manufacturing [25]. It also covers digital twin technology. It also argues for the importance of digital twins in today's environment.

b: SIMULATION AND ARTIFICIAL INTELLIGENCE. In previous papers, we explored a number of important elements of using digital twins with simulation and AI-based manufacturing processes. Kuehn argues for and analyzes a hypothesis that uses "virtual clones" and machine learning techniques to boost industrial efficiency. This theory supports it. In order to highlight the many different goals and ideas that can be achieved by introducing a digital twin into a manufacturing process, the author organizes important components of a manufacturing process into categories. This helps explain how digital twins are used in manufacturing. As a result of this advancement, companies may simulate, model, and improve their production processes digitally. This guarantees improved product quality and production efficiency. Min *et al.* [36] comprehensively identified the main enabling technologies, like Kuehn. The authors focus on AI-based digital twin solutions, particularly machine learning, and analyze their capabilities. The authors of this study [36] use an industrial IoT petrochemical plant to evaluate the pros and cons of using machine learning and digital twins in the petrochemical sector. This helps them to weigh the merits and downsides of utilising these two technologies in the petrochemical business. Case studies are utilized to complete the evaluation. The digital twin and algorithms are facility-specific, therefore they can only be used in petrochemical plants, which was discovered as the case study was being conducted to be a negative. This means they do not apply to other production plants. Transfer learning lets users build solutions for a variety of industrial processes by finding commonality in several algorithms. Transfer learning helps find these parallels. Manufacturing's defect detection topic has two fascinating articles. Jain *et al.* use a digital twin to simulate distributed solar system failure diagnosis (PV). As a result of recent developments in the field of digital twins, the crew was able to construct a digital twin that can precisely analyze flaws in PV energy units in real time [16]. This was possible because they created a digital twin that could do the job. An AI-based approach for problem diagnostics in a smart manufacturing setting is provided by Xu *et al.* [24]. Xu *et al.* note that training data is scarce for developing accurate AI algorithms for novel industrial processes. These issues are highlighted for context. This is needed to create lifelike digital twins. In order to learn, identify issues, and generate training data simultaneously, a digital twin is used in the model's front running. This is done to make the model more efficient. This is done to lessen the chance and severity of any potential complications. The next step of the algorithm will take the training data from the previous step to perform transfer learning. Digital twins and transfer learning can make a more accurate flaw diagnostic system [24]. The authors provide a practical case study in an automotive production context to test and assess the idea. The case study takes place in an automotive production environment. This study examines how useful the idea is in different circumstances. Point c will discuss this further.

c: SYSTEM DESIGN AND DEVELOPMENT. The design and development of industrial processes and systems is another significant and necessary area where digital twins can be used. One of the most well-known and significant domains in the world is this one. Shangguan and colleagues deploy digital twins and introduce cyber-physical systems (CPS) in their paper [18]. This method benefits from both trends. The authors introduce a hierarchical digital twin framework (HDTM) for the design and development of dynamic CPSs in a smart manufacturing environment. This framework can be used to create and develop dynamic CPSs in a smart manufacturing setting. This HDTM can be used in smart manufacturing to develop and build dynamic CPSs. The digital twin is used throughout the CPS design process to reuse and test actual data on a virtual twin. In the past, this was impossible. A case study shows the benefits of a digital twin. This shows the benefits of a digital twin. Owing to the fact that it allows the authors to use the digital twin in a real-time predictive design scenario and train for large-scale system adjustments, these benefits are likely to be observed more clearly in this context than in a similar situation where it was not possible. Their virtues are highlighted. Framework software is crucial to industrial robot design. Howard [12] offers a study that examines the digital twin trend, its uses, and its suitability for smart manufacturing. The purpose of this study was to examine the digital twin trend. A digital twin was used for "virtual validation" [12]. This was done quite well. This was the main goal, and it was accomplished. The authors of each of these papers address challenges raised by digital twin, analytics, and smart manufacturing thought leaders and visionaries. In addition to providing high-level system design, Karadeniz *et al.* and Chhetri *et al.* present contextual papers and case studies on the development of digital twin systems for manufacturing processes. The last post used a case study to examine how Industry 4.0 and the

Internet of Things are helping AR and VR trends. This study examines how augmented and virtual reality trends are benefiting from these improvements (virtual reality). The authors discuss a concept that, similar to the Internet of Things, creates a "digital twin" of various culinary products (equipment and processes related to food and cooking). They refer to this idea as "eGastronomic stuff." Gastronomical processes need IoT-connected physical sensors to collect data for a digital twin. The authors explain how a digital twin can assist in monitoring and maintaining the operation of "eGastronomic" processes, specifically an ice cream maker, by using an ice cream machine as a case study. The authors discuss how a digital twin can monitor and sustain eGastronomic processes. The authors utilize the operation of an ice cream maker as a case study to show how valuable a digital twin may be. The ice cream machine is utilized as a case study by writers. According to a more recent study [34] by Chhetri and colleagues, the emergence of smart industrial design and development can be ascribed to digitalization [34].

The authors offer a strategy and conduct a case study to create a digital twin of a manufacturing process using the Internet of Things. This is done by leveraging the Internet of Things. This approach collects and stores data streams using Internet of Things sensors. These data streams are then used to infer side channel states like acoustics, power, and magnetic output of a process. These states can be deduced from data streams. All of these are examples of outcomes from procedures. They can locate a problem and analyze production concerns. A case study of a fused-deposition modeling (FDM) system validates the team's work. This is the first system of its kind to do anomaly detection using this format with such high accuracy. In their work, Mandolla *et al.* [2] mention blockchain technology as another trend in smart manufacturing. They do so in the context of the study. Data can be recorded on multiple workstations at once using a growing list of blocks kept in a decentralized ledger. Distributed ledgers enable this. A cryptographic hash links blocks. This links preceding blocks. These hashes provide timestamps and information about the block's transactions. Each block has a unique hash. Blockchain technology's rapid growth is tied to its widespread use in applications like bitcoin exchanges [5] Blockchain was initially used to record digital currency transactions. Mandolla *et al.* [2] demonstrate a blockchain use with a digital twin. The study shows how the authors linked the two. This is done inside the case study framework. This work focuses on metal additive manufacturing of aircraft components. Mandolla and his team create a digital duplicate of the technique as part of their inquiry. They also address conceptually how blockchain technology and a digital twin can safeguard and oversee the process. [2]. Bilberg and Malik [1] are investigating how robotics might be adjusted for production. This study examines product manufacture. They describe an event-driven digital twin that works with a robot on an assembly line. Together, the two machines do their jobs. This digital copy was a cooperative effort. The other way combines the physical and virtual tasks to produce a real-time, skills-based robotic production line that accurately distributes tasks to humans and robots based on their optimal productivity levels. In order to construct a real-time, skills-based robotic production line, this method blends the physical task with the virtual task [1]. This technique creates a real-time, skills-based robotic assembly line that can produce more. Both articles note the difficulties of seamless integration, which is essential for assembly lines and digital twins. Both papers offer case studies to highlight the benefits of digitalization [1, 2].

d: ENERGY-EFFICIENT MANUFACTURING. It is expected that the manufacturing sector will monitor and reduce its energy use. Due to this assumption, innovative energy-efficient production line solutions are urgently needed. Lu *et al.* [35] and Mawson and Hughes [28] present energy-efficient manufacturing systems and architecture. This is because it benefits the local ecology. Yet, a more efficient production process would cut costs, increase earnings, and enable future expenditures. Lu and his colleagues' work on an energy-aware digital twin model architecture using MCLoud was presented to the audience. The public was shown this art. These factors make it easier to build an Industry 4.0 environment where CPSs and digital twins may automatically self-configure and monitor manufacturing techniques. In this case, saving energy through efficiency measures is very crucial. Mawson and Hughes [28] evaluate Industry 4.0's effects on manufacturing process automation, connectivity, and adaptability. This assessment was made possible by Mawson and Hughes [28] using Industry 4.0. This research provides a detailed overview and a case study using methods and frameworks from machine process energy consumption research. Both of these contributions were made feasible by this research. This research enabled both of these contributions. Mawson and his colleagues mention several simulation tools, but none of them have multi-level integration. To develop an accurate model, start with material flows and go on to resource flows throughout the manufacturing process. This is needed to model reality accurately. These challenges show how digital twins, virtual reality, and augmented reality will simplify research in the future.

3.8 Open Research

The penultimate section of this review consists of a brief discussion of open research issues for digital twins and an analysis of the previously examined literature for obstacles that will be addressed by future study. Both of these sections are presented below. In addition to that, the next and last part of the review will be introduced after this section.

3.8.1 An Explanation of the Function Played by the Digital Partner in the Manufacturing Process

The concept of digital twins that are completely included into the production process has been discussed in the corpus of scholarly work; nevertheless, the commercial world has not yet put this concept into action. In the part that came before this one, which was labeled "Section 3.7," we discussed and looked into a variety of publications that are connected to the manufacturing industry. This section is this one's predecessor. The publications in question contained things like reviews and case studies in addition to fundamental ideas. The more in-depth advancements of something that is known as a "digital twin" have been the topic of discussion in a great number of articles and books that have been published. We went over several instances of case studies that were conducted within the industry. In Section 3.7.6, subsection B, we went over even more specific information regarding those examples. A summary of the conversations that took place is presented in the following paragraphs. There is a body of work that has been done in the field of study, but not all of the publications combine all of the elements that make up a digital twin. Not only does the modeling of a digital twin include the physical and virtual features, but it also includes the modeling of the data, connectivity, and service components. Modeling and scaling are two processes that are required in order to generate generic digital twins, and both of these processes are absolutely important. Examples of sensor-based modeling of various stages of a digital twin can be found in the works of Zheng *et al.* [8], Schleich *et al.* [35], and Shangguan *et al.* [18]. Instead of concentrating solely on the hierarchical modelling of a digital twin's virtual and physical modeling, there is an immediate need to establish a generic model for a complete digital twin. This is because hierarchical modeling is only one aspect of a digital twin. Hierarchical modeling is just one component of what is known as a "digital twin," so this is why. When it comes to the application of data fusion and digital twins in an industrial context, there has been less research undertaken on the topic overall. However, the literature addressing data fusion is another subject that has undergone substantial investigation throughout many disciplines of science. This is the case despite the fact that the body of literature pertaining to data fusion is another subject that has been the focus of considerable investigation. Along the same lines, the utilization of digital twin modeling allows for the potential implementation of data fusion as an additional method that can be used in the creation of generic models. In the research that has been done, the possibility of predictive maintenance combining data fusion with digital twins has been investigated [7]. This is an area of research that has the potential to result in significant financial gain. This area of investigation has the potential to be very fruitful in the long run. The modeling of digital twins, which combine virtual and physical data, is currently deficient in research. The phrase "cyber-physical" fusion, which is strongly tied to this phenomenon and can also be used to describe the same thing, is also closely connected to this phenomenon.

In order to mimic data fusion by utilizing digital twins, standardized methods are required; unfortunately, there are currently no approaches that fulfil this requirement. Reference [7] addresses various applications of data fusion that can be used for digital twins in the manufacturing industry. These applications can be used to create more accurate simulations of physical products. In addition to this, they present a number of the possible challenges that may appear following the implementation of data fusion. Concerns about the connection, as well as possible dangers to the system's safety, are two examples of these potential obstacles to overcome. Because it can be applied to the health and manufacturing processes of plants ranging in size from small to large [24], and because it is the result of years of research into the maintenance of machines, the term "prognostics and health management" (PHM) is primarily used in an industrial setting [24]. As a result of the development of digital twins, both issue diagnosis and predictive maintenance for industrial processes have become more viable [10]. PHM provides a platform on which research can be conducted in these two subfields. By the application of the technology of digital twins, PHM is being advocated as a potential avenue for research in the aforementioned fields. The digital twin not only makes it possible for machines to potentially communicate and work together, but it also grants the ability to model processes, which makes it possible to get that much closer to the goal of more precise manufacturing. In other words, the digital twin makes it possible for machines to potentially communicate and work together. In the research that was just reviewed above, seven different authors give data from a variety of case studies on the application of digital twins. These case studies cover a range of diverse topics. Just now, an analysis of this research was carried out. The question of how digital twins may be utilized to boost the efficiency of industrial processes is investigated in all three publications and case studies. This is done in order to offer an overview of [1], [36], and [2]. In [1], the author describes how the utilization of digital twins can be applied to increase the efficiency of various industrial operations. The previously stated reference [1], explores the advantages that can be gained by implementing a digital twin into an assembly line. In article [36], which Min and his colleagues wrote, they investigate the ways in which the application of digital twins might make the production process more effective in the petrochemical industry. Specifically, they focus on the ways in which digital twins could improve the efficiency of the process of converting raw materials into finished products. The work that was presented and discussed can be found in the reference that was cited [2], and it focuses on the ways in which digital twins and blockchain technology can be utilized to support and improve the aircraft manufacturing company. The first three publications are case studies that investigate optimization, whilst the remaining four publications [28] are focused with the modeling of digital twins of various sorts of systems. These papers were discussed in the section that came before this one. Both [28] and [18] investigate the use of hierarchical modeling frameworks for the development of digital twins; [28] uses a case study to investigate how to develop digital twin models utilizing a multi-leveled framework, and [18] looks into the use of a hierarchical modeling framework for the development of digital twins. Reference [28] uses a multi-leveled framework to investigate how to develop digital twin models; [18] uses a multi-leveled framework to investigate how to develop digital twin models. In order to assess how Internet of Things sensors can be utilized to model and maintain a digital twin, a case study is analyzed here. The application of a hierarchical modeling framework to the creation of digital twins is the topic of discussion in this research study. Karandiz *et al.* give a case study on the 3D modeling of an ice cream maker in their most recent paper [37] in order to evaluate the potential uses of a digital twin. The purpose of the study is to model the machine in three dimensions.

3.8.2 Duality of the Digital in Healthcare

In the lines that follow, we will discuss the research questions that have not yet been answered regarding digital twins in the context of the medical field. During a few of the tests that the researchers have carried out, there has been considerable discussion over the possibility of applying the digital twin technology to individuals. One example of this would be a digital twin of a person that was able to keep track of their day-to-day well-being as well as their health. This would pave the way for the development of the concept of a human twin that would be able to replicate the impact that various lifestyle choices might have, both favorably and negatively, on the actual person. This would be a huge step forward in the field of regenerative medicine. The significant open research takes the form of modeling and removing the obstacles that stand in the way of correctly modeling a human body as a digital twin. You should be able to locate the findings of this study on the website. This difficulty can be summed up in a single sentence as the fact that there are no standardized modeling methodologies for digital twins [41]. PHM is utilized in the manufacturing industry in a manner that is analogous to the way that the digital twin is utilized to monitor and control the health of individuals in a manner that is analogous to the way that PHM is utilized in the manufacturing industry. This idea is a fascinating area of research. It is conceivable to put the digital twin to use in a manner that is comparable to that of PHM by combining data analytics with the purpose of ensuring that patients are in excellent health. This would allow the digital twin to be used in a manner that would be comparable to that of PHM. This could include day-to-day medical treatment in addition to ongoing worry about one's health. The findings of the research that are presented in reference [33] are freely accessible to the general public and can be obtained by anyone interested in doing so. It combines data from the past with data received in real time at the present moment in order to create digital twin simulations of surgical procedures and healthcare in general. This is done for the goal of developing digital twin simulations. Using the patient's virtual twin, the objective is to identify potential dangers in advance of their becoming manifest in the patient's condition so that appropriate action can be taken. The field of data fusion is one of the areas of research that is being done. There is a need for research in this area for a number of reasons, one of which is that [33] and [38] note the demand for research in effectively dealing with data that has been acquired and processed for a digital twin. This is primarily owing to the fact that it deals with sensitive patient data via the virtual and physical digital twin, and that appropriate interaction and convergence are required for greater confidence and use; as a consequence, there is a need for additional research in this field. An additional fascinating area of investigation is the study of medical and surgical treatments that can be carried out at a distance. As a result of the fact that digital twins make it possible for doctors to do pre-operative tests from a remote place, there is reason to be hopeful about the possibility of reducing the danger of sustaining an accident or losing one's life as a direct result of undergoing surgical procedures. Another recommendation that was made by Laaki and colleagues [33] was for there to be open research done on network-supported remote surgery. This idea coincides with the advancements that 5G would make possible for mobile networked surgery, which is another field that digital twins will investigate in the near future in the context of the healthcare sector [33, 41]. The recent developments in data fusion, modeling, remote surgery, and the application of digital twins in healthcare have made it far less difficult to carry out research in more specialized fields. In references [33, 40, 41], and [38], the continued issue of the safety of the data collection and processing for a digital twin is discussed in each of the respective references. When dealing with sensitive data obtained from a healthcare context, this worry is especially significant because there is a possibility that the data could be stolen or otherwise compromised. The work that will be done in the future will primarily focus on guaranteeing the privacy of the data that is utilized in the process of creating a digital twin. This is because the work that will be done in the future will be done.

3.8.3 The Digital Twins that Complement Smart Cities and Smart Cities Themselves

The final area of interest pertaining to this topic is proposal c, which discusses the unresolved research challenges that are associated with digital twins in connection to smart cities. According to the review that was discussed in Part 3.7, which can be found above, this sector receives almost the same amount of academic research that the healthcare sector does. At this point in time, the papers cover a large portion of this wider and more all-encompassing concept of a smart city. The scope of these articles ranges from city-wide digital twins for smart cities to more specific topics like digital twins for traffic management [23] and livestock management [16] in addition to writings on renewable energy [19]. It was one of the subjects that was discussed whether or not it would be possible to create a large town or metropolis that is entirely connected to the internet, as well as whether or not the digital twin might be used on a more extensive scale. An intriguing idea that has been looked into by reference [3] is whether or not the development and management of smart cities could benefit from the use of three-dimensional modeling. When the digital twin is paired with the physical infrastructure that already exists, this potential becomes apparent in the shape of an intriguing scenario. The application of data analytics, such as predictive analytics applied to a digital twin with the purpose of constructing smart cities, is an example of a gap in knowledge that needs to be filled by additional study. In these areas, there is a need for additional research. It is impossible to have a smart city that does not have its own digital doppelganger that is standardized. This is a need of the highest importance. Having said that, this does call for additional research to be conducted in a variety of different fields, including the interactions of data fusion, the physical and virtual exchange of data, and the building of a digital twin. These are just some of the many questions that need to be answered through further research. The significance of modeling in a smart city is emphasized in references [23] and [42], which make a plea for more generic models that take into consideration all of the component components of a smart city. Both of these references highlight the importance of modeling in a smart city. Both of these references emphasize how important modeling is in the development of a smart city. The significance of modeling in the context of a smart city is highlighted by this example. Chen [23] proposes integrating traffic management without further modeling or development. The idea can be applied immediately. This standard will ensure smart buildings and smart traffic are compatible to maximize efficiency. Manufacturing digital twins are creating new smart city opportunities [3, 23].

(1) DATA MODELS. The first set of findings addresses the digital twin paradigm and its architecture. The published research lacks unified models, a general digital twin architecture, and a consensus on how to develop a digital twin system. Digital twins' best use is also debated. A fundamental digital twin system can be implemented by creating a digital twin design paradigm. This method involves digitizing a physical object. This paradigm would guide digital twin creation.

(2) HETEROGENEOUS SYSTEMS. The digital twin's heterogeneous environment is connected to massive distributed networks, so academics in this field strive to learn how to interact with such systems. Compare and contrast the many existing digital twin systems to see how they handle smaller and larger diversified digital twin environments. Comparing and contrasting the many digital twin systems can do this.

(3) ARTIFICIAL INTELLIGENCE. Due to digitalization, artificial intelligence has become one of the leaders and facilitators for growth and flexibility in the process of allowing digital twins, and it is quickly becoming the main component of such systems. Artificial intelligence has led the growth and flexibility of digital twins. Artificial intelligence has emerged as a leader and facilitator for growth and flexibility in digital twins. AI may explain this. The unexplored research areas that aim to determine how AI, machine learning,

and deep learning algorithms may affect digital twin technology invention and deployment. These fields are unexplored. As demand for digital twin technology rises, Industry 4.0 principles have improved how people live, work, and communicate, creating new academic research opportunities. Technology has also changed human behavior [44]. AI allows industrial case studies on predictive maintenance and system usability. AI allows these case studies. These case studies will prepare for future ones.

(4) SECURITY. Blockchain technology has created new data privacy options. This creates these new opportunities. These new data confidentiality methods focus on digital twin protection solutions.

(5) DATA EXCHANGE. As connectivity has improved intelligence and security, digital twin systems and their technology have proliferated. Connectivity benefits these. Increased connectivity allows devices to share more data. To determine if the digital twin can thrive when devices share data, more research is needed. Data integration for small IoT devices and large heterogeneous systems is a challenge. To fill a space. This open space should realize the Internet of Things. This investigation is public. Given the limited data exchange, smaller digital twins must be created faster.

(6) IoT. Most digital twin research focuses on the Internet of Things. Retrofitting sensors ensures reliable data exchange. If possible, this can be done. Internet of Things and digital twin researchers have yet to explore edge computing.

3.8.4 Difficulties in the Process of Free Research

As discussed in the next section, solving open digital twin research questions is difficult. (1) Multidisciplinarity is one of the biggest challenges during planning and creation. The collaboration of many researchers from different fields produces them. This may help breakthroughs, but it may also hurt. Because many fields have different goals, research goes in different directions, slowing results. (2) This paper highlights the lack of standardization in many new and developing technologies, another hurdle the digital twin must overcome before it can become a reality. This will distinguish digital twin initiatives. The digital twin technology is being slowed by the wide variety of definitions and lack of standardisation. (3) Global advancements of digital twins and the technologies that enable them have many benefits and drawbacks. Goals that are too distant or unattainable will slow the adoption and development of new technologies and opinions [45]. Industry 4.0 and digitalization have accelerated technology development at different rates. This rapid expansion presents many challenges, particularly in connectivity and data exchange. The digital twin may not work with Internet of Things-connected auxiliary systems [46]. Due to its different rates, its expansion may pose security risks [47, 48].

3.9 Conclusion

Digital twin use has changed in recent years. More publications and industry leaders' investment in digital twin technology have helped this movement. Without the rise of AI, IoT, and IIoT, which are key enablers for digital twins, they will never be possible. As shown by the vast amount of literature in this area, most digital twin research is industrial-focused. Most of the literature was analyzed. Manufacturing has more papers on digital twins than smart cities and healthcare, indicating gaps in the study for both industries. Digital twins are becoming increasingly dependent on artificial intelligence, and one of the most important research directions is the use of many algorithmic methodologies. Articles discuss small-scale effects of AI and digital twins. Exciting and necessary future research will study ways to build on successful artificial intelligence and digital twin efforts. The lack of standardization and widespread misperceptions about digital twins are significant findings.

Standardization is necessary to ensure that future advancements are digital twins and not misinterpretations. Addressing issues is the only way. Digital healthcare twins and smart cities are two other growing industries. The study contributes to a categorical evaluation that includes manufacturing, healthcare, and smart cities. The researchers' digital twin development is explained in each subject's section. The paper also identifies challenges and crucial enabling technologies that will aid future work. The research also shows that the term "digital twin" has not changed much since 2012. Many studies mistake digital twins for models and shadows. The literature contains few examples of smaller digital twin projects, but none of larger ones. Lack of domain knowledge about scaling up larger digital twins is a contributing factor. Predictive maintenance and machine health are growing fields in manufacturing articles about digital twins. Several publications have studied using digital twins to predict human behaviour. Digital twins use analogous concepts to track patients' health statuses. The study also discusses data fusion research and remote surgery advances. This is mostly due to healthcare data's sensitivity. There has not been much research on smart cities, but it may be possible to analyze digital twins for traffic management systems and smart city improvements. This is an improvement. This assessment lays the groundwork for future work in digital twins, which is still in its infancy and mostly focused on manufacturing. This work lays the groundwork for future field research by various researchers.

References

[1] G. L. Knapp, T. Mukherjee, J. S. Zuback, H. L. Wei, T. A. Palmer, A. De, and T. DebRoy, "Building blocks for a digital twin of additive manufacturing," *Acta Mater.*, vol. 135, pp. 390–399, Aug. 2017.

[2] A. Bilberg, and A. A. Malik, "Digital twin driven human–robot collaborative assembly," *CIRP Ann.*, vol. 68, no. 1, pp. 499–502, 2019.

[3] T. Ruohomaki, E. Airaksinen, P. Huuska, O. Kesaniemi, M. Martikka, and J. Suomisto, "Smart city platform enabling digital twin," in *Proc. Int. Conf. Intell. Syst. (IS)*, Sep. 2018, pp. 155–161.
[4] J. Liu, X. Du, H. Zhou, X. Liu, L. ei Li, and F. Feng, "A digital twin-based approach for dynamic clamping and positioning of the flexible tooling system," *Procedia CIRP*, vol. 80, pp. 746–749, Jan. 2019.
[5] C. Mandolla, A. M. Petruzzelli, G. Percoco, and A. Urbinati, "Building a digital twin for additive manufacturing through the exploitation of blockchain: A case analysis of the aircraft industry," *Comput. Ind.*, vol. 109, pp. 134–152, Aug. 2019.
[6] N. Mohammadi, and J. E. Taylor, "Smart city digital twins," in *Proc. IEEE Symp. Ser. Comput. Intell. (SSCI)*, Nov. 2017, pp. 1–5.
[7] M. Grieves, "Digital twin: Manufacturing excellence through virtual factory replication," NASA, Washington, DC, USA, White Paper 1, 2014.
[8] E. Glaessgen, and D. Stargel, "The digital twin paradigm for future NASA and U.S. Air force vehicles," in *Proc. 53rd AIAA/ASME/ASCE/AHS/ASC Struct., Struct. Dyn. Mater. Conf. 20th AIAA/ASME/AHS Adapt. Struct. Conf. 14th AIAA*, Apr. 2012, pp. 1818.
[9] Y. Chen, "Integrated and intelligent manufacturing: Perspectives and enablers," *Engineering*, vol. 3, no. 5, pp. 588–595, Oct. 2017.
[10] D. Wu, C. Jennings, J. Terpenny, and S. Kumara, "Cloud-based machine learning for predictive analytics: Tool wear prediction in milling," in *Proc. IEEE Int. Conf. Big Data (Big Data)*, Dec. 2016, pp. 2062–2069.
[11] Z. Liu, N. Meyendorf, and N. Mrad, "The role of data fusion in predictive maintenance using digital twin," in *Proc. Annu. Rev. Prog. Quant. Nondestruct. Eval.*, Provo, UT, USA, 2018, Art. no. 020023.
[12] S. E. Bibri, and J. Krogstie, "The core enabling technologies of big data analytics and context-aware computing for smart sustainable cities: A review and synthesis," *J. Big Data*, vol. 4, no. 1, pp. 38, Dec. 2017.
[13] S. R. Chhetri, S. Faezi, A. Canedo, and M. A. A. Faruque, "QUILT: Quality inference from living digital twins in IoT-enabled manufacturing systems," in *Proc. Int. Conf. Internet Things Design Implement.*, Apr. 2019, pp. 237–248.
[14] Y. Zheng, S. Yang, and H. Cheng, "An application framework of digital twin and its case study," *J. Ambient Intell. Humanized Comput.*, vol. 10, no. 3, pp. 1141–1153, Jun. 2018.
[15] R. Vrabič, J. A. Erkoyuncu, P. Butala, and R. Roy, "Digital twins: Understanding the added value of integrated models for through-life engineering services," *Procedia Manuf.*, vol. 16, pp. 139–146, Jan. 2018.
[16] A. Madni, C. Madni, and S. Lucero, "Leveraging digital twin technology in model-based systems engineering," *Systems*, vol. 7, no. 1, pp. 7, Jan. 2019.
[17] W. Kritzinger, M. Karner, G. Traar, J. Henjes, and W. Sihn, "Digital twin in manufacturing: A categorical literature review and classification," *IFACPapersOnLine*, vol. 51, no. 11, pp. 1016–1022, 2018.
[18] D. Howard, "The digital twin: Virtual validation in electronics development and design," in *Proc. Pan Pacific Microelectron. Symp. (Pan Pacific)*, Feb. 2019, pp. 1–9.
[19] A. Coraddu, L. Oneto, F. Baldi, F. Cipollini, M. Atlar, and S. Savio, "Datadriven ship digital twin for estimating the speed loss caused by the marine fouling," *Ocean Eng.*, vol. 186, Aug. 2019, Art. no. 106063.

[20] J. David, A. Lobov, and M. Lanz, "Leveraging digital twins for assisted learning of flexible manufacturing systems," in *Proc. IEEE 16th Int. Conf. Ind. Informat. (INDIN)*, Jul. 2018, pp. 529–535.

[21] T. DebRoy, W. Zhang, J. Turner, and S. S. Babu, "Building digital twins of 3D printing machines," *Scripta Mater.*, vol. 135, pp. 119–124, Jul. 2017.

[22] S.-K. Jo, D.-H. Park, H. Park, and S.-H. Kim, "Smart livestock farms using digital twin: Feasibility study," in *Proc. Int. Conf. Inf. Commun. Technol. Converg. (ICTC)*, Oct. 2018, pp. 1461–1463.

[23] R. He, G. Chen, C. Dong, S. Sun, and X. Shen, "Data-driven digital twin technology for optimized control in process systems," *ISA Trans.*, vol. 95, pp. 221–234, Dec. 2019.

[24] D. Burnett, J. Thorp, D. Richards, K. Gorkovenko, and D. Murray-Rust, "Digital twins as a resource for design research," in *Proc. 8th ACM Int. Symp. Pervasive Displays*, Jun. 2019, pp. 1–2.

[25] K. Sivalingam, M. Sepulveda, M. Spring, and P. Davies, "A review and methodology development for remaining useful life prediction of offshore fixed and floating wind turbine power converter with digital twin technology perspective," in *Proc. 2nd Int. Conf. Green Energy Appl. (ICGEA*), Mar. 2018, pp. 197–204.

[26] H. Pargmann, D. Euhausen, and R. Faber, "Intelligent big data processing for wind farm monitoring and analysis based on cloud-technologies and digital twins: A quantitative approach," in *Proc. IEEE 3rd Int. Conf. Cloud Comput. Big Data Anal. (ICCCBDA)*, Apr. 2018, pp. 233–237.

[27] C. Brosinsky, D. Westermann, and R. Krebs, "Recent and prospective developments in power system control centers: Adapting the digital twin technology for application in power system control centers," in *Proc. IEEE Int. Energy Conf. (ENERGYCON)*, Jun. 2018, pp. 1–6.

[28] Y. Xu, Y. Sun, X. Liu, and Y. Zheng, "A Digital-Twin-Assisted fault diagnosis using deep transfer learning," *IEEE Access*, vol. 7, pp. 19990–19999, 2019.

[29] Q. Qi, and F. Tao, "Digital twin and big data towards smart manufacturing and industry 4.0: 360 degree comparison," *IEEE Access*, vol. 6, pp. 3585–3593, 2018.

[30] M. Kritzler, J. Hodges, D. Yu, K. Garcia, H. Shukla, and F. Michahelles, "Digital companion for industry," in *Proc. Companion World Wide Web Conf.*, San Francisco, CA, USA, ACM Press, 2019, pp. 663–667.

[31] Y. Umeda, J. Ota, F. Kojima, M. Saito, H. Matsuzawa, T. Sukekawa, A. Takeuchi, K. Makida, and S. Shirafuji, "Development of an education program for digital manufacturing system engineers based on 'digital triplet' concept," *Procedia Manuf.*, vol. 31, pp. 363–369, Jan. 2019.

[32] V. J. Mawson, and B. R. Hughes, "The development of modelling tools to improve energy efficiency in manufacturing processes and systems," *J. Manuf. Syst.*, vol. 51, pp. 95–105, Apr. 2019.

[33] F. Longo, L. Nicoletti, and A. Padovano, "Ubiquitous knowledge empowers the smart factory: The impacts of a service-oriented digital twin on enterprises' performance," *Annu. Rev. Control*, vol. 47, pp. 221–236, Jan. 2019.

[34] M. Lohtander, N. Ahonen, M. Lanz, J. Ratava, and J. Kaakkunen, "Micro manufacturing unit and the corresponding 3D-model for the digital twin," *Procedia Manuf.*, vol. 25, pp. 55–61, Jan. 2018.

[35] H. Laaki, Y. Miche, and K. Tammi, "Prototyping a digital twin for real time remote control over mobile networks: Application of remote surgery," *IEEE Access*, vol. 7, pp. 20325–20336, 2019.
[36] H. Zhang, Q. Liu, X. Chen, D. Zhang, and J. Leng, "A digital twin-based approach for designing and multi-objective optimization of hollow glass production line," *IEEE Access*, vol. 5, pp. 26901–26911, 2017.
[37] B. Schleich, N. Anwer, L. Mathieu, and S. Wartzack, "Shaping the digital twin for design and production engineering," *CIRP Ann.*, vol. 66, no. 1, pp. 141–144, 2017.
[38] Q. Min, Y. Lu, Z. Liu, C. Su, and B. Wang, "Machine learning based digital twin framework for production optimization in petrochemical industry," *Int. J. Inf. Manage.*, vol. 49, pp. 502–519, Dec. 2019.
[39] M. Joordens, and M. Jamshidi, "On the development of robot fish swarms in virtual reality with digital twins," in *Proc. 13th Annu. Conf. Syst. Syst. Eng. (SoSE)*, Jun. 2018, pp. 411–416.
[40] Y. Liu, L. Zhang, Y. Yang, L. Zhou, L. Ren, F. Wang, R. Liu, Z. Pang, and M. J. Deen, "A novel cloud-based framework for the elderly healthcare services using digital twin," *IEEE Access*, vol. 7, pp. 49088–49101, 2019.
[41] S. Gahlot, S. R. N. Reddy, and D. Kumar, "Review of smart health monitoring approaches with survey analysis and proposed framework," *IEEE Internet Things J.*, vol. 6, no. 2, pp. 2116–2127, Apr. 2019.
[42] A. El Saddik, "Digital twins: The convergence of multimedia technologies," *IEEE MultimediaMag.*, vol. 25, no. 2, pp. 87–92, Apr. 2018.
[43] D. Ross, "Digital twinning [information technology virtual reality]," *Eng. Technol.*, vol. 11, pp. 44–45, May 2016.
[44] R. Magargle, L. Johnson, P. Mandloi, P. Davoudabadi, O. Kesarkar, S. Krishnaswamy, J. Batteh, and A. Pitchaikani, "A simulation-based digital twin for model-driven health monitoring and predictive maintenance of an automotive braking system," in *Proc. 12th Int. Modelica Conf., Prague, Czech Republic*, Jul. 2017, pp. 35–46.
[45] D. Petrik, and G. Herzwurm, "IIoT ecosystem development through boundary resources: A siemens MindSphere case study," in *Proc. 2nd ACM SIGSOFT Int. Workshop Software-Intensive Bus., Start-ups, Platforms, Ecosyst.* IWSiB, 2019, pp. 1–6.
[46] K. E. Barkwell, A. Cuzzocrea, C. K. Leung, A. A. Ocran, J. M. Sanderson, J. A. Stewart, and B. H. Wodi, "Big data visualisation and visual analytics for music data mining," in *Proc. 22nd Int. Conf. Inf. Visualisation (IV)*, Jul. 2018, pp. 235–240.
[47] I. Singh, J. Gupta, R. Kumar, S. Sriramulu, A. Daniel, and N. Partheeban, (2022). "A Model For Identifying Fake News In Social Media", *Applications of Computational Methods in Manufacturing and Product Design*, Springer Nature Singapore. www.springerprofessional.de/en/a-model-for-identifying-fake-news-in-social-media/20368666
[48] S. Premkumar, and A. N. Sigappi, "Processing capacity-based decision mechanism edge computing model for IoT applications," *Computational Intelligence*, vol. 39, no. 4, pp. 532–553, Aug. 2023. doi: 10.1111/coin.12541

Chapter 4

Twin Technology

Exploring Types and Applications of Digital Twins

S. Dinesh Krishnan, K.V. Mahalakshmi, Dyagala Naga Sudha, V. Sathya Priya, and A. Daniel

4.1 Introduction

Twin technology, also known as digital twin technology, has emerged as a transformative concept in various industries, revolutionizing the way we understand, monitor, and optimize physical objects, systems, and processes. By creating a virtual replica or digital twin of a physical entity, organizations can unlock a multitude of benefits and gain deeper insights into its behavior, performance, and potential improvements. The idea behind twin technology is to bridge the gap between the physical and digital realms, enabling a dynamic and interactive representation of the real-world object or system. Through the integration of cutting-edge technologies like artificial intelligence (AI), the Internet of Things (IoT), data analytics, and modeling, the digital twin can replicate the physical counterpart's characteristics, allowing for real-time monitoring, analysis, and simulation.

The digital twin serves as a powerful tool that facilitates a deeper understanding of complex systems and enables organizations to make data-driven decisions. By capturing and analyzing real-time data from sensors, actuators, and other sources, the digital twin provides a comprehensive view of the physical entity's performance, enabling proactive maintenance, optimization, and predictive analytics. One of the fundamental aspects of twin technology is its ability to capture not only the physical attributes of the object or system but also the contextual information associated with it. This contextual information includes historical data, environmental factors,

DOI: 10.1201/9781003469612-4

operational parameters, and even human interaction, providing a holistic representation of the physical entity [1].

The applications of twin technology are vast and diverse, spanning across industries. In manufacturing, digital twins facilitate product design, process optimization, and predictive maintenance. In healthcare, they enable personalized medicine, surgical planning, and patient-specific treatment. In transportation, digital twins optimize traffic flow, enable predictive maintenance of vehicles, and enhance autonomous systems. In smart cities, digital twins optimize energy consumption, manage infrastructure, and support urban planning [8].

As twin technology continues to evolve and mature, new opportunities and challenges arise. Advancements in AI, machine learning, and connectivity enable even more sophisticated digital twin capabilities. However, ensuring data security, privacy, and interoperability remains a critical consideration [6].

In this chapter, we will explore the various types of twin technology, delve into their characteristics and applications, and provide real-world examples and conceptual diagrams to illustrate the practical implementation and benefits. By understanding the potential of twin technology, we can unlock new possibilities for innovation, optimization, and problem-solving across industries.

4.2 Definition of Twin Technology

Twin technology, also known as digital twin technology, refers to the practice of creating a virtual replica or digital representation of a physical object, system, or process. It involves leveraging advanced technologies such as AI, the IoT, data analytics, and modeling to capture real-time data and simulate the behavior, characteristics, and performance of the physical counterpart.

The digital twin is designed to mirror the physical entity in a virtual environment, allowing for monitoring, analysis, and optimization of its operations. It acts as a bridge between the physical and digital worlds, enabling organizations to gain insights, make informed decisions, and improve overall efficiency. The digital twin incorporates a wide range of information, including geometry, physical properties, sensor data, operational parameters, and historical records. By continuously updating and synchronizing with the physical entity, the digital twin can provide a real-time representation that can be used for various purposes, such as predictive maintenance, performance optimization, simulations, and experimentation.

The concept of twin technology originated from the idea of improving understanding, control, and management of complex systems by creating a virtual counterpart. It has gained significant traction in industries, such as manufacturing, healthcare, transportation, and smart cities, where the ability to replicate and analyze real-world objects and processes in a virtual environment offers numerous benefits, including reduced costs, enhanced decision-making, and improved operational efficiency.

4.3 Purpose and Significance of Creating Digital Twins

The purpose and significance of creating digital twins are manifold, offering numerous benefits and driving advancements in various industries. Here are some key purposes and significances of digital twins:

1. *Enhanced Understanding:* digital twins provide a comprehensive and dynamic representation of physical entities, enabling a deeper understanding of their behavior, performance, and interactions. By capturing and analyzing real-time data, organizations can gain insights into complex systems that were previously difficult to obtain, leading to improved knowledge and informed decision-making.
2. *Improved Design and Development:* digital twins facilitate iterative design and development processes by allowing virtual simulations and testing. Engineers and designers can experiment with different configurations, materials, and parameters in the virtual environment, reducing costs and time associated with physical prototypes. This enables faster innovation, improved product quality, and increased efficiency.
3. *Real-time Monitoring and Control:* by continuously synchronizing with the physical counterpart, digital twins enable real-time monitoring of various parameters and performance metrics. This real-time feedback allows organizations to detect anomalies, predict failures, and optimize operations promptly. It enables proactive maintenance, reduces downtime, and enhances operational efficiency.
4. *Predictive Analytics and Maintenance:* digital twins leverage historical data and advanced analytics algorithms to predict and prevent failures or performance degradation. By analyzing patterns and trends, organizations can implement predictive maintenance strategies, optimizing maintenance schedules, and reducing costs. This approach minimizes unplanned downtime, maximizes asset utilization, and extends the lifespan of equipment.
5. *Optimization and Efficiency:* digital twins provide a platform for optimizing processes, systems, and resources. Through simulations and scenario modeling, organizations can identify bottlenecks, optimize workflows, and improve resource allocation. This results in enhanced efficiency, reduced waste, and increased productivity.
6. *Personalized Services and Solutions:* in sectors like healthcare, digital twins enable personalized medicine and treatment planning. By creating virtual replicas of patients or body parts, healthcare professionals can simulate procedures, predict outcomes, and tailor treatments to individual needs. This leads to improved patient care, optimized treatment plans, and better healthcare outcomes.

7. *Risk Mitigation and Decision Support:* digital twins provide a virtual environment for testing and experimentation without affecting the physical system. Organizations can simulate different scenarios, evaluate risks, and test alternative strategies before implementing them in the real world. This reduces the potential for costly errors, supports robust decision-making, and enables organizations to navigate complex environments more effectively.
8. *Collaboration and Communication:* digital twins serve as a shared platform for collaboration and communication among stakeholders involved in the lifecycle of a physical entity. Designers, engineers, operators, and other stakeholders can access and contribute to the digital twin, facilitating knowledge sharing, problem-solving, and cross-functional collaboration.

In summary, the creation of digital twins offers significant benefits across industries, ranging from improved understanding and optimized operations to predictive maintenance, personalized solutions, and collaborative decision-making. By leveraging the power of digital twins, organizations can drive innovation, increase efficiency, and unlock new opportunities for growth and competitiveness.

4.4 Understanding Digital Twins

4.4.1 Definition and Characteristics of Digital Twins

Definition: a digital twin is a virtual replica or representation of a physical object, system, or process. It is a digital counterpart that mimics the physical entity's characteristics, behavior, and interactions in a virtual environment. The digital twin is created by capturing and integrating real-time data from sensors, IoT devices, and other sources, allowing for monitoring, analysis, and simulation of the physical counterpart.

4.4.1.1 Characteristics

1. Real-Time Synchronization: digital twins are continuously synchronized with their physical counterparts, capturing and reflecting real-time data and changes. This synchronization ensures that the digital twin remains an accurate representation of the physical entity, allowing for real-time monitoring and analysis.
2. Data Integration: digital twins incorporate various data sources, including sensor data, operational parameters, historical records, and environmental factors. This integration enables a comprehensive understanding of the physical entity and facilitates analysis and decision-making based on a wide range of data inputs.

3. Virtual Modeling and Simulation: Ddigital twins utilize virtual modeling and simulation techniques to replicate the physical entity's structure, behavior, and performance. This allows for testing different scenarios, predicting outcomes, and optimizing operations without impacting the physical counterpart.
4. Predictive Analytics: digital twins leverage historical and real-time data to perform predictive analytics. By analyzing patterns, trends, and anomalies, they can predict potential issues, failures, or performance degradation. This enables proactive maintenance, optimized operations, and improved decision-making.
5. Interconnectivity: digital twins can be interconnected with other digital twins or systems, forming a network of interconnected entities. This interconnectivity enables the exchange of data, information, and insights, facilitating collaborative decision-making and optimization across multiple entities or systems.
6. Lifecycle Representation: digital twins span the entire lifecycle of a physical entity, from design and development to operation and maintenance. They capture information and insights throughout the lifecycle, enabling improvements, optimization, and decision-making at every stage.
7. Feedback Loop: digital twins provide a feedback loop between the physical and digital realms. The data and insights generated by the digital twin can be used to optimize and improve the physical entity's performance, while real-time data from the physical counterpart enhances the accuracy and relevance of the digital twin.
8. Visualization and User Interface: digital twins often incorporate visualization tools and user interfaces to present data, simulations, and insights in a user-friendly and accessible manner. This allows stakeholders to interact with and understand the digital twin, facilitating effective communication and decision-making.

Overall, digital twins combine real-time data, virtual modeling, and simulation, predictive analytics, and interconnectivity to create a virtual representation that mirrors the physical entity's characteristics and behavior. These characteristics enable organizations to gain deeper insights, make informed decisions, and optimize operations in various industries.

4.4.2 Components and Architecture of Digital Twins

Digital twins consist of several key components and follow a specific architecture to ensure their effectiveness and functionality. Here are the essential components and architectural elements of digital twins:

1. Physical Entity: the physical entity is the actual object, system, or process that the digital twin represents. It can be a machine, a manufacturing plant, a healthcare device, a transportation system, or any other physical entity of interest.
2. Sensors and Data Sources: sensors and data sources play a crucial role in capturing real-time data from the physical entity. These sensors can include temperature sensors, pressure sensors, accelerometers, cameras, RFID tags, and various other types of sensors depending on the nature of the physical entity and the data required.
3. Data Acquisition: the data acquisition component involves the collection, aggregation, and preprocessing of data from sensors and other sources. This component ensures the availability of accurate and reliable data for the digital twin.
4. Data Integration and Storage: the data integration and storage component involves integrating and consolidating data from various sources into a central repository. This allows for efficient data management and provides a comprehensive dataset for analysis and simulation.
5. Modeling and Simulation: the modeling and simulation component is responsible for creating the virtual representation of the physical entity. It involves developing mathematical models, algorithms, and simulation techniques that replicate the behavior, characteristics, and interactions of the physical entity.
6. Analytics and Algorithms: analytics and algorithms form a critical component of digital twins. They analyze the collected data, perform calculations, identify patterns, and generate insights. This component enables predictive analytics, anomaly detection, optimization, and decision support.
7. Visualization and User Interface: the visualization and user interface component provides a user-friendly interface to interact with the digital twin. It includes graphical representations, dashboards, and data visualization tools that present the data, simulations, and insights in a comprehensible manner.
8. Communication and Connectivity: the communication and connectivity component allows for seamless data exchange between the digital twin and the physical entity. It involves networking protocols, connectivity options, and integration with other systems or digital twins in the ecosystem.
9. Control and Actuation: in some cases, digital twins include a control and actuation component that enables the digital twin to influence the behavior or operation of the physical entity. This can involve sending commands or instructions based on the insights and simulations generated by the digital twin.
10. Feedback Loop: the feedback loop is an integral part of the digital twin architecture. It involves the continuous exchange of data and insights between the physical entity and the digital twin. Real-time data from the physical entity enhances the accuracy and relevance of the digital twin, while insights and optimizations generated by the digital twin can be fed back to improve the performance of the physical entity.

The architecture of digital twins can vary depending on the specific application, industry, and complexity of the physical entity. However, these components form the foundational elements that enable the creation, functionality, and effectiveness of digital twins [2].

4.4.3 Data Acquisition and Integration Processes

Data acquisition and integration are crucial steps in creating and maintaining digital twins. These processes involve capturing data from various sources, ensuring its quality and consistency, and integrating it into a centralized repository for analysis and simulation. Here are the key steps involved in data acquisition and integration for digital twins:

1. Sensor Deployment: the first step is to deploy sensors and data sources on the physical entity to capture relevant data. Sensors can be embedded within the entity or placed in its proximity to monitor parameters such as temperature, pressure, vibration, motion, or other variables specific to the entity's characteristics and behavior.
2. Data Collection: sensors and data sources continuously collect data from the physical entity. This data can be in the form of real-time sensor readings, operational parameters, status updates, or any other information that provides insights into the entity's behavior and performance.
3. Data Preprocessing: raw data collected from sensors often requires preprocessing to ensure its quality, consistency, and compatibility with the digital twin. This preprocessing stage involves filtering out noise, correcting errors, aligning timestamps, and converting data into a standardized format suitable for integration and analysis.
4. Data Validation: the acquired data undergoes validation processes to ensure its accuracy and reliability. Validation techniques may involve comparing sensor readings against predefined thresholds, cross-referencing data with historical records, or using statistical methods to identify outliers or anomalies.
5. Data Transformation: in some cases, data transformation is required to align the collected data with the digital twin's modeling and simulation requirements. This may involve converting data into specific units, scaling values, or transforming data into suitable mathematical representations for simulation purposes.
6. Data Integration: once the data is preprocessed and validated, it is integrated into a centralized repository or data management system. This integration process brings together data from various sources, such as different sensors or subsystems, into a unified dataset for analysis and simulation.
7. Data Mapping and Correlation: during the integration process, data from different sources needs to be mapped and correlated to establish

relationships and connections between variables. This enables a comprehensive understanding of how different parameters or subsystems influence each other and contributes to the overall behavior of the digital twin.

8. Data Storage and Management: the integrated data is stored in a centralized data repository or a database system designed for efficient storage, retrieval, and management of large datasets. This repository serves as the foundation for analysis, simulation, and visualization of the digital twin.
9. Data Security and Privacy: data security and privacy measures should be implemented to protect the collected and integrated data. This includes encryption, access control, and adherence to data protection regulations to ensure the confidentiality and integrity of the data.
10. Continuous Data Updating: digital twins rely on continuous data acquisition and integration to remain up-to-date and accurately reflect the physical entity's real-time behavior. Therefore, the data acquisition and integration processes need to be ongoing, ensuring that new data is captured, validated, and integrated into the digital twin in a timely manner.

By following these data acquisition and integration processes, organizations can ensure that their digital twins are built on accurate, reliable, and integrated data, enabling effective analysis, simulation, and decision-making for improved understanding and optimization of the physical entity.

4.4.4 Virtual Modeling and Simulation Techniques

Virtual modeling and simulation techniques are fundamental components of digital twin technology. They enable the creation of a virtual representation of the physical entity and facilitate various analyses, predictions, and optimizations. Here are some common virtual modeling and simulation techniques used in digital twin technology:

1. Three-dimensional (3D) Modeling: 3D modeling involves creating a virtual representation of the physical entity using computer-aided design (CAD) or other modeling tools. It allows for visualizing the shape, structure, and components of the entity in a three-dimensional space.
2. Physics-Based Modeling: physics-based modeling focuses on replicating the physical behavior and interactions of the entity. It uses mathematical equations and physical laws to simulate the entity's mechanics, dynamics, thermodynamics, and other relevant physical phenomena.
3. Finite Element Analysis (FEA): FEA is a simulation technique used to analyze the structural integrity and performance of complex systems. It divides the virtual model into smaller elements and calculates the stresses, strains, and deformations based on the material properties and applied loads.

4. Computational Fluid Dynamics (CFD): CFD is employed to simulate fluid flow and heat transfer in systems such as HVAC, pipelines, or aerodynamics. It models fluid properties, boundary conditions, and geometric configurations to analyze flow patterns, pressure distributions, and thermal behavior.
5. Multi-body Dynamics: multi-body dynamics simulates the motion and interactions of multiple interconnected bodies or components within a system. It accounts for forces, collisions, and constraints to analyze the system's dynamics and predict its behavior under different conditions.
6. Discrete Event Simulation (DES): DES is used to model systems with discrete events and processes, such as manufacturing systems, supply chains, or transportation networks. It simulates the sequence of events, interactions, and decision-making processes to evaluate system performance and optimize operations.
7. Agent-Based Modeling (ABM): ABM focuses on modeling the behavior and interactions of individual agents or entities within a system. Each agent follows predefined rules and behaviors, allowing for the simulation of complex emergent phenomena and understanding system-level dynamics.
8. Machine Learning and Artificial Intelligence: machine learning and AI techniques are increasingly integrated into digital twin models. They enable the digital twin to learn from data, make predictions, identify patterns, and optimize operations based on real-time or historical information.
9. Real-Time Data Integration: virtual modeling and simulation techniques are enhanced by integrating real-time data from sensors and other sources. This integration enables the digital twin to reflect the current state and behavior of the physical entity accurately.
10. Scenario Testing and Optimization: virtual modeling and simulation allow for testing different scenarios and optimizing system performance. By altering parameters, inputs, or conditions, various what-if analyses and optimizations can be performed to identify optimal settings or improve performance under different circumstances.

These virtual modeling and simulation techniques empower digital twins to replicate the physical entity's behavior, predict future outcomes, optimize operations, and support decision-making. They enable organizations to explore different possibilities, evaluate performance under various conditions, and improve the overall efficiency, reliability, and sustainability of the physical entity.

4.5 Types of Digital Twins

4.5.1 Product Digital Twins

Overview of Product Digital Twins: product digital twins are virtual replicas of physical products that represent their design, characteristics, and behavior. They enable manufacturers to simulate and analyze product performance, optimize design iterations, and enhance the entire product lifecycle. Product digital twins combine real-time data, simulations, and analytics to provide insights into product behavior, usage patterns, and maintenance requirements.

4.5.1.1 Applications in Product Design, Development, and Optimization

1. Design Validation and Iteration: product digital twins allow designers to test and validate design concepts virtually. They can simulate how different design variations perform under various conditions, identify potential issues or improvements, and optimize the design before physical prototyping.
2. Performance Optimization: digital twins enable the optimization of product performance by simulating its behavior under different operating conditions. Manufacturers can analyze and fine-tune parameters, materials, or components to improve product efficiency, durability, and functionality.
3. Predictive Maintenance: by integrating sensors with the product digital twin, manufacturers can monitor real-time data on component health, usage patterns, and environmental conditions. This allows for predictive maintenance, as anomalies or potential failures can be detected early, reducing downtime and improving reliability.
4. Product Lifecycle Management: product digital twins support the entire product lifecycle management process. They help track and manage product changes, updates, and versions, ensuring efficient collaboration among different teams, suppliers, and stakeholders.
5. Customer Experience Enhancement: digital twins enable manufacturers to gain insights into how customers use and interact with the product. By analyzing data from connected products, manufacturers can identify usage patterns, customer preferences, and areas for improvement, leading to enhanced customer experiences and personalized product offerings.

4.5.1.2 Case Studies and Examples

1. Rolls-Royce: Rolls-Royce utilizes digital twins for its aircraft engines. By creating digital replicas of engines, they can monitor real-time data on engine performance, fuel consumption, and maintenance requirements. This data enables predictive maintenance and optimization of engine operations.
2. General Electric (GE): GE uses digital twins for its gas turbines. By combining sensor data with virtual models, they can simulate turbine behavior, analyze performance data, and predict maintenance needs. This helps optimize operations, reduce downtime, and enhance turbine efficiency.

3. Siemens: Siemens employs product digital twins for industrial equipment and machinery. They create virtual models that represent the physical machines and use real-time data to monitor performance, identify potential issues, and optimize maintenance schedules.
4. Procter & Gamble (P&G): P&G uses digital twins for its consumer products, such as washing machines and detergent dispensers. They simulate product behavior, usage patterns, and performance in different scenarios, allowing for product optimization and personalized product recommendations.
5. Automotive Industry: automotive manufacturers utilize digital twins to simulate vehicle performance, safety features, and fuel efficiency. They can test different design configurations, optimize vehicle aerodynamics, and predict crash performance, enabling faster and safer vehicle development.

These examples illustrate how product digital twins are employed across various industries to improve product design, development, and optimization. By leveraging virtual replicas and real-time data, manufacturers can enhance product performance, reduce costs, and deliver innovative and customer-centric solutions.

4.5.2 Process Digital Twins

Introduction to Process Digital Twins: Process digital twins are virtual representations of manufacturing and operational processes. They simulate and analyze the behavior, performance, and interactions of equipment, materials, and workflows in real-time. Process digital twins enable organizations to monitor, optimize, and streamline their processes, leading to improved efficiency, productivity, and quality [3].

4.5.2.1 Implementation in Manufacturing and Operational Processes

1. Production Line Optimization: process digital twins are used to optimize production lines by simulating the flow of materials, machines, and workers. They help identify bottlenecks, optimize production schedules, and improve overall throughput and efficiency.
2. Quality Control and Defect Detection: digital twins can monitor quality control parameters and analyze real-time data to identify defects or anomalies in the production process. By comparing expected and actual outputs, organizations can take corrective actions and enhance product quality.
3. Supply Chain Management: process digital twins enable real-time monitoring of supply chain processes, such as inventory management, logistics, and demand forecasting. They provide insights into the movement of goods, identify potential delays or disruptions, and optimize inventory levels [10].

4. Energy and Resource Optimization: digital twins help optimize energy consumption and resource utilization in manufacturing processes. By simulating different operational scenarios and analyzing data, organizations can identify energy-intensive processes, optimize resource allocation, and reduce waste.
5. Maintenance Planning and Predictive Analytics: process digital twins integrate real-time data from sensors and equipment to monitor machine health, performance, and maintenance requirements. This enables predictive maintenance planning, reduces downtime, and increases equipment lifespan.

4.5.2.2 Real-Time Monitoring, Analysis, and Optimization

1. Real-Time Data Integration: process digital twins integrate data from various sources, including sensors, equipment, and control systems, to provide real-time insights into process behavior and performance.
2. Monitoring and Visualization: digital twins offer real-time monitoring and visualization capabilities, allowing operators and managers to track process variables, performance metrics, and KPIs. Visual representations, such as dashboards or 3D models, facilitate easy interpretation of data.
3. Analytics and Optimization: process digital twins employ advanced analytics techniques, such as machine learning and optimization algorithms, to analyze real-time data and identify patterns, anomalies, or optimization opportunities. This enables organizations to make data-driven decisions and optimize process parameters for improved performance.
4. What-If Analysis: digital twins allow for what-if scenarios and simulations to understand the impact of changes or modifications to process parameters. Organizations can evaluate different scenarios, predict outcomes, and optimize processes before implementing changes in the physical environment.

4.5.2.3 Case Studies and Practical Examples

1. Airbus: Airbus utilizes process digital twins to optimize aircraft assembly processes. By simulating assembly workflows and analyzing real-time data, they improve efficiency, reduce errors, and enhance collaboration among workers and machines.
2. BMW: BMW employs process digital twins in their manufacturing plants to optimize production processes. They simulate and analyze assembly line workflows, material flow, and logistics to improve productivity, reduce costs, and enhance quality control.
3. Shell: Shell utilizes process digital twins in their oil and gas refineries to monitor and optimize process parameters, energy consumption, and equipment performance. This improves operational efficiency, reduces downtime, and enhances safety.

4. Tesla: Tesla uses process digital twins to optimize their manufacturing processes for electric vehicles. They simulate production line operations, material flow, and worker interactions to enhance production efficiency, quality control, and customization capabilities.
5. Coca-Cola: Coca-Cola implements process digital twins to monitor and optimize their beverage production processes. They analyze real-time data on ingredient quantities, mixing parameters, and bottling operations to ensure consistent product quality and production efficiency.

These case studies highlight how process digital twins are implemented in various industries to optimize manufacturing and operational processes. By leveraging real-time data, simulations, and analytics, organizations can achieve higher productivity,

4.5.3 System Digital Twins

Overview of System Digital Twins: system digital twins represent complex systems composed of interconnected subsystems, such as smart cities, healthcare systems, or supply chains. They provide a holistic view of the system's behavior, interactions, and performance. System digital twins integrate data from various sources, enable real-time monitoring and analysis, and facilitate decision-making for optimizing system-level efficiency, resilience, and sustainability.

4.5.3.1 Applications in Complex Systems

1. Smart Cities: system digital twins are used to monitor and optimize various aspects of smart cities, including energy management, traffic flow, waste management, and public services. They enable real-time monitoring of city infrastructure, predict and respond to events, and optimize resource allocation.
2. Healthcare Systems: in healthcare, system digital twins are employed to monitor patient health, optimize resource allocation, and improve operational efficiency. They integrate data from medical devices, electronic health records, and patient monitoring systems to facilitate real-time decision-making and personalized healthcare.
3. Supply Chains: system digital twins help optimize supply chain operations by monitoring and analyzing data from various stages of the supply chain, including procurement, production, logistics, and distribution. They enable real-time visibility, demand forecasting, and inventory optimization, leading to improved efficiency and customer satisfaction [10].

4.5.3.2 Integration of Subsystems and Data Sources

1. Data Integration: system digital twins integrate data from diverse subsystems and data sources, including sensors, IoT devices, databases, and external data

feeds. This integration ensures a comprehensive view of the system's behavior and enables effective analysis and decision-making.
2. Interoperability: system digital twins require interoperability between subsystems to exchange data and enable seamless communication. Standardized data formats, protocols, and interfaces are essential for integrating subsystems and ensuring data compatibility.
3. Model Integration: in complex systems, different subsystems may have their own modeling approaches or simulation tools. System digital twins require the integration of these individual models to create a unified representation of the entire system. This integration may involve data mapping, synchronization, and coordination of simulation processes.

4.5.3.3 Case Studies and Real-World Examples

1. Barcelona Smart City: Barcelona implemented a system digital twin to monitor and optimize various aspects of the city, including energy consumption, traffic flow, waste management, and public safety. The digital twin integrates data from sensors, IoT devices, and city infrastructure to provide real-time insights and enable data-driven decision-making.
2. Mayo Clinic: The Mayo Clinic utilizes a system digital twin in their healthcare system to monitor patient health, optimize resource allocation, and improve patient outcomes. The digital twin integrates data from medical devices, electronic health records, and patient monitoring systems to provide a comprehensive view of patient health and support personalized care.
3. Walmart: Walmart employs a system digital twin to optimize its supply chain operations. The digital twin integrates data from various stages of the supply chain, including inventory levels, transportation routes, and customer demand. This enables real-time visibility, demand forecasting, and efficient inventory management.
4. Singapore Port: The Port of Singapore utilizes a system digital twin to optimize port operations and logistics. The digital twin integrates data from various subsystems, including vessel tracking, cargo handling, and terminal operations. It provides real-time insights for efficient resource allocation, congestion management, and improved port performance.
5. Siemens Digital Rail: Siemens implemented a system digital twin for rail infrastructure, enabling real-time monitoring and optimization of train operations, maintenance, and passenger flow. The digital twin integrates data from signaling systems, train sensors, and passenger information systems to enhance operational efficiency and passenger experience.

These examples demonstrate how system digital twins are applied in complex systems such as smart cities, healthcare, and supply chains. By integrating subsystems, data sources, and simulation models, system digital twins provide a comprehensive understanding of system behavior, facilitate real-time monitoring and optimization, and support data-driven decision- making.

4.6 Digital Twin Technologies and Enablers

4.6.1 Overview of Technologies Enabling Digital Twin Implementation

Digital twin implementation relies on several key technologies that enable the creation, integration, and analysis of digital twins. Some of these technologies include the following:

1. Internet of Things (IoT): IoT devices and sensors play a crucial role in collecting real-time data from physical assets and environments. They enable the monitoring of various parameters, such as temperature, pressure, vibration, and location, which are essential for creating accurate digital representations.
2. Artificial Intelligence (AI): AI techniques, including machine learning, deep learning, and cognitive computing, enable advanced analytics and decision-making capabilities. AI algorithms can analyze large volumes of data generated by digital twins, detect patterns, predict behavior, and optimize performance.
3. Data Analytics: data analytics tools and techniques are employed to process and derive insights from the vast amount of data collected by digital twins. They help identify trends, anomalies, correlations, and patterns, providing valuable information for optimization and decision-making.
4. Cloud Computing: cloud platforms provide scalable and flexible infrastructure for storing, processing, and analyzing digital twin data. Cloud computing enables easy access, collaboration, and integration of data and models from different sources, making digital twins more accessible and manageable.
5. Augmented reality (AR) and virtual reality (VR): AR and VR technologies enhance the visualization and interaction capabilities of digital twins. They enable users to immerse themselves in the virtual environment, visualize data overlays on physical objects, and simulate scenarios for training, maintenance, and design purposes. Figure 4.1 refers to the visualization and interaction of digital twins on AR and VR.

4.6.2 Sensor Technologies and Data Collection Methods

Sensors play a critical role in collecting data from physical assets and environments to feed into digital twins. Various sensor technologies and data collection methods are used, including the following:

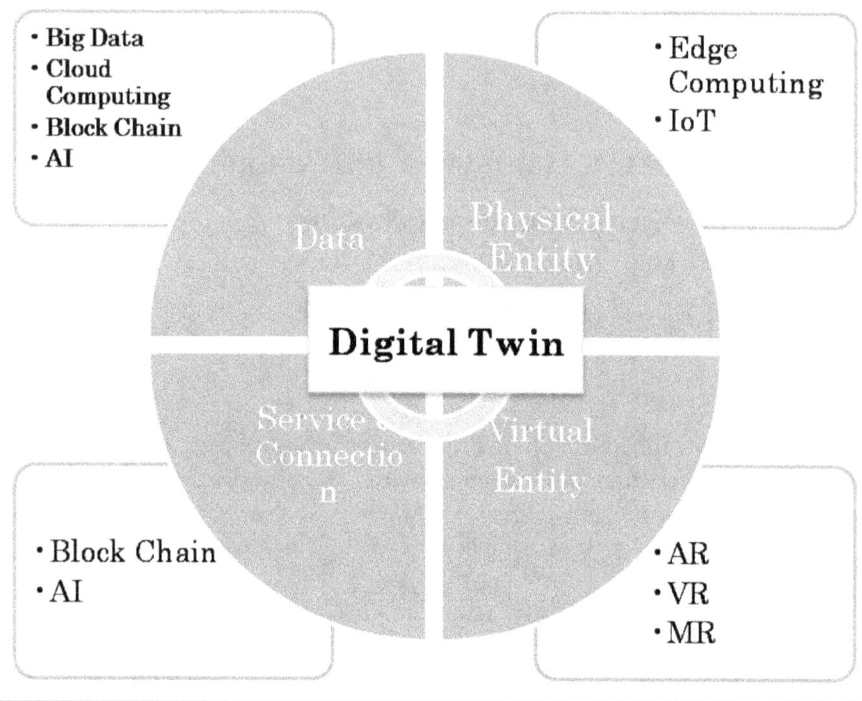

Figure 4.1 Augmented reality (AR) and virtual reality (VR).

1. Internet of Things (IoT) Sensors: IoT sensors, such as temperature sensors, humidity sensors, accelerometers, and pressure sensors, are widely used for monitoring physical parameters in real-time. They provide continuous data streams that contribute to the creation of accurate and up-to-date digital twins.
2. Imaging and Vision Sensors: cameras and imaging sensors capture visual information about physical assets, allowing for visual inspection, object recognition, and tracking. This visual data can be integrated into digital twin models for enhanced monitoring and analysis.
3. RFID and Barcode Technologies: RFID (radio frequency identification) and barcode technologies are utilized for asset tracking and identification. They enable automatic data capture, facilitating the integration of asset information into digital twin systems.
4. GPS and Location Sensors: GPS (global positioning system) and location sensors provide accurate positioning information. They are used to track the location and movement of assets, vehicles, and personnel, enabling spatial awareness in digital twin models.
5. Wireless Communication Technologies: wireless communication technologies, such as Wi-Fi, Bluetooth, and cellular networks, enable the transmission

of sensor data to centralized systems. They facilitate the real-time transfer of data from sensors to digital twin platforms.

4.6.3 Modeling and Simulation Tools

Modeling and simulation tools are essential for creating accurate virtual representations within digital twin technology. Some commonly used tools include the following:

1. Computer-Aided Design (CAD) Software: CAD software allows for the creation of detailed 3D models of physical assets, components, and environments. These models serve as the foundation for digital twins, providing a visual representation of the physical entity.
2. Finite Element Analysis (FEA) Software: FEA software is used for simulating the structural and mechanical behavior of physical assets. It enables engineers to analyze stress, strain, and deformation characteristics, helping to validate and refine the digital twin models.
3. Computational Fluid Dynamics (CFD) Software: CFD software is employed for simulating fluid flow, heat transfer, and other fluid-related phenomena. It is used to analyze and optimize the performance of systems such as HVAC (heating, ventilation, and air conditioning), fluid pipelines, and energy systems.
4. Discrete Event Simulation (DES) Software: DES software is utilized for simulating dynamic systems and processes. It enables the modeling and analysis of complex workflows, material flows, and operational processes within digital twins.
5. Data-Driven Modeling and Machine Learning Tools: data-driven modeling techniques, such as machine learning algorithms and statistical analysis tools, are used to develop predictive models based on historical data. These models can be integrated into digital twins to simulate and predict future behavior.

4.6.4 Integration Platforms and Data Management Systems

Integration platforms and data management systems are critical for aggregating, storing, and managing the vast amounts of data generated by digital twins. They provide the infrastructure for seamless integration and analysis of data from different sources. Some key components include the following:

1. Data Integration and Middleware Platforms: these platforms enable the integration of data from various sensors, devices, and systems into a unified digital twin environment. They handle data preprocessing, data transformation, and data flow management.

2. Database Management Systems (DBMSs): DBMSs provide storage and retrieval capabilities for digital twin data. They ensure data integrity, security, and efficient query processing to support real-time analytics and decision-making.
3. Big Data Platforms: big data platforms, such as Apache Hadoop and Apache Spark, are used to handle large volumes of data generated by digital twins. These platforms provide distributed storage and processing capabilities, enabling scalable and parallel data analysis.
4. Cloud-based Platforms: cloud platforms, such as Amazon Web Services (AWS) and Microsoft Azure, offer infrastructure and services for hosting digital twin environments. They provide scalability, flexibility, and accessibility for managing digital twin data and applications.
5. Visualization and Collaboration Tools: visualization and collaboration tools enable users to interact with digital twins, visualize data, and collaborate with other stakeholders. These tools facilitate data interpretation, decision-making, and knowledge sharing within digital twin environments.

These technologies and enablers collectively support the implementation and operation of digital twins by providing the necessary tools and infrastructure for data collection, modeling, simulation, integration, and analysis.

4.6.5 Manufacturing and Industrial Applications

Manufacturing and industrial sectors have been at the forefront of adopting digital twin technology to enhance efficiency, productivity, and overall performance. Here are some key applications of digital twins in manufacturing and industrial settings [4]:

1. Manufacturing Process Optimization: digital twins are used to optimize manufacturing processes by simulating and analyzing various parameters, such as equipment settings, material flow, and production schedules. They enable real-time monitoring and analysis of process variables, allowing for continuous improvement and optimization. Digital twins can identify bottlenecks, inefficiencies, and quality issues, helping manufacturers streamline operations and achieve higher productivity.
2. Predictive Maintenance and Quality Control: digital twins facilitate predictive maintenance by monitoring equipment health, performance, and usage patterns in real-time. By integrating data from sensors and other sources, digital twins can detect anomalies, predict failures, and recommend maintenance actions. This proactive approach helps manufacturers reduce downtime, extend equipment lifespan, and optimize maintenance schedules.

Additionally, digital twins enable real-time quality control by analyzing process variables, identifying deviations, and triggering alerts or adjustments to maintain product quality [5].

3. Supply Chain Optimization and Management: digital twins are utilized to optimize supply chain operations, including procurement, inventory management, logistics, and demand forecasting. By integrating data from various stages of the supply chain, digital twins provide real-time visibility, facilitate demand-supply matching, and optimize inventory levels. They enable scenario analysis, allowing manufacturers to evaluate different supply chain strategies and optimize resource allocation. Digital twins help reduce costs, minimize stockouts, improve delivery performance, and enhance overall supply chain efficiency [10].
4. Product Lifecycle Management: digital twins support the entire product lifecycle, from design and development to manufacturing, operation, and maintenance. They enable virtual prototyping and simulation, allowing manufacturers to optimize product designs, assess performance, and validate functionality before physical production. During the operational phase, digital twins provide real-time insights into product performance, usage patterns, and maintenance needs. This data-driven approach helps manufacturers enhance product reliability, optimize maintenance schedules, and deliver improved customer experiences.
5. Human–Machine Collaboration and Training: digital twins facilitate human–machine collaboration by providing virtual environments for training, simulation, and decision support. They enable operators and technicians to interact with virtual representations of equipment and processes, gaining hands-on experience and practicing troubleshooting scenarios. Digital twins also support remote collaboration, allowing experts to provide guidance and support from a distance. This capability enhances training effectiveness, reduces downtime, and improves operational efficiency.
6. Energy Efficiency and Sustainability: digital twins help manufacturers optimize energy consumption, reduce waste, and improve sustainability. By monitoring and analyzing energy usage patterns in real-time, digital twins identify energy-intensive processes and equipment, enabling manufacturers to implement energy-saving measures and optimize resource allocation. This contributes to cost savings, environmental sustainability, and compliance with energy efficiency regulations.

These manufacturing and industrial applications of digital twins demonstrate their potential to transform operations, enhance productivity, and drive continuous improvement in various aspects of the manufacturing process, supply chain management, and product lifecycle management.

4.6.6 Healthcare and Medical Applications

Digital twin technology has immense potential to revolutionize healthcare and medical fields by enabling personalized medicine, improving treatment planning, facilitating surgical simulations, and enabling remote patient monitoring. Here are some key applications of digital twins in healthcare and medicine:

1. Adoption of Digital Twins in Healthcare: digital twins are increasingly being adopted in healthcare settings to create virtual replicas of individual patients or specific organs/systems. These replicas capture and simulate the physiological characteristics, behaviors, and responses of patients, allowing for personalized and patient-specific care.
2. Personalized Medicine and Patient-Specific Treatment Planning: digital twins enable personalized medicine by integrating patient data, including genetic information, medical records, and imaging data, into a comprehensive virtual model. This allows healthcare professionals to simulate and predict treatment outcomes, optimize drug dosages, and customize treatment plans based on the individual patient's characteristics. Digital twins can help identify the most effective treatment options, reduce adverse effects, and improve patient outcomes.
3. Surgical Simulations: digital twins are used to create virtual surgical simulations that replicate the anatomy and dynamics of the patient. Surgeons can use these simulations to plan and practice complex procedures, evaluate different surgical techniques, and anticipate potential challenges or complications. This enhances surgical precision, reduces risks, and improves patient safety.
4. Remote Patient Monitoring: digital twins enable remote patient monitoring by capturing real-time data from wearable devices, sensors, and other monitoring technologies. This data is integrated into the digital twin, allowing healthcare providers to remotely monitor patients' vital signs, disease progression, and response to treatment. Remote patient monitoring improves patient engagement, enables early detection of abnormalities or worsening conditions, and reduces the need for frequent hospital visits.
5. Predictive Analytics and Early Disease Detection: digital twins combined with advanced analytics and machine learning techniques can analyze patient data to identify patterns, detect early signs of diseases, and predict disease progression. This enables early intervention, timely treatment, and improved patient outcomes. Digital twins can assist in predicting the efficacy of specific interventions or treatments for individual patients based on their unique characteristics and medical history.
6. Virtual Clinical Trials: digital twins are employed in virtual clinical trials to simulate and test the effectiveness and safety of new drugs or medical devices. Virtual trials reduce the need for traditional, time-consuming, and costly clinical trials, accelerating the development and approval of new treatments.

Digital twins enable researchers to simulate the behavior of drugs or devices in a virtual patient population, facilitating better understanding of efficacy, side effects, and optimal usage.

7. Rehabilitation and Prosthetics: digital twins can aid in rehabilitation by creating virtual models that replicate patients' movements, biomechanics, and neuromuscular activities. This allows for the development of personalized rehabilitation programs and the optimization of prosthetic devices. Digital twins can simulate the interaction between the patient and the prosthesis, optimizing fit, comfort, and functionality.

These healthcare and medical applications of digital twins demonstrate their potential to improve patient outcomes, enable personalized care, enhance treatment planning, and support medical research and development. Digital twins have the capability to transform healthcare by leveraging data-driven insights and virtual simulations to drive innovation and improve patient-centric care delivery.

4.6.7 Transportation and Infrastructure Applications

Digital twin technology is increasingly being adopted in transportation and infrastructure sectors to optimize operations, enhance safety, and improve the efficiency of transportation systems. Figure 4.2 refers to transportation and infrastructure sectors to optimize its operations. Here are some key applications of digital twins in transportation and infrastructure:

1. Traffic Management and Optimization: digital twins enable real-time monitoring and analysis of traffic flow, congestion patterns, and transportation infrastructure conditions. By integrating data from various sources, such as sensors, cameras, and GPS systems, digital twins provide insights into traffic conditions, enabling transportation authorities to optimize signal timings, manage traffic flow, and reduce congestion. They facilitate predictive analytics to anticipate traffic patterns and plan efficient routes for vehicles, thereby enhancing overall transportation efficiency.
2 Intelligent Transportation Systems (ITSs): digital twins support the implementation of ITSs, which integrate various technologies to improve safety, efficiency, and sustainability in transportation. Digital twins facilitate real-time monitoring of vehicle movements, traffic conditions, and infrastructure status, enabling dynamic traffic management, incident detection, and emergency response. They contribute to the development of adaptive traffic control systems, smart intersections, and real-time traveler information systems.
3. Infrastructure Planning and Management in Smart Cities: digital twins play a crucial role in the planning and management of transportation infrastructure

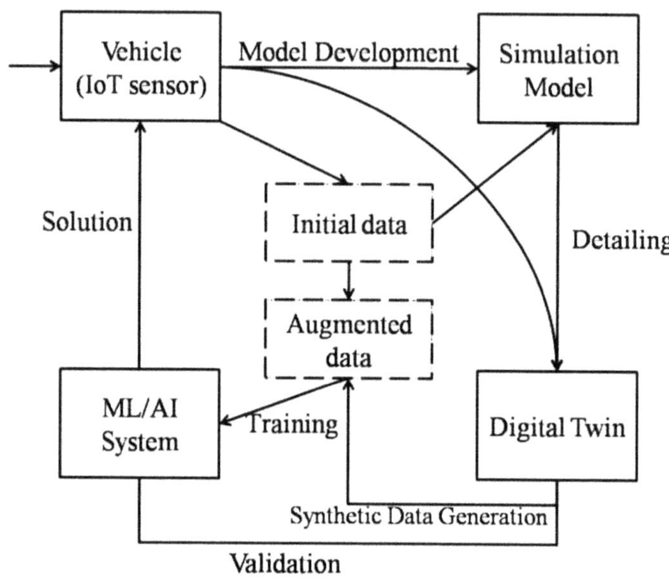

Figure 4.2 Transportation and infrastructure applications.

in smart cities. They provide a virtual representation of the entire transportation network, including roads, bridges, public transportation systems, and parking facilities. Digital twins enable urban planners and policymakers to simulate and evaluate different scenarios, optimize infrastructure designs, and assess the impact of proposed changes. They support the development of sustainable transportation systems, efficient land use, and improved connectivity within smart cities [7].

4. Asset Management and Maintenance: digital twins facilitate asset management and maintenance in transportation infrastructure, such as bridges, tunnels, and railways. By capturing real-time data from sensors, digital twins monitor the condition, performance, and structural health of assets. This data-driven approach enables predictive maintenance, identifies potential issues, and helps prioritize maintenance activities. Digital twins also support asset lifecycle management by providing insights into the aging process, deterioration patterns, and the need for repairs or replacements.
5. Virtual Simulations for Transportation Planning: digital twins enable virtual simulations for transportation planning, allowing stakeholders to assess the impact of new infrastructure projects, changes in transportation policies, or modifications to existing systems. Through virtual simulations, planners can evaluate traffic flow, pedestrian movement, and the effectiveness of proposed interventions before implementing them in the real world. This helps

optimize resource allocation, improve transportation accessibility, and minimize disruptions during construction or infrastructure upgrades.

6. Public Transportation Optimization: digital twins contribute to the optimization of public transportation systems, including buses, trains, and metros. By integrating real-time data on vehicle locations, passenger flow, and scheduling, digital twins facilitate efficient route planning, dynamic scheduling, and demand-responsive transportation services. This improves the reliability, capacity, and overall user experience of public transportation, leading to increased ridership and reduced congestion on roads.

These transportation and infrastructure applications of digital twins demonstrate their potential to enhance the efficiency, safety, and sustainability of transportation systems. By leveraging real-time data, predictive analytics, and virtual simulations, digital twins support better decision-making, optimized resource allocation, and improved transportation experiences for both individuals and communities.

4.6.8 Benefits and Impact of Digital Twins

Digital twin technology offers numerous benefits and has a significant impact across various industries and domains. Here are some key benefits and impacts of digital twins:

1. Improving Operational Efficiency and Productivity: digital twins enable organizations to optimize their operations and enhance productivity. By providing real-time insights into processes, assets, and systems, digital twins help identify inefficiencies, bottlenecks, and areas for improvement. They enable data-driven decision-making, process optimization, and resource allocation, resulting in improved operational efficiency, reduced costs, and increased productivity.
2. Enhancing Decision-Making through Real-Time Insights: digital twins provide real-time data and analytics, enabling organizations to make informed decisions based on accurate and up-to-date information. They offer a holistic view of operations, enabling stakeholders to monitor performance, track key metrics, and identify patterns or anomalies. This real-time visibility empowers decision-makers to respond quickly to changes, mitigate risks, and seize opportunities, leading to improved outcomes and competitive advantage.
3. Enabling Predictive Maintenance and Reducing Downtime: one of the significant advantages of digital twins is their ability to facilitate predictive maintenance. By continuously monitoring and analyzing data from sensors and other sources, digital twins can detect early signs of equipment failure, performance degradation, or maintenance needs. This proactive approach allows organizations to schedule maintenance activities, replace parts, or take

corrective actions before failures occur. As a result, downtime is reduced, equipment lifespan is extended, and maintenance costs are optimized.

4. Facilitating Innovation and Product Development: digital twins serve as virtual environments for innovation and product development. They enable virtual prototyping, simulation, and testing, allowing organizations to evaluate design alternatives, assess performance, and optimize products or processes before physical production. Digital twins facilitate collaboration and knowledge sharing among teams, enabling iterative improvements, accelerating time-to-market, and reducing development costs.
5. Supporting Sustainable Practices and Resource Optimization: digital twins contribute to sustainability efforts by optimizing resource usage and reducing waste. By analyzing data on energy consumption, material flow, and environmental factors, digital twins help identify opportunities for efficiency improvements and sustainable practices. Organizations can optimize energy usage, reduce emissions, and minimize resource wastage through data-driven insights provided by digital twins.
6. Enhancing Safety and Risk Management: digital twins enable organizations to enhance safety and risk management by simulating and evaluating various scenarios. They can assess the impact of potential risks, such as equipment failures, natural disasters, or security breaches, and develop mitigation strategies accordingly. Digital twins also support training and simulation exercises for emergency response, allowing organizations to improve preparedness and minimize the impact of adverse events.
7. Enabling Remote Monitoring and Collaboration: digital twins facilitate remote monitoring and collaboration, allowing stakeholders to access and interact with virtual representations of assets, processes, or systems. This capability is particularly valuable in scenarios where physical access is challenging or costly, such as offshore installations, remote locations, or hazardous environments. Remote monitoring and collaboration supported by digital twins improve efficiency, reduce travel costs, and enable experts to provide remote guidance or support.
8. Driving Continuous Improvement and Innovation: digital twins foster a culture of continuous improvement and innovation within organizations. By providing insights into performance metrics, customer feedback, and market trends, digital twins enable organizations to identify areas for improvement, develop new products or services, and adapt to changing customer demands. They support data-driven innovation, allowing organizations to stay ahead of the competition and respond effectively to market dynamics.

The adoption of digital twins brings about transformative benefits across industries, including improved operational efficiency, enhanced decision-making, predictive maintenance, innovation facilitation, sustainability, safety, and remote collaboration. These benefits drive organizational growth, competitiveness, and agility in an increasingly data-driven and interconnected world.

4.6.9 Challenges and Future Directions

While digital twin technology offers significant benefits, there are several challenges that need to be addressed to realize its full potential. Additionally, there are emerging trends and future directions that are shaping the evolution of digital twin technology. Here are some key challenges and future directions in digital twin technology:

1. *Data Privacy and Security Considerations:* as digital twins rely on vast amounts of data from various sources, ensuring data privacy and security becomes crucial. Organizations must implement robust security measures to protect sensitive data from unauthorized access, breaches, or cyber-attacks. Data anonymization techniques, encryption, access controls, and compliance with privacy regulations are essential to address these challenges.
2. *Integration Challenges and Interoperability Issues:* digital twins often require the integration of multiple data sources, systems, and platforms. Ensuring seamless data integration and interoperability among diverse technologies and data formats can be challenging. Organizations need to develop standardized protocols, data models, and application programming interfaces (APIs) to facilitate smooth integration and interoperability across different components of the digital twin ecosystem.
3. *Scalability and Complexity Management:* as digital twins become more comprehensive and encompass larger systems or networks, managing scalability and complexity becomes a challenge. Handling vast amounts of real-time data, performing complex simulations, and ensuring real-time responsiveness can strain computational resources. Advancements in cloud computing, edge computing, and high-performance computing can address scalability and complexity challenges, allowing digital twins to handle larger-scale systems effectively.
4. Emerging Trends and Future Directions in Digital Twin Technology:
    1. Digital Twins in Cyber-Physical Systems: the integration of digital twins with cyber-physical systems, such as the Internet of Things (IoT), robotics, and automation, will enable more seamless and dynamic interactions between the physical and virtual worlds, enhancing system performance and efficiency.
    2. AI and Machine Learning in Digital Twins: the incorporation of AI and machine learning techniques in digital twins will enable advanced analytics, predictive capabilities, and autonomous decision-making. AI-powered digital twins can learn from historical data, adapt to changing conditions, and optimize system performance in real-time.
    3. Digital Twins for Sustainability and Circular Economy: digital twins can play a crucial role in driving sustainability initiatives and supporting the circular economy. By analyzing resource consumption, waste generation, and environmental impact, digital twins can identify opportunities for resource optimization, waste reduction, and sustainable practices.

4. Digital Twins for Social Infrastructure: the application of digital twins is expanding beyond traditional industries into social infrastructure domains such as smart cities, healthcare systems, and public utilities. Digital twins can help optimize resource allocation, improve service delivery, and enhance the quality of life for citizens.
5. Digital Twins for Personalized Services: digital twins are increasingly being utilized to create personalized services and experiences. By capturing and analyzing individual preferences, behaviors, and characteristics, digital twins can tailor products, services, and recommendations to meet specific customer needs.
6. Blockchain for Trust and Transparency: blockchain technology can enhance the trust and transparency of digital twins by providing immutable records and secure data sharing. Blockchain-based digital twins enable secure data transactions, provenance tracking, and decentralized control, fostering trust among stakeholders.
7. Digital Twins for Resilience and Risk Management: digital twins can support resilience planning and risk management by simulating and evaluating the impact of potential disruptions or disasters. They enable organizations to develop proactive strategies, assess vulnerabilities, and enhance resilience in critical infrastructure and complex systems. Addressing the challenges and leveraging these emerging trends will drive the future development and adoption of digital twin technology. As technology advances, digital twins will become more sophisticated, interconnected, and instrumental in driving innovation, efficiency, and sustainability across various industries and domains.

4.7 Conclusion

1. *Recap of Key Points Discussed*: in this paper, we have explored digital twin technology and its various aspects. We started by defining digital twins and discussing their characteristics, including the ability to create a virtual representation of a physical entity and the synchronization of data between the physical and digital realms. We then delved into the components and architecture of digital twins, highlighting the importance of data acquisition and integration processes, as well as virtual modeling and simulation techniques.

   We explored three main types of digital twins: product digital twins, process digital twins, and system digital twins. For each type, we discussed their overview, applications, and provided case studies and examples to illustrate their practical use in different industries. We also examined the enabling technologies and tools that support digital twin implementation, such as IoT, AI, data analytics, sensor technologies, modeling and simulation tools, and integration platforms.

Furthermore, we explored the benefits and impact of digital twins, including improved operational efficiency, enhanced decision-making, predictive maintenance, and innovation facilitation. We discussed how digital twins contribute to various industries, such as manufacturing, healthcare, transportation, and infrastructure, by optimizing processes, enabling personalized medicine, and enhancing traffic management and infrastructure planning.

2. *Summary of the Benefits and Applications of Digital Twin Technology:* digital twin technology offers numerous benefits and has a wide range of applications across industries. It improves operational efficiency and productivity by providing real-time insights and facilitating data-driven decision-making. Digital twins enable predictive maintenance, reducing downtime and optimizing resource allocation. They also foster innovation and product development through virtual prototyping and simulation.

   The applications of digital twins are diverse and impactful. They optimize manufacturing processes, enable personalized medicine and surgical simulations in healthcare, and support traffic management and infrastructure planning in transportation and smart cities. Digital twins also find applications in areas such as supply chain optimization, asset management, and public transportation optimization [9].
3. *Final Thoughts on the Future Prospects of Digital Twin Technology*: digital twin technology is rapidly evolving and holds immense potential for the future. As technology advances, challenges such as data privacy and security, integration, scalability, and complexity management will be addressed. The emerging trends in digital twin technology, including its integration with AI, machine learning, blockchain, and its application in social infrastructure, personalized services, and resilience planning, indicate a promising future.

   The future prospects of digital twin technology are bright, as organizations recognize its transformative capabilities. The ability to create virtual replicas of physical entities and systems will continue to revolutionize industries, enhance operational efficiency, and drive innovation. With ongoing advancements in technology and increased adoption, digital twins will play a pivotal role in shaping the future of industries, infrastructure, and society as a whole.

   In conclusion, digital twin technology has emerged as a powerful tool for optimizing processes, enabling data-driven decision-making, and driving innovation. Its benefits and applications are wide-ranging, spanning various industries and domains. By leveraging digital twins, organizations can achieve improved efficiency, enhanced productivity, and better outcomes. As we move forward, digital twin technology will continue to evolve, opening up new possibilities and revolutionizing the way we design, operate, and optimize physical systems.

References

[1] Chen, X., Xu, X., Zhao, Y., & Wu, D. (2020). Digital twin enabled proactive maintenance & performance optimization for complex systems. *Annual Reviews in Control*, 50, 348–359.
[2] Dong, L., Xu, X., Xu, X., & Wang, L. (2020). Digital twin technology in construction industry: A review. *Automation in Construction*, 116, 103208.
[3] Hu, Y., Shen, W., Zhang, X., Liu, S., & Qi, J. (2020). Digital twin-driven manufacturing quality prediction and control: Framework and applications. *Journal of Manufacturing Systems*, 56(Part 2), 502–513.
[4] Liu, S., Yang, Y., & Yan, X. (2020). Digital twin-driven smart manufacturing: A survey. *Journal of Industrial Information Integration*, 17, 100120.
[5] Liao, L., Li, Y., & Fan, Z. (2020). Digital twin-enabled predictive maintenance for manufacturing systems: A review. *Journal of Manufacturing Systems*, 56(Part 2), 552–566.
[6] Lu, Y., & Li, G. (2020). Digital twin-driven smart manufacturing: Conception, benefits & challenges. *Journal of Manufacturing Systems*, 56(Part 2), 465–476.
[7] Tao, F., Cheng, J., Qi, Q., Zhang, M., Zhang, H., & Sui, F. (2020). Digital twin-driven sustainable manufacturing: Framework, approaches & applications. *Journal of Cleaner Production*, 272, 122632.
[8] Tao, F., Zhang, M., & Nee, A. Y. C. (2018). Digital twin-driven product design, manufacturing & service with big data. *International Journal of Advanced Manufacturing Technology*, 94(9–12), 3563–3576.
[9] Srinivasan, R., B Ganesh, H. S., & Parthasarathy, R. (2020). A review on digital twin technology: Architecture, benefits, & applications. *International Journal of Advanced Trends in Computer Science and Engineering*, 9(1), 3917–3924.
[10] Wei, G., Zhang, J., Lu, Y., & Cui, H. (2021). Digital twin technology & its application in smart logistics. *Discrete Dynamics in Nature and Society*, 2021, 8835296.

Chapter 5

The Convergence of Data Analytics, Digital Twins, and the IoT/IIoT

A New Era of Data-Driven Decision Making

S. Geerthik, N. Vel Murugesh Kumar, P.V. Gopirajan, and B. Jaison

5.1 Introduction

By allowing infrastructures and gadgets to become linked to the World Wide Web, interact, and collaborate on facts, the Internet of Things (IoT) has fundamentally altered how we inhabit our environments and work. The Industrial Internet of Things (IIoT) has taken this revolution further, providing new ways of automating and optimizing industrial processes, from production to maintenance (Fuller et al., 2020). The rise of the IoT and IIoT has also led to an explosion in data generation, providing organizations with a wealth of information about their products, processes, and customers.

Data analytics has emerged as a vital tool for making sense of this data, providing organizations with insights that can help drive innovation, improve efficiency, and enhance the customer experience. However, the immense data amount and intricacy generated by the IoT and IIoT can make it challenging to extract meaningful insights. Digital twin technology (DTT) can help with this scenario by giving

DOI: 10.1201/9781003469612-5

organizations a cybernetic representation of physical processes and systems, thereby allowing them to look deeper into that link underlying facts and what is happening in them.

The world quickly evolves into a complex network of connected systems and devices. As the IoT (Elazhary, 2019) and IIoT (Boobalan et al., 2022; Corallo et al., 2022; Sengupta et al., 2020) continue to grow, the need for connecting them to data analytics and DTT (Mauro & Kana, 2023; Semeraro et al., 2023a; Somers et al., 2022) is becoming increasingly important. However, a crucial part of the process is still missing: combining the data from these emerging technologies with the physical world. That is where digital twins come in. With the development of DTT, companies can now use these technologies to create smarter, connected systems that are more efficient, reliable, and secure (Tao et al., 2022).

Consequently, the rapport between the analysis of data, IoT, and IIoT with DTT is crucial, offering businesses a new approach to optimize operations, improve decision-making, and ultimately enhance the experience of customers (Zhu et al., 2022a). This book will investigate this connection in-depth, looking into DTT, applications, and tools and how they affect the IoT, IIoT, and data analytics (Fernandes et al., 2022).

This book chapter overviews the various technologies and how they are used to create more intelligent and efficient systems and processes. It discusses data analytics, digital twins, IIoT, and obstacles and opportunities for the future associated with these technological advances. Through a combination of practical examples, illustrations, and case studies, readers will gain a comprehensive understanding of the link between all these technologies. This book chapter also offers perspectives on the potential destiny of these breakthroughs and how they could be applied to develop more innovative, more effective procedures and frameworks.

5.2 Digital Twins: The Future of Data-Driven Decision-Making in the IoT and IIoT

5.2.1 IoT and IIoT

Two technologies that allow for connectivity between systems, networks, and devices are IoT and IIoT (Singh et al., 2019). IoT connects standard systems and objects to the internet to communicate with one another and with people. IIoT refers to the internet's integration of industrial machines and procedures, which enables remote monitoring and control.

The IoT changes how we work, reside, and engage with technological devices. For example, smart homes allow homeowners to control lighting, temperature, and security from their smartphones, while connected cars can provide real-time traffic updates and navigation assistance.

The IIoT refers to using sensors, actuators and other devices to connect and oversee machinery and processes. Its primary purpose is gathering and analyzing data to enhance efficiency and productivity. Manufacturing machinery and operations are tracked and administered through sensors in the IIoT. This enables real-time data collection and analysis, increasing efficiency, improved safety, and reduced downtime. One of the key benefits of the IIoT is improved asset management. By monitoring machines and equipment in real-time, maintenance crews can quickly identify potential problems and respond before they become significant issues, reducing downtime and increasing overall efficiency.

Furthermore, it is possible to enhance product quality and manufacturing procedures using data gathered by IIoT equipment (Zhu et al., 2022a). Another benefit of the IIoT is increased safety. By monitoring industrial processes in real-time, organizations can identify potential safety hazards and take action to prevent incidents. For example, connected sensors can detect gas leaks and alert workers, reducing the risk of accidents and environmental damage.

The IIoT also enables remote monitoring and control, improving the efficiency and productivity of industrial operations. For example, maintenance crews can remotely monitor equipment and diagnose problems, reducing the need for physical inspections. Additionally, remote control of machines and equipment can reduce the time required to perform tasks and improve overall efficiency. In order to benefit from the growing IIoT, businesses must implement strong cybersecurity measures to safeguard their connected devices and networks from online attacks. To prevent unauthorized accessibility of critical information and systems. This incorporates computer firewalls, data encryption, along additional safety protocols.

The critical differences between IoT and IIoT are as follows:

- IoT is broader and encompasses consumer devices, while IIoT is explicitly focused on industrial applications.
- IIoT often involves more robust and reliable technology, as it is used in mission-critical industrial processes.
- The data generated by IIoT is often more complex and requires specialized analysis than IoT data.

5.2.2 Data Analytics

The act of gathering, arranging, and analyzing data to find patterns and themes is known as data analytics (Kaluzny, 2021; Nyoman Kutha Krisnawijaya et al., 2022). Data analytics helps organizations make data-based decisions, identify problems and opportunities, and optimize operations (Akhter & Sofi, 2022). Data analytics has become a critical tool in many organizations and industries as they face the challenge of managing and processing several information points. Data analytics allows organizations to leverage the power of big data to gain insights into customer behavior, operational performance, and market trends. Data analytics can be applied to various industries and applications, including marketing, healthcare, finance, retail, and government. In marketing, data analytics enables organizations to understand customer preferences and buying patterns, tailoring their marketing campaigns and products to meet customer needs. By examining client information to spot anomalies and anticipate future client needs, data analytics concerning healthcare can enhance the safety of patients. In finance, data analytics can help organizations reduce risk and increase profits by analyzing market trends and customer behavior.

Also, the regeneration of artificial intelligence (AI) protocol is significantly influenced by data analytics (Bhatti et al., 2021a; Huynh-The et al., 2023; Shao et al., 2023a). Implementing such techniques, machines can now learn from data and build forecasts as well as choices on it. This can increase productivity and judgment precision across various markets and uses. The process of data analytics typically begins with data collection and preparation. This involves data collection from heterogeneous databases, spreadsheets, and weblogs and preparing it for analysis by cleaning, transforming, and integrating the data (Ghenai et al., 2022). The next step involves data exploration, where data analysts use visualizations and statistical techniques for pattern recognition and data trend analysis. The next step is building predictive models, where the data is used to train algorithms that can develop forecasts based on fresh information. Finally, the analysis results are communicated to decision-makers, who use the insights to inform their decisions.

Data analysis, methods like data visualization, and machine learning platforms are just a few of the technologies and techniques utilized in data analytics. These technologies allow data professionals to swiftly and effectively handle and analyze enormous volumes of data, allowing them to learn new knowledge and speed up decision-making. Data analytics is a rapidly growing field, and demand for data analysts is anticipating massive growth. As organizations collect and store more data, the need for skilled professionals to analyze and interpret that data increases.

5.2.3 Relations between IoT, IIoT, and Data Analytics

IoT and IIoT provide enormous volumes of information that get examined to learn new knowledge and enhance services. The significance of data analytics is vital IoT and IIoT applications. In IoT, data analytics can be used to process and make sense of the data generated by connected devices, such as consumer devices like smart home appliances, wearables, and connected vehicles. This data can be used to improve user experiences and optimize device performance. In IIoT, data analytics is even more critical as it helps to monitor, control, and optimize industrial processes in real-time. For example, data generated from connected machines and sensors can be analyzed to identify maintenance needs, improve production efficiency, and increase safety. In IoT and IIoT, data analytics can help identify patterns, trends, and anomalies in the data generated by connected devices, leading to improved decision-making and better outcomes. Therefore, IoT, IIoT, and data analytics are closely interrelated and dependent on each connected device and the IoT.

In the IIoT, edge analytics describes the procedure of gathering, examining, and responding to data produced by networked appliances and systems at the network's edge instead of the data center or cloud. This approach enables real-time decision-making and control, latency minimization, and increasing industrial production pipeline. Edge analytics can be used in various industrial applications, such as predictive maintenance, process optimization, and quality control. By analyzing data close to the source, industrial organizations can gain valuable insights into their operations to improve efficiency, productivity, and market sustainability.

5.3 Digital Twins: The Future of Industrial Manufacturing

Digital twin is a platform to connect physical assets to virtual models. This technology uses data collected from physical assets to create a virtual replica, allowing organizations real-time performance monitoring. IoT and IIoT enable organizations to manage data from physical assets in real-time, which can then be used for data analysis. DTT combines this data with simulations and models to create a virtual model. This virtual model helps to analyze the asset's performance and predict future performance.

Combining these technologies gives organizations a powerful tool to optimize operations (Ji et al., 2022; Sifat et al., 2023). By monitoring and analyzing data in real-time, organizations can identify problems quickly and make the necessary adjustments to improve processes. This technology also allows organizations to simulate new scenarios and test the effects of changes before they are implemented. Figure 5.1 depicts the digital twin model's workflow. Its initial phase includes the development of a virtual model for the current working physical model, the application of machine learning (ML) algorithms to find a solution to complex problems, compare the predictions attained by ML, and improve the prediction accuracy by regular analysis (Gong et al., 2022).

Monitoring, simulating, and optimizing the operating characteristics of a physical asset or system are the goals of a DTT. The following phases commonly make up a DTT's workflow.

Data collection: data collection from numerous devices, including sensors, machinery, and control mechanisms, is the initial phase. The actual asset and the system are represented virtually using this data. The gathered data is subsequently sorted, cleansed, and converted to make it ready for analysis. Isolated cases must be eliminated, numerical data must be filled in, and the data must be formatted to be analyzed.

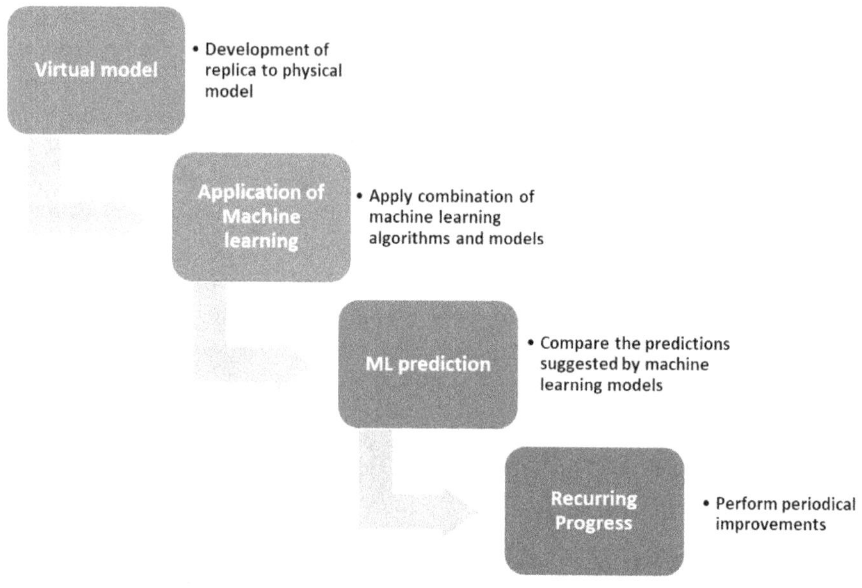

Figure 5.1 Workflow of a digital twin system.

Modeling: digital twins will be established with the real-time data collected. Computer-aided design (CAD) tools or simulation software shall be used to develop a virtual model, including information about the asset's physical and operational characteristics. The prepared data is then explored to gain insights into the data's patterns, trends, and relationships. This can be done using statistical methods, visualization techniques, or machine learning algorithms.

Monitoring: DTT allows continuous surveillance of the performance and conduct of real-time items. This can be achieved by comparing the synchronous output with that of the DTT model, which can provide insights into how the asset performs and help identify potential problems.

Simulation: the application of DTT will aid in testing and improving the performance of a tangible asset or system by simulating various situations and circumstances. This helps see possible issues before they arise and enables decision-making on improving efficiency.

Optimization: the behavior of an actual item or technology can be enhanced with the data produced by the DTT. This may involve adjusting parameters, controlling processes, or implementing new technologies to improve performance (Semeraro et al., 2023b).

Decision-making: operational decision-making shall be influenced by the DTT's insights and information.

Continuous improvement: DTT provides a continuing and sustaining activity (Jiang et al., 2022).

This workflow provides a continuous monitoring, simulation, and optimization loop, which can improve physical asset performance.

Digital twin modeling helps pretend and analyze the physical system's conduct (Ezhilmathi & Samy, 2022). The process involves creating a digitized system and then using software tools and algorithms to demonstrate system sturdiness.

The DTT is typically built using data from sensors and other sources and engineering knowledge and expertise about the system being modeled. Following that, the model may simulate exactly what is happening and will respond to various events, like alterations in the environment, unexpected failures of equipment, and system overhauls.

The simulation results can be used to identify potential problems and optimize system performance. For example, digital twin modeling and simulation in manufacturing can optimize production lines, reduce downtime, and increase overall efficiency. In healthcare, digital twin modeling and simulation can be used to optimize patient care and improve health outcomes.

5.3.1 Benefits of Linking Data Analytics with DTT

Linking data analytics with DTT has several benefits as shown in Figure 5.2, including the following:

Improved Decision Making: data analytics provides the way for digital twin by providing complex systems, processes, and operations. Complex DTTs provides the support for realistic information for achieving better operational efficiency and cost saving.

Predictive Maintenance: organizations may forecast when infrastructure is likely to break by integrating data mining and DTTs. This enables them to schedule maintenance and repairs in advance, minimizing downtime and maximizing productivity.

Increased Productivity: by integrating data analytics and DTT, organizations shall witness the progress, predict bottlenecks, inefficiencies, and take steps to improve productivity.

Enhanced Customer Experience: this DTT helps to simulate the customer experience, allowing organizations to test new products, services, and customer interactions before they are launched, leading to a more customer-centric approach.

Better Collaboration: the combination of data analytics and DTT enables organizations to collaborate with internal and external stakeholders, improving communication and collaboration across the supply chain.

Improved Supply Chain Management: by integrating data analytics and DTT, organizations can better manage their supply chain, reducing the risk of disruptions, increasing efficiency, and improving customer satisfaction.

Enhanced Safety: by integrating data analytics and DTT, organizations can witness progress, identify potential safety risks, and take steps to prevent accidents and improve safety.

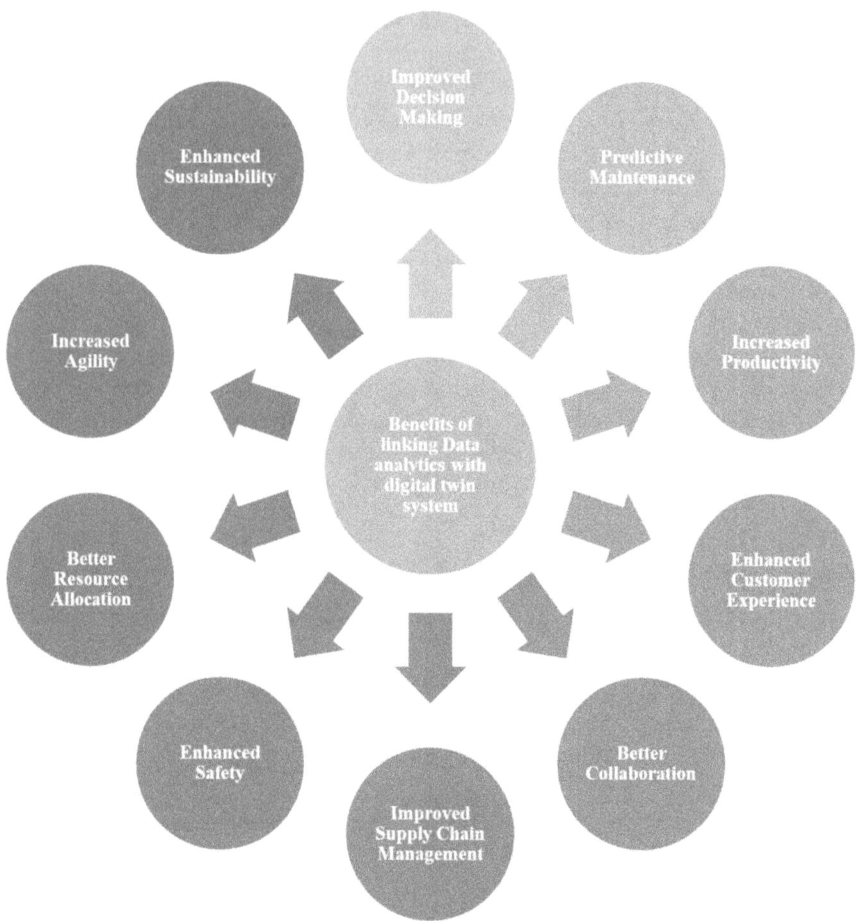

Figure 5.2 Benefits of linking data analytics with digital twin.

Better Resource Allocation: by integrating data analytics and DTT, organizations can optimize resource allocation, reducing waste and increasing efficiency.

Increased Agility: by integrating data analytics and DTT, organizations can immediately react to market requirements, trends, and operational challenges, increasing their agility and competitiveness.

Enhanced Sustainability: by integrating data analytics and DTT, organizations can monitor and analyze resource usage, emissions, and other sustainability indicators, allowing them to make informed decisions about their sustainability practices.

It is possible that digital twins could become the "next big thing" in analytics, as they offer many potential benefits and opportunities. Digital twins provide digitized monitoring and analysis of the system, which can provide valuable insights and improve decision-making.

In manufacturing, healthcare, and transportation industries, digital twins have already demonstrated their ability to improve operational efficiency, reduce downtime, and increase productivity. As technology advances and more organizations adopt digital twins, they may become increasingly widespread and widely used for data analysis and decision-making.

Furthermore, since DTT is a relatively new technology, some obstacles and disadvantages must be considered and resolved. These include the cost of implementation and maintenance, the difficulties involved in establishing a digital twin, and issues with anonymity and safety.

Digital twins and predictive analytics are complementary methodologies that provide synchronous insights and support decision-making. Machine language-based predictions and mathematical models provide the opportunity to evaluate data and formulate forecasts for the future.

By combining digital twins with predictive analytics, organizations can use the digital twin's digitized data to make informed predictions about future events, such as equipment failures, demand fluctuations, and resource utilization.

In the manufacturing sector, a digital twin could be employed to track a manufacturing line's operation in real-time, and predictive analytics can be used to evaluate the data and foresee whenever machinery is most likely to break down. This information can then be used to schedule maintenance and reduce downtime.

Digital twins shall observe patients' vital signs continuously. This monitoring gives an idea to classify whether the patient needs immediate attention. This information can then be used to provide proactive care and reduce the risk of adverse events.

5.3.2 Unlocking the Value of IoT and IIoT Data with Digital Twins

Connecting these technologies is a powerful combination that allows organizations to collect, interpret, manage, and store up-to-date physical assets, making proactive decisions, improving operational efficiency, and reducing costs. This technology creates a digitized infrastructure, which can be updated instantaneously through sensors, connected devices, and analytics. Organizations can use this data to gain insight into their operations, identify potential problems and take corrective action. By allowing businesses to make digital duplicates of real assets, tools, and procedures, this technique provides a single point of reference for data analysis, prediction, and decision-making (Foth et al., 2020). It provides a comprehensive view of physical assets and their performance, enabling predictive maintenance, optimization, and performance management. Digital twin also enables predictive analytics, machine learning, and artificial intelligence, leading to more efficient operations and better customer experiences. By connecting IoT, IIoT, and data analytics, businesses can create virtual representations of their physical systems, enabling them to visualize how those systems perform in real-time and make data-driven decisions. This can help enterprises to optimize their operations and stay ahead of the competition. Additionally, DTT can visualize and simulate the impact of proposed changs, further optimizing processes and reducing costs.

5.4 Ensuring the Interoperability of IoT, IIoT, Data Analytics, and Digital Twins

Digital twin transforms how organizations manage and monitor their physical assets and systems. Digital twins allow organizations to simulate and analyze their real-time behavior, performance, and interactions by creating a virtual representation of physical objects and systems. However, integrating DTT with others can be challenging and requires significant technical expertise and planning. As shown in Figure 5.3, the following are some key challenges organizations must overcome to integrate these technologies successfully.

Technical Challenges: implementing DTT requires significant technical expertise, including data collection and storage capability and analyze substantial

Figure 5.3 Digital twin technology integration challenges.

real-time data. This can challenge organizations without a strong data management infrastructure or skilled workforce. To overcome these challenges, organizations must invest in the right tools and technologies, including IoT devices, IIoT platforms, and data analytics software, and have a skilled workforce, including data scientists and engineers, who can analyze and interpret the data generated by digital twins.

Data Quality: ensuring that the data collected from IoT and IIoT devices is accurate, complete, and relevant is vigorous for DTTs realization. However, collecting high-quality data from these devices can be challenging, especially in real-world environments where conditions are constantly changing. To overcome these challenges, organizations must implement robust data quality control measures.

Integration Challenges: it might be difficult to integrate DTT into current systems and processes, particularly for businesses with legacy systems and procedures. This requires significant planning and coordination to collect and analyze data consistently and accurately. Organizations must take a holistic approach to DTT to overcome these challenges, incorporating it into their overall digital strategy and ensuring that it integrates with other key technologies and systems.

Cybersecurity Threats: one of the critical challenges associated with DTT is cybersecurity. As more and more devices and systems become connected, the risk of cyber-attacks, such as hacking, malware, and ransomware, increases. These attacks can compromise sensitive data and systems, leading to data breaches, intellectual property theft, and operations disruption. Enterprises must implement robust security protocols to safeguard their connected gadgets, and techniques can reduce or eliminate the risk, including firewalls, encryption, and access restrictions.

Data Privacy Threats: the collection and analysis of data from IoT and IIoT devices raise privacy concerns, as it can be used to monitor individuals, track their behavior, and collect sensitive personal information. Companies must implement robust privacy policies and data protection measures to ensure data is used ethically and responsibly. This includes data minimization, retention policies, and secure data storage and transfer methods.

Cost and Scalability: implementing DTT can be expensive, especially for organizations that need to upgrade their data management infrastructure and invest in new technologies and tools. Additionally, companies may find it hard to update their resources as DTT evolves and expands. Organizations must carefully plan their implementation to overcome these challenges and ensure they have the support needed to continue investing in and changing their DTT over time.

Human Error: while DTT offers significant benefits, it also introduces new risks, including potential human error. For example, if the digital twin is not configured correctly, it may provide inaccurate information or insights, leading to incorrect decisions or actions.

Lack of standardization: DTT lags in standardization, and organizations must find ways to work together to establish common standards and protocols. This will help ensure that DTT is widely adopted and that organizations can work together effectively to solve common challenges.

Technical skills gap: there is currently a shortage of technical skills in the field of DTT, and organizations need to find ways to upskill their employees and ensure that they have the skills required to implement DTT solutions effectively.

5.4.1 Examples of Successful Integration of IoT Technologies with DTT

Manufacturing: DTT maintains connected equipment devices' performance, accurately simulates their behavior, and optimizes production. This can help manufacturers identify and address inefficiencies and reduce costs. By connecting IoT devices to data warehouses, manufacturers can analyze and monitor their production process in real-time and identify improvement areas (Liu et al., 2023; Shao et al., 2023b; Zhu et al., 2022b).

Smart Cities: digital twin shall help in monitoring data of a city's infrastructure, allowing city planners and administrators to make informed decisions on how to lead sophisticated life for its citizens (Gericke et al., 2019; Huang et al., 2022a; Wang et al., 2023). For instance, they can use the data to optimize traffic flow and reduce emissions. Smart cities shall depend on integrated DTT to monitor traffic, optimize public services, and improve citizen experience. By connecting devices to the city's systems, data necessitated for urban planning would be easy and reduce traffic congestion.

Digital Twin in Automotive: in the automotive industry, a digital twin can track performance data from connected cars, such as fuel efficiency, engine temperature, and tire pressure (Tavakolibasti et al., 2023; Zhang et al., 2022). Collected data provides support to inform decisions related to vehicle maintenance and repair and optimize the driving experience.

Healthcare: digital twins can provide real-time patient health status updates to healthcare providers (Ahmed et al., 2022; Maleki et al., 2022). By combining facts from various devices and sensors, the digital twin model helps diagnose and treat illnesses before they become more serious.

Figure 5.4 illustrates the applications of the DTT. It is widely used in various general-purpose applications to improve workflow performance.

5.4.2 Companies Implemented Digital Twins with Supporting Technologies

Philips Healthcare: Philips Healthcare uses a digital twin to help healthcare providers create comprehensive models of their physical assets. This includes tracking patient and equipment data, analyzing data to gain insights, and predicting future outcomes.

GE Digital: GE Digital uses digital twinning to create virtual replicas of real-world machines and systems. This allows them to predict and optimize machine

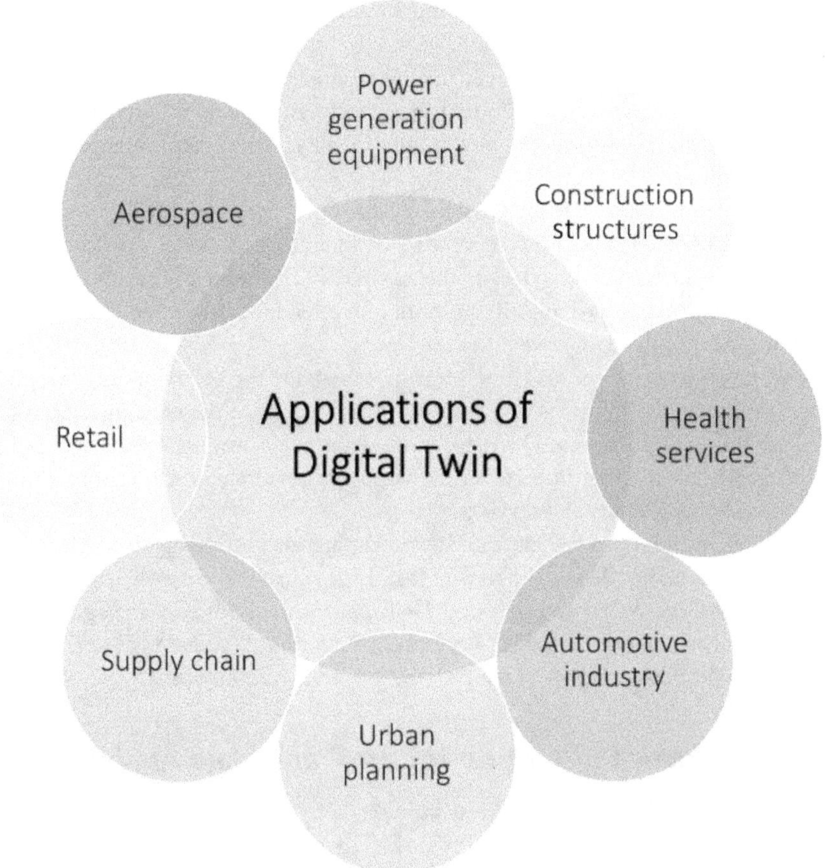

Figure 5.4 Real-world examples of digital twins in use.

performance, track real-time machine data, and create a digital twin of entire factories.

Siemens: Siemens uses a digital twin to interconnect merchandise with the data their customers and partners generate. They can collect and analyze data from multiple sources to gain insights and optimize their products and services.

Microsoft: Microsoft enhances its IoT offerings with a digital twin. Developers may attach IoT gadgets according to their cloud services and build electronic copies of real-world devices. As a result, they can continuously oversee, track, analyze, and enhance the performance of their IoT systems.

IoT, IIoT, and data analytics technologies work together to enable the development of DTT in businesses. The collected data is processed and analyzed using ML-based IIoT skills, which offer knowledge about how things work and the function of what is happening or item (Balaji et al., 2020).

Digital twin benefits the real-time physical systems by simulating and analyzing the data. This enables organizations to test and optimize their products and processes before they are deployed in the real world, reducing the risk of failure and improving efficiency. To boost productivity and decrease interruptions, firms might, for instance, utilize DTT to mimic their factories, spot obstacles, and optimize workflows.

The enhancement of upkeep and fix is yet another benefit of DTTs. Organizations recognize possible issues and take steps to avoid failures by eyeing how their physical structures behave instantaneously. For example, DTT helps airline systems monitor aircraft performance and identify potential maintenance issues, reducing the risk of downtime and improving overall efficiency.

By giving enterprises real-time information about the behavior and efficiency of physical items, digital twins also help businesses enhance the consumer experience. For example, producers use DTTs for monitoring their products' performance in the field, providing insights into customer usage and preferences and enabling them to improve their products and services.

Organizations need to have a strong and secured real-time database management infrastructure to implement digital twins. They must also have the right tools and technologies, including IoT devices, IIoT platforms, and data analytics software. Additionally, they need a skilled workforce, including data scientists and engineers, who can analyze and interpret the data generated by digital twins.

5.4.2 Influence of DTT on IoT, IIoT, and Data Analytics

The influence of DTTs on these technologies is immense. Digital twin provides virtual models for IoT and IIoT networks. These virtual models can monitor and control required physical objects' behavior. This data helps to improve operational efficiency and reduce costs and future risks. Digital twin also enables organizations to leverage data from their physical environment for analytics. By having a virtual representation of their background, organizations shall analyze the collected data to identify any patterns or correlations and any potential issues that need to be addressed. This data can then be used to improve processes or products and make decisions around future investments.

Furthermore, DTT can help organizations better manage their IoT and IIoT devices. By creating a virtual model of these devices, organizations can better manage them and keep track of their usage and any potential issues that may arise. This information can then be used to improve systems, processes, and products, as well as to identify any potential vulnerabilities. DTTs can be implemented to enhance data analysis. Integrating physical and virtual model data makes gaining more significant insights into the thing's performance possible. This data can be used to develop better predictive models, optimize maintenance and operations, and identify opportunities for improvement. Additionally, the data gathered from DTTs, with the help of ML and AI applications, allows for additional automated decision support.

DTT significantly impacts the IoT, IIoT, and data analytics. By creating a virtual replica of physical objects, digital twins allow for a more comprehensive and detailed analysis of data collected from IoT and IIoT devices. This provides data analytics to more accurately identify patterns, trends, and correlations in the data, providing insights that can be used to improve business processes, products, services, and operations. Data collected from DTTs can also power ML algorithms, allowing organizations to explore additional functions. This data can also be used to optimize processes and boost operational efficiency.

Finally, digital twins benefit the development of new stuff and skills and improve customer experiences. Organizations can rapidly develop, test, and launch new products and services by leveraging the data from digital twins. This, in turn, can help organizations remain competitive in the market and increase their profits. Integrating DTTs with IoT and IIoT provides revolutionary results (Bhatti et al., 2021b). Providing more significant insights into physical objects' performance can enable businesses to make smarter decisions, improve operational efficiency, and reduce costs.

5.4.3 Integration in Digital Twins

DTTs are integrated to other systems, technologies, and data resources to provide a complete picture of a modeled actual system. This integration can be done in several ways, including the following:

API Integration: digital twins can be integrated with other systems and technologies through APIs (application programming interfaces), which allow data to be exchanged between systems in a standardized format.

Middleware Integration: middleware can integrate digital twins with other systems and technologies. Middleware acts as a bridge between systems, allowing data to be exchanged and processed between different systems.

Data Exchange: DTTs can give immediate feedback and data concerning the simulated material system by integrating with supplementary data sources, such as sensing. Present-day information is provided by using this data for updating the digital twin.

Modeling and Simulation: digital twins can be integrated with simulating tools to achieve a precise, comprehensive physical system.

Cloud Integration: digital twins can be integrated with cloud-based technologies, such as cloud storage and cloud computing, to provide scalable and accessible data storage and processing capabilities.

Edge Computing Integration: digital twins can be integrated with edge computing technologies to reduce latency and increase digital twin accuracy.

Integrating digital twins into other systems and technologies is crucial for providing a comprehensive view of the modeled physical system and making informed decisions based on the digital twin. The specific integration method depends on the requirements and constraints of the modeled system and the technology used.

5.4.4 Threats of Linking these Technologies

Integrating IoT, IIoT, and DTTs is witnessing enormous benefits for organizations. Companies can simulate and analyze their behavior, performance, and interactions in actual instances by creating simulated systems. However, as with any new technology, this trend has significant risks and challenges. Specific essential threats should be tackled to ensure digital twin's safe and effective use.

Cybersecurity Threats: as more and more devices and systems become connected, the risk of cyber-attacks, such as hacking, malware, and ransomware, increases. These attacks can compromise sensitive data and systems, leading to data breaches, intellectual property theft, and operations disruption. To mitigate these risks, robust cybersecurity firewalls must be installed.

Data Privacy Threats: data collected from IoT and IIoT devices raises privacy concerns, as it can be used to monitor individuals, track their behavior, and collect sensitive personal information. Companies must implement robust privacy policies and data protection measures for data safety. This includes data minimization, retention policies, and secure data storage and transfer methods.

Technical Challenges: implementing DTT requires significant technical expertise, including a large volume of real-time data storage and smooth retrieval. This can challenge organizations without a robust data management infrastructure or skilled workforce. To overcome these challenges, companies must invest in the right tools and technologies, including IoT devices, IIoT platforms, and data analytics software, and have a skilled workforce, including data scientists and engineers, who can analyze and interpret the data generated by digital twins.

Cost and Scalability: implementing DTT can be expensive, especially for organizations that need to upgrade their data management infrastructure and invest in new technologies and tools. Additionally, as DTT evolves and expands, it can be difficult for companies to track updates. To overcome these challenges, companies must carefully plan their implementation and ensure they have the resources and support needed to continue investing in and evolving their DTT over time.

Integration Challenges: integrating DTT with existing systems and processes can be challenging, especially for organizations with legacy systems and processes (Balaji et al., 2020; Jiang et al., 2022). To overcome these challenges, companies must take a holistic approach to DTT, incorporating it into their digital strategy and ensuring it integrates with other key technologies and systems.

Human Error: while DTT offers significant benefits, it also introduces new risks, including potential human error. For example, if the digital twin is not configured correctly, it may provide inaccurate information or insights, leading to incorrect decisions or actions. To mitigate this risk, organizations must have robust processes and procedures to ensure that digital twins are configured correctly and that the data they provide is accurate and reliable.

Data Quality: Another challenge associated with DTT is data quality. They ensure that the data collected from IoT and IIoT devices is accurate, complete, and relevant to specific applications. However, collecting high-quality data from these devices can be challenging, especially in real-world environments where conditions are constantly changing.

5.5 Summary

In today's rapidly advancing technology landscape, the connection between IoT, IIoT, and data analytics with DTT has become increasingly important. This interconnection enables real-time data analysis from their physical assets. This strengthens the organization's decision-making probability, optimizes operations, improves efficiency, and enhances competitiveness.

The first part of this chapter explores the concept of IoT and IIoT and how these technologies are being used to collect data from physical assets and systems. IoT provides an internet-enabled interconnected device that can transfer and receive data and perform actions, while IIoT refers to the application of industrial IoT technology. These technologies allow organizations to create repository of physical assets and systems, including their behavior, performance, and interactions.

The next part of the chapter focuses on the importance of data analytics in IoT and IIoT approaches. IoT and IIoT real-time organization data analytics allow organizations to analyze and interpret results from the data (Gharaibeh et al., 2017). Organizations can gain valuable insights into their physical assets and systems using advanced data analytics techniques, including their behavior, performance, and interactions. This information can be used to optimize operations, improve efficiency, and enhance decision-making.

The third part of the chapter introduces DTT. Digital twins allow the virtual representation of physical assets and systems of the organizations. (Huang et al., 2022a). This simulation is only as accurate and valuable as the data used to create it, which is where data analytics comes in. By using data analytics with DTT, organizations can gain valuable insights into their physical assets and systems, allowing them to make informed decisions, enhance predictive maintenance, improve resource utilization, and monitor their real-time systems (Sengupta et al., 2021).

The final part of the chapter explores some of the challenges associated with integrating these technologies. Ensuring the data collected from IoT and IIoT devices is accurate, complete, and relevant is critical to the success of data analytics with DTT. However, collecting high-quality data from these devices can be challenging, especially in real-world environments where conditions are constantly changing. Integrating data analytics with DTT can also be complex, especially for organizations with legacy systems and processes. Additionally, cybersecurity threats are a key challenge associated with data analytics with digital twins, as the risk of cyber-attacks, such as hacking, malware, and ransomware, increases with the growing number of connected devices and systems.

Finally, connecting IoT, IIoT, and data analytics with DTTs can revolutionize how organizations run. Organizations may acquire essential insights and arrive at knowledgeable choices by analyzing real-time data from physical assets, optimizing operations, boosting efficacy, and increasing competitiveness. Whilst linking IoT, IIoT, and data analytics with DTTs presents obstacles, they may be solved by taking a holistic approach and incorporating these technologies into an organization's entire digital planning process.

Bibliography

Ahmed, I., Ahmad, M., & Jeon, G. (2022). Integrating digital twins and deep learning for medical image analysis in the era of COVID-19. *Virtual Reality & Intelligent Hardware, 4*(4), 292–305. https://doi.org/10.1016/J.VRIH.2022.03.002

Akhter, R., & Sofi, S. A. (2022). Precision agriculture using IoT data analytics and machine learning. *Journal of King Saud University – Computer and Information Sciences, 34*(8), 5602–5618. https://doi.org/10.1016/J.JKSUCI.2021.05.013

Balaji, V. R., Maheshwaran, S., Rajesh Babu, M., Kowsigan, M., Prabhu, E., & Venkatachalam, K. (2020). Combining statistical models using modified spectral subtraction method for embedded system. *Microprocessors and Microsystems, 73*, 102957. https://doi.org/10.1016/J.MICPRO.2019.102957

Bhatti, G., Mohan, H., & Raja Singh, R. (2021a). Towards the future of smart electric vehicles: Digital twin technology. *Renewable and Sustainable Energy Reviews, 141*, 110801. https://doi.org/10.1016/J.RSER.2021.110801

Bhatti, G., Mohan, H., & Raja Singh, R. (2021b). Towards the future of smart electric vehicles: Digital twin technology. *Renewable and Sustainable Energy Reviews, 141*, 110801. https://doi.org/10.1016/J.RSER.2021.110801

Boobalan, P., Ramu, S. P., Pham, Q. V., Dev, K., Pandya, S., Maddikunta, P. K. R., Gadekallu, T. R., & Huynh-The, T. (2022). Fusion of federated learning and industrial Internet of Things: A survey. *Computer Networks, 212*, 109048. https://doi.org/10.1016/J.COMNET.2022.109048

Corallo, A., Lazoi, M., Lezzi, M., & Luperto, A. (2022). Cybersecurity awareness in the context of the Industrial Internet of Things: A systematic literature review. *Computers in Industry, 137*, 103614. https://doi.org/10.1016/J.COMPIND.2022.103614

Elazhary, H. (2019). Internet of Things (IoT), mobile cloud, cloudlet, mobile IoT, IoT cloud, fog, mobile edge, and edge emerging computing paradigms: Disambiguation and research directions. *Journal of Network and Computer Applications, 128*(June 2018), 105–140. https://doi.org/10.1016/j.jnca.2018.10.021

Ezhilmathi, S., & Samy, S. S. (2022). Evaluating Performance Metrics in Classifying Bitcoin Mixing Services Using Decision Tree Algorithm. *2022 International Conference on Electronic Systems and Intelligent Computing (ICESIC)*, 304–310. https://doi.org/10.1109/ICESIC53714.2022.9783490

Fernandes, S., João, D., Cardoso, B., Martins, M., & Carvalho, E. (2022). Digital twin concept developing on an electrical distribution system—An application case. *Energies, 15*, 2836. https://doi.org/10.3390/en15082836

Foth, M., Anastasiu, I., Mann, M., & Mitchell, P. (2020). From Automation to Autonomy: Technological Sovereignty for Better Data Care in Smart Cities. In *Advances in 21st Century Human Settlements*. https://doi.org/10.1007/978-981-15-8670-5_13

Fuller, A., Fan, Z., Day, C., & Barlow, C. (2020). Digital twin: Enabling technologies, challenges and open research. *IEEE Access, 8*, 108952–108971. https://doi.org/10.1109/ACCESS.2020.2998358

Gericke, G. A., Kuriakose, R. B., Vermaak, H. J., & Mardsen, O. (2019). Design of Digital Twins for Optimization of a Water Bottling Plant. *IECON 2019 – 45th Annual Conference of the IEEE Industrial Electronics Society, 1*, 5204–5210. https://doi.org/10.1109/IECON.2019.8926880

Gharaibeh, A., Salahuddin, M. A., Hussini, S. J., Khreishah, A., Khalil, I., Guizani, M., & Al-Fuqaha, A. (2017). Smart cities: A survey on data management, security, and enabling technologies. *IEEE Communications Surveys & Tutorials, 19*(4), 2456–2501. https://doi.org/10.1109/COMST.2017.2736886

Ghenai, C., Husein, L. A., al Nahlawi, M., Hamid, A. K., & Bettayeb, M. (2022). Recent trends of digital twin technologies in the energy sector: A comprehensive review. *Sustainable Energy Technologies and Assessments, 54*, 102837. https://doi.org/10.1016/J.SETA.2022.102837

Gong, H., Cheng, S., Chen, Z., Li, Q., Quilodrán-Casas, C., Xiao, D., & Arcucci, R. (2022). An efficient digital twin based on machine learning SVD autoencoder and generalised latent assimilation for nuclear reactor physics. *Annals of Nuclear Energy, 179*, 109431. https://doi.org/10.1016/J.ANUCENE.2022.109431

Huang, W., Zhang, Y., & Zeng, W. (2022a). Development and application of digital twin technology for integrated regional energy systems in smart cities. *Sustainable Computing: Informatics and Systems, 36*, 100781. https://doi.org/10.1016/J.SUSCOM.2022.100781

Huynh-The, T., Pham, Q. V., Pham, X. Q., Nguyen, T. T., Han, Z., & Kim, D. S. (2023). Artificial intelligence for the metaverse: A survey. *Engineering Applications of Artificial Intelligence, 117*, 105581. https://doi.org/10.1016/J.ENGAPPAI.2022.105581

Ji, T., Huang, H., & Xu, X. (2022). Digital twin technology—A bibliometric study of top research articles based on local citation score. *Journal of Manufacturing Systems, 64*, 390–408. https://doi.org/10.1016/J.JMSY.2022.06.016

Jiang, T., Zhou, J., Zhao, J., Wang, M., & Zhang, S. (2022). A multi-dimensional cognitive framework for cognitive manufacturing based on OAR model. *Journal of Manufacturing Systems, 65*, 469–485. https://doi.org/10.1016/J.JMSY.2022.09.019

Kaluzny, B. L. (2021). Data analytics in military human performance: Getting in the game: Summary of a keynote address. *Journal of Science and Medicine in Sport, 24*(10), 970–974. https://doi.org/10.1016/J.JSAMS.2021.04.003

Liu, S., Lu, Y., Li, J., Shen, X., Sun, X., & Bao, J. (2023). A blockchain-based interactive approach between digital twin-based manufacturing systems. *Computers & Industrial Engineering, 175*, 108827. https://doi.org/10.1016/J.CIE.2022.108827

Maleki, S., Jazdi, N., & Ashtari, B. (2022). Intelligent digital twin in health sector: Realization of a software-service for requirements- and model-based-systems-engineering. *IFAC-PapersOnLine, 55*(19), 79–84. https://doi.org/10.1016/J.IFACOL.2022.09.187

Mauro, F., & Kana, A. A. (2023). Digital twin for ship life-cycle: A critical systematic review. *Ocean Engineering, 269*, 113479. https://doi.org/10.1016/J.OCEANENG.2022.113479

Nyoman Kutha Krisnawijaya, N., Tekinerdogan, B., Catal, C., & Tol, R. van der. (2022). Data analytics platforms for agricultural systems: A systematic literature review. *Computers and Electronics in Agriculture, 195*, 106813. https://doi.org/10.1016/J.COMPAG.2022.106813

Semeraro, C., Olabi, A. G., Aljaghoub, H., Alami, A. H., al Radi, M., Dassisti, M., & Abdelkareem, M. A. (2023a). Digital twin application in energy storage: Trends and challenges. *Journal of Energy Storage, 58*, 106347. https://doi.org/10.1016/J.EST.2022.106347

Semeraro, C., Olabi, A. G., Aljaghoub, H., Alami, A. H., al Radi, M., Dassisti, M., & Abdelkareem, M. A. (2023b). Digital twin application in energy storage: Trends and challenges. *Journal of Energy Storage, 58*, 106347. https://doi.org/10.1016/J.EST.2022.106347

Sengupta, J., Ruj, S., & Bit, S. D. (2021). A Secure fog-based architecture for industrial Internet of Things and industry 4.0. *IEEE Transactions on Industrial Informatics, 17*(4), 2316–2324. https://doi.org/10.1109/TII.2020.2998105

Sengupta, J., Ruj, S., & das Bit, S. (2020). A comprehensive survey on attacks, security issues and blockchain solutions for IoT and IIoT. *Journal of Network and Computer Applications, 149*, 102481. https://doi.org/10.1016/J.JNCA.2019.102481

Shao, G., Hightower, J., & Schindel, W. (2023a). Credibility consideration for digital twins in manufacturing. *Manufacturing Letters, 35*, 24–28. https://doi.org/10.1016/J.MFGLET.2022.11.009

Shao, G., Hightower, J., & Schindel, W. (2023b). Credibility consideration for digital twins in manufacturing. *Manufacturing Letters, 35*, 24–28. https://doi.org/10.1016/J.MFGLET.2022.11.009

Sifat, Md. M. H., Choudhury, S. M., Das, S. K., Ahamed, Md. H., Muyeen, S. M., Hasan, Md. M., Ali, Md. F., Tasneem, Z., Islam, Md. M., Islam, Md. R., Badal, Md. F. R., Abhi, S. H., Sarker, S. K., & Das, P. (2023). Towards electric digital twin grid: Technology and framework review. *Energy and AI, 11*, 100213. https://doi.org/10.1016/J.EGYAI.2022.100213

Singh, I., Centea, D., & Elbestawi, M. (2019). IoT, IIoT and cyber-physical systems integration in the SEPT learning factory. *Procedia Manufacturing, 31*, 116–122. https://doi.org/10.1016/J.PROMFG.2019.03.019

Somers, R. J., Douthwaite, J. A., Wagg, D. J., Walkinshaw, N., & Hierons, R. M. (2022). Digital-twin-based testing for cyber–physical systems: A systematic literature review. *Information and Software Technology*, 107145. https://doi.org/10.1016/J.INFSOF.2022.107145

Tao, F., Xiao, B., Qi, Q., Cheng, J., & Ji, P. (2022). Digital twin modeling. *Journal of Manufacturing Systems, 64*, 372–389. https://doi.org/10.1016/J.JMSY.2022.06.015

Tavakolibasti, M., Meszmer, P., Böttger, G., Kettelgerdes, M., Elger, G., Erdogan, H., Seshaditya, A., & Wunderle, B. (2023). Thermo-mechanical-optical coupling within a digital twin development for automotive LiDAR. *Microelectronics Reliability, 141*, 114871. https://doi.org/10.1016/J.MICROREL.2022.114871

Wang, W., He, F., Li, Y., Tang, S., Li, X., Xia, J., & Lv, Z. (2023). Data information processing of traffic digital twins in smart cities using edge intelligent federation learning. *Information Processing & Management, 60*(2), 103171. https://doi.org/10.1016/J.IPM.2022.103171

Zhang, Q., Shen, S., Li, H., Cao, W., Tang, W., Jiang, J., Deng, M., Zhang, Y., Gu, B., Wu, K., Zhang, K., & Liu, S. (2022). Digital twin-driven intelligent production line for automotive MEMS pressure sensors. *Advanced Engineering Informatics, 54*, 101779. https://doi.org/10.1016/J.AEI.2022.101779

Zhu, Q., Huang, S., Wang, G., Moghaddam, S. K., Lu, Y., & Yan, Y. (2022a). Dynamic reconfiguration optimization of intelligent manufacturing system with human-robot collaboration based on digital twin. *Journal of Manufacturing Systems, 65*, 330–338. https://doi.org/10.1016/J.JMSY.2022.09.021

Zhu, Q., Huang, S., Wang, G., Moghaddam, S. K., Lu, Y., & Yan, Y. (2022b). Dynamic reconfiguration optimization of intelligent manufacturing system with human-robot collaboration based on digital twin. *Journal of Manufacturing Systems, 65*, 330–338. https://doi.org/10.1016/J.JMSY.2022.09.021

Chapter 6

Simulation Strategies for Analyzing Data

S.S. Darly and D. Kadhiravan

6.1 Defining Data

Data refers to any collection of information that can be analyzed and interpreted to gain insights or knowledge. It can be in various forms, including text, numbers, images, videos, or any other type of digital information. In today's digital age, data is generated at an unprecedented rate by various sources such as social media, sensors, and IoT devices.

There are two main types of data: qualitative and quantitative data.

Qualitative Data: qualitative data is non-numerical data that is descriptive in nature. It provides information about the qualities or characteristics of a particular phenomenon, such as attitudes, beliefs, opinions, and behaviors. Qualitative data is typically collected through methods, such as interviews, observations, and focus groups.

Qualitative data can be analyzed through various techniques, including content analysis, grounded theory, and ethnography. These methods involve examining the data for patterns, themes, and relationships to gain insights and understanding about the phenomenon being studied.

Examples of qualitative data include the following:

1. Interview transcripts: transcripts of interviews with participants, which provide detailed information about their experiences, opinions, and attitudes.

 DOI: 10.1201/9781003469612-6

2. Field notes: notes taken during observation of a particular phenomenon, which describe behaviors, interactions, and other aspects of the situation.
3. Open-ended survey responses: survey responses that allow participants to provide detailed information about their experiences or opinions, rather than selecting from a list of predetermined options.
4. Focus group transcripts: transcripts of discussions in a group setting, which provide information about the attitudes and opinions of the participants.

Qualitative data is often used in social science research to explore complex phenomena and gain insights into human behavior. It is also useful in fields such as marketing, where understanding consumer attitudes and preferences is essential for developing effective advertising and marketing strategies [1]. Overall, qualitative data provides rich, detailed information that can help researchers gain a deeper understanding of the phenomenon being studied.

Quantitative Data: quantitative data is numerical data that is used for statistical analysis. It provides information about the quantities, measurements, or counts of a particular phenomenon. Quantitative data is often collected through methods such as experiments, surveys, and observational studies.

Quantitative data can be analyzed through various statistical techniques, including descriptive statistics, inferential statistics, and regression analysis. These methods involve using mathematical and statistical tools to analyze the data and draw conclusions about the phenomenon being studied.

Examples of quantitative data include the following:

1. Height measurements: measurements of the height of individuals in a sample, which can be used to determine the average height of the population.
2. Test scores: scores on a standardized test, which can be used to determine the performance of students or the effectiveness of a particular educational program.
3. Sales figures: data on the sales of a particular product or service, which can be used to analyze market trends and make business decisions.
4. Survey responses: responses to closed-ended survey questions that provide numerical data, such as age, income, or rating scales.

Quantitative data is often used in scientific research to test hypotheses and make predictions about the phenomenon being studied. It is also useful in fields, such as finance, where numerical data is essential for making informed investment decisions. Overall, quantitative data provides precise, measurable information that can be used to make objective decisions and draw reliable conclusions.

Data can also be categorized as structured and unstructured data.

Structured Data: structured data refers to data that is organized in a specific format or structure, such as a table, database, or spreadsheet. Structured data is typically stored in a consistent format, with defined fields and categories that make it easy to search, sort, and analyze the data. Structured data can be easily processed and analyzed using software tools such as databases, statistical software, and programming languages like Python and SQL. These tools can be used to extract, manipulate, and analyze the data to gain insights and make informed decisions.

Examples of structured data include the following:

1. Customer data: data on customer demographics, purchase history, and behavior, which can be used to develop targeted marketing campaigns.
2. Financial data: data on financial transactions, budgets, and expenses, which can be used to analyze company performance and make financial decisions.
3. Inventory data: data on inventory levels, sales, and restocking, which can be used to manage inventory and optimize supply chain operations.
4. Employee data: data on employee demographics, job titles, and performance, which can be used to track employee productivity and make HR decisions.

Structured data is commonly used in business and scientific research to make data-driven decisions. It is also used in fields such as healthcare, where electronic health records are used to store patient data and facilitate medical research. Overall, structured data provides a reliable and organized way to store and analyze data, making it easier to extract insights and make informed decisions.

Unstructured Data: unstructured data refers to data that does not have a predefined structure or format. It typically includes data in the form of text, images, audio, and video, as well as social media posts, email messages, and other types of content. Unstructured data is often generated in large volumes and can be difficult to process and analyze using traditional software tools.

Examples of unstructured data include the following:

1. Social media posts: posts on social media platforms such as X (formerly Twitter), Facebook, and Instagram, which can include text, images, and videos.
2. Email messages: email messages that contain unstructured text and attachments, such as images or documents.
3. Customer feedback: feedback from customers in the form of comments, reviews, and ratings, which can be difficult to categorize and analyze.
4. Audio and video files: audio and video recordings, which can include speech, music, and other sounds.

Sensor data: data from sensors that measure environmental factors such as temperature, humidity, and air quality. Unstructured data is increasingly being used in fields such as natural language processing, computer vision, and machine learning, where algorithms can be used to analyze and interpret the data. Software tools such as text analytics, image recognition, and speech recognition can be used to extract insights from unstructured data, allowing organizations to make more informed decisions. Overall, unstructured data provides a wealth of information that can be used to gain insights and make informed decisions, but it requires specialized tools and techniques to process and analyze.

In addition to these categories, there are also big data and small data.

Big Data: big data refers to extremely large and complex data sets that are beyond the ability of traditional data processing tools to manage, process, and analyze within a reasonable timeframe. The term "big" is relative and can vary depending on the context, but it generally refers to data sets that are too large and too diverse to be analyzed using traditional data processing techniques.

Big data is typically characterized by the three "Vs": volume, velocity, and variety. Volume refers to the sheer amount of data, which can be in the petabyte or exabyte range. Velocity refers to the speed at which data is generated and needs to be processed. Variety refers to the different types of data that make up a big data set, including structured data (such as data in databases), semi-structured data (such as data in XML or JSON formats), and unstructured data (such as text, images, and videos).

Big data can be generated from a variety of sources, including social media, IoT devices, transactional systems, and scientific experiments. Big data is often used to identify patterns, trends, and insights that can help organizations make better decisions and improve their operations [2]. To manage and analyze big data, organizations typically use specialized software tools and platforms that can handle the scale and complexity of big data sets.

Small Data: small data refers to datasets that are relatively small in size, typically ranging from a few hundred to a few thousand data points. In contrast, big data refers to datasets that are extremely large in size, often in the range of terabytes or even petabytes. Small data may be easier to manage and analyze than big data because it requires less computational power and storage space. However, small data may still be useful in certain contexts, such as when analyzing data from a single source or when working with data that is difficult to collect or expensive to obtain. Overall, the definition of small data can vary depending on the specific field or application, but it generally refers to datasets that are manageable for analysis and do not require advanced data processing techniques or tools.

6.2 Introduction to Data Analysis

Data plays an essential role in many fields, including business, science, healthcare, and government. With the advancements in technology and the growing availability of data, the importance of data in decision-making and gaining insights has become even more critical. Proper collection, cleaning, and analysis of data can lead to better decision-making, improved efficiency, and increased productivity in organizations.

Data analysis is the process of examining and interpreting data using statistical and computational methods to extract meaningful insights and draw conclusions. It is an essential part of many fields, including business, science, social science, and healthcare, among others [5]. In this introduction to data analysis, we will discuss the basics of the process, the tools and techniques used, and the importance of data analysis.

The Data Analysis Process: the data analysis process involves several steps, including the following:

Data Collection: data collection is the process of gathering information and data from various sources, which can be used for analysis and decision-making. The process of data collection typically involves the following steps:

1. Define the research question: the first step in data collection is to define the research question or problem that needs to be addressed. This involves identifying the purpose of the data collection, the specific variables to be measured, and the population or sample to be studied.
2. Select the data collection method: the next step is to select the appropriate data collection method. The choice of method will depend on the research question, the type of data to be collected, and the resources available. Common data collection methods include surveys, interviews, observations, experiments, and secondary data analysis.
3. Design the data collection instrument: the data collection instrument is the tool used to collect the data, such as a questionnaire, interview guide, or observation protocol. The instrument should be designed to ensure that the data collected is reliable and valid, and that it measures the variables of interest.
4. Pilot test the instrument: before collecting data from the entire sample or population, it is important to pilot test the data collection instrument with a small group of participants. This will help to identify any problems with the instrument and make necessary revisions.
5. Collect the data: the next step is to collect the data from the sample or population. The data should be collected in a systematic and standardized manner to ensure that it is accurate and reliable.
6. Clean and organize the data: after collecting the data, it is important to clean and organize it to ensure that it is accurate and usable. This involves checking for missing or inconsistent data, and ensuring that the data is in the correct format for analysis.

7. Analyze the data: the final step in data collection is to analyze the data using appropriate statistical or analytical techniques. The results of the analysis can be used to draw conclusions and make informed decisions.

Overall, data collection is a critical component of the research process, and it is important to follow best practices to ensure that the data collected is accurate, reliable, and valid.

Data Cleaning: data cleaning, also known as data cleansing, is the process of identifying and correcting or removing errors, inconsistencies, and inaccuracies in data. The goal of data cleaning is to ensure that the data is accurate, consistent, and usable for analysis and decision-making.

The process of data cleaning typically involves the following steps:

1. Identify missing data: the first step in data cleaning is to identify any missing data, such as incomplete records or fields with missing values.
2. Handle missing data: once missing data has been identified, the next step is to handle it. This may involve imputing missing values using statistical methods or removing records with missing data.
3. Identify and handle duplicates: duplicate data can occur when multiple records have the same values for all or some of the variables. Identifying and handling duplicates is important to ensure that the data is not skewed and analysis results are accurate.
4. Check for outliers: outliers are data points that are significantly different from the rest of the data. Identifying and handling outliers is important to ensure that the data is not skewed and analysis results are accurate.
5. Standardize data: standardizing data involves ensuring that the data is in a consistent format and follows the same conventions, such as using the same units of measurement.
6. Correct errors and inconsistencies: finally, data cleaning involves identifying and correcting errors and inconsistencies in the data. This may involve reviewing individual records or using automated tools to detect and correct errors.

Data cleaning is a critical step in the data analysis process. By ensuring that the data is accurate, consistent, and usable, data cleaning helps to improve the reliability and validity of analysis results.

Data Exploration: data exploration is the process of analyzing and visualizing data to understand its characteristics and patterns. The goal of data exploration is to gain insights and identify relationships or trends in the data that can be used to inform further analysis or decision-making.

Data exploration typically involves the following steps:

1. Summarize data: the first step in data exploration is to summarize the data using descriptive statistics such as mean, median, and standard deviation. This helps to understand the distribution and variability of the data.
2. Visualize data: visualization is a powerful tool for exploring data. Data can be visualized using charts and graphs, such as scatter plots, histograms, and box plots, to identify patterns and relationships in the data.
3. Identify outliers: outliers are data points that are significantly different from the rest of the data. Identifying outliers is important to ensure that the data is not skewed and analysis results are accurate.
4. Identify correlations: correlations are relationships between variables in the data. Identifying correlations can help to understand how variables are related and may suggest possible causal relationships.
5. Identify trends: trends are patterns that emerge over time or across different variables. Identifying trends can help to understand how the data is changing and may suggest possible future outcomes.

Data exploration is a critical step in the data analysis process. By gaining insights into the data, data exploration helps to inform further analysis and decision-making. It can also help to identify potential issues or limitations with the data, which can be addressed through data cleaning or other data preparation techniques. Overall, data exploration is an important tool for unlocking the value of data and making informed decisions.

Data Analysis: in this step, various statistical and computational methods are used to analyze the data. The goal is to extract meaningful insights and draw conclusions from the data.

Interpretation and Communication: finally, the insights and conclusions drawn from the data need to be interpreted and communicated effectively to the stakeholders. This involves presenting the findings in a clear and concise manner, using visualizations, tables, and charts.

Data analysis can involve a wide range of techniques and methods, depending on the type of data and the questions being asked. Some common techniques used in data analysis include the below:

Descriptive statistics: descriptive statistics involve summarizing and visualizing data to understand its characteristics and patterns. This can include measures of central tendency (such as mean, median, and mode), measures of variability (such as standard deviation and range), and graphical representations (such as histograms and scatter plots).

Inferential statistics: inferential statistics involve using sample data to make inferences about a larger population. This can include techniques such as hypothesis testing and regression analysis.

Data mining: data mining involves using machine learning and other techniques to identify patterns and relationships in data. This can include clustering, classification, and association rule mining.

Text analysis: text analysis involves using natural language processing and other techniques to analyze unstructured text data, such as social media posts or customer reviews. This can include sentiment analysis, topic modeling, and text classification.

Visualization: visualization involves using charts, graphs, and other visual representations to communicate data insights. This can include static or interactive visualizations, such as heat maps or network diagrams.

6.3 Tools and Techniques Used in Data Analysis

Data analysis is the process of inspecting, cleaning, transforming, and modeling data to discover useful information, draw conclusions, and support decision-making [4]. Here are some commonly used tools and techniques in data analysis:

Spreadsheet tools: spreadsheet tools like Microsoft Excel or Google Sheets are commonly used for data analysis. They can perform basic calculations, create charts and graphs, and apply statistical functions.

Statistical software: statistical software like SPSS, SAS, and R are commonly used in data analysis. They provide more advanced statistical analysis tools, including regression analysis, ANOVA, and cluster analysis.

Data visualization tools: data visualization tools like Tableau, PowerBI, and QlikView are used to create interactive charts, graphs, and dashboards that make it easy to understand and communicate insights from data.

Data cleaning tools: data cleaning tools like OpenRefine and Trifacta are used to clean, transform and preprocess data to prepare it for analysis.

Machine learning tools: machine learning tools like TensorFlow and Scikit-learn are used to build predictive models and uncover patterns in data.

Text analytics tools: text analytics tools like RapidMiner and IBM Watson are used to analyze text data, including sentiment analysis, topic modeling, and text classification.

Big data processing tools: big data processing tools like Apache Hadoop and Apache Spark are used to process and analyze large volumes of data quickly and efficiently.

Data mining techniques: data mining techniques like association rule mining, clustering, and classification are used to identify patterns and relationships in data.

Exploratory data analysis techniques: exploratory data analysis techniques like histograms, scatter plots, and box plots are used to visualize and explore data.

Time series analysis techniques: time series analysis techniques like forecasting and trend analysis are used to analyze time-based data.

Making predictions: data analysis can be used to make predictions based on historical data, which can help organizations plan for the future.

Improving efficiency: data analysis can help identify inefficiencies and areas for improvement, which can help organizations optimize their operations.

Improving decision-making: data analysis provides insights and information that can help inform decision-making, leading to better outcomes.

Improving customer satisfaction: data analysis can help organizations better understand their customers' needs and preferences, leading to improved products and services.

6.3.1 Conclusion

Data analysis is an essential process for extracting meaningful insights and making informed decisions. It involves several steps, including data collection, cleaning, exploration, analysis, interpretation, and communication. There are many tools and techniques used in data analysis, including statistical analysis, machine learning, data visualization, text mining, and data mining. Data analysis is important for identifying patterns and trends, making predictions, improving efficiency, improving decision-making, and improving customer satisfaction.

6.4 Data Analysis with Simulation Strategy

Simulation strategy is a method of modeling complex systems or processes using computer software in order to understand and analyze their behavior. Simulation involves creating a virtual environment that replicates the real-world conditions and variables of the system or process being studied [7]. Figure 6.1 represents process flow for generalized simulation strategy for a generalized problem statement.

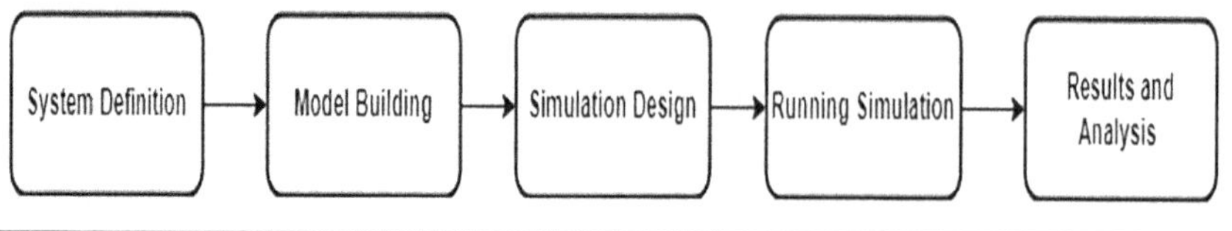

Figure 6.1 Process flow representation for simulation strategy.

A simulation strategy typically involves several key steps, including the following:

(i) Defining the system: this involves identifying the components, variables, and parameters of the system to be simulated.
(ii) Building the model: this involves developing a mathematical or computational model that represents the behavior of the system.
(iii) Designing the simulation: this involves selecting the appropriate simulation software and programming the model to simulate the behavior of the system.
(iv) Running the simulation: this involves executing the simulation and collecting data on the behavior of the system.
(v) Analyzing the results: this involves interpreting the data collected from the simulation to draw conclusions about the behavior of the system.

Simulation strategies are used in many fields, including engineering, science, finance, healthcare, and social sciences. They are particularly useful for analyzing complex systems or processes that are difficult to understand or observe in the real world, such as weather patterns, stock market fluctuations, or the spread of diseases [10]. By simulating these systems, researchers and decision-makers can better understand their behavior and make informed decisions based on the insights gained.

Analyzing data using simulation strategies involves using computer software to create a virtual environment that replicates the real-world conditions and variables of the system or process being studied [9]. The simulated environment allows researchers to study the behavior of the system or process under different conditions and to observe how it responds to different inputs or changes.

Simulation strategies can be used to analyze data in a variety of ways, including the following:

Predictive modeling: predictive modeling is the process of using statistical techniques and machine learning algorithms to analyze historical data and make predictions about future outcomes. The goal of predictive modeling is to identify patterns and relationships in the data that can be used to make accurate predictions about future

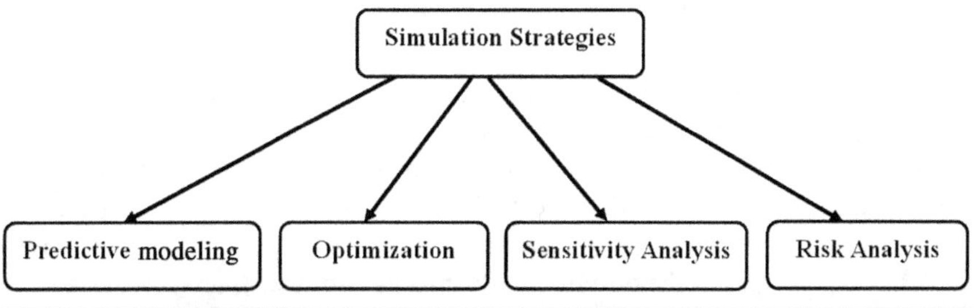

Figure 6.2 Types of simulation strategies.

events. Predictive modeling can be applied to a wide range of fields, including finance, healthcare, marketing, and sports. For example, in finance, predictive modeling can be used to forecast stock prices, identify investment opportunities, and manage risk. In healthcare, it can be used to predict patient outcomes, identify patients who are at high risk for developing certain conditions, and optimize treatment plans.

The success of predictive modeling depends on several factors, including the quality of the data used for training, the selection of appropriate features and algorithms, and the ability to test and validate the model. Properly executed, predictive modeling can provide valuable insights and help businesses and organizations make better decisions.

Optimization: optimization refers to the process of finding the best solution to a problem within a given set of constraints. In many fields, including engineering, computer science, economics, and operations research, optimization plays a crucial role in decision-making and problem-solving. Optimization problems can be formulated in many different ways, but they all involve finding the best solution among a set of possible options. For example, in engineering, optimization can be used to design more efficient systems or structures, while in economics, optimization can be used to maximize profits or minimize costs.

The process of optimization typically involves defining the objective function, which is the function that needs to be optimized, and the constraints, which are the limitations on the solution. Optimization algorithms then search for the solution that maximizes or minimizes the objective function, subject to the given constraints. There are many different optimization algorithms, ranging from simple methods like gradient descent to more complex methods like genetic algorithms and simulated annealing. The choice of algorithm depends on the specific problem being solved and the characteristics of the objective function and constraints. Optimization is a powerful tool that can be used to solve many different types of problems, from designing more efficient systems to improving decision-making in complex scenarios.

Sensitivity analysis: sensitivity analysis is a technique used to evaluate the effect of changes in input variables or parameters on the output of a model or system. It is commonly used in decision-making and risk management to assess the robustness and reliability of a model or system. The purpose of sensitivity analysis is to identify which variables or parameters have the greatest impact on the output of the model or system. By varying these variables or parameters and observing the resulting changes in the output, analysts can determine which variables or parameters are most critical to the performance of the system or model [3]. Sensitivity analysis can be performed in many different ways, depending on the type of model or system being analyzed. In some cases, it may involve simply varying one input variable at a time and observing the resulting changes in the output. In other cases, more complex methods such as Monte Carlo simulation may be used to generate a range of possible outcomes based on varying multiple input variables simultaneously.

Sensitivity analysis can be used in a wide range of fields, including finance, engineering, and healthcare. For example, in finance, sensitivity analysis can be used to evaluate the impact of changes in interest rates or market conditions on the performance of a portfolio, while in engineering, it can be used to assess the robustness of a design to variations in materials or operating conditions.

Risk analysis: risk analysis is the process of identifying, assessing, and managing potential risks or uncertainties that may affect a project, investment, or business decision. It involves evaluating the likelihood and potential impact of risks and developing strategies to mitigate or manage them. The goal of risk analysis is to minimize potential losses or negative consequences and maximize opportunities by making informed decisions based on a thorough understanding of potential risks and uncertainties.

There are several steps involved in risk analysis, including the following:

(i) Risk identification: identifying potential risks or uncertainties that may affect the project or decision.
(ii) Risk assessment: evaluating the likelihood and potential impact of each identified risk.
(iii) Risk management: developing strategies to mitigate or manage risks, including risk avoidance, risk transfer, risk reduction, and risk acceptance.
(iv) Risk monitoring: continuously monitoring and reassessing risks throughout the project or decision-making process.

Risk analysis is used in a variety of fields, including finance, engineering, healthcare, and project management. For example, in finance, risk analysis can be used to assess the risk of investments and develop strategies to minimize potential losses, while in engineering, it can be used to identify potential safety hazards and develop mitigation strategies. In healthcare, risk analysis can be used to assess the potential risks associated with new treatments or procedures and develop strategies to minimize those risks.

Simulation strategies can be particularly useful in fields such as engineering, finance, healthcare, and social sciences, where complex systems or processes are involved. By simulating these systems, researchers and decision-makers can gain insights into their behavior and make informed decisions based on the results. When analyzing data, simulation strategies can help you gain insights into how the data is distributed, how it behaves under different conditions, and how it might change over time. Here are some common simulation strategies for analyzing data:

1. *Monte Carlo simulation*: this is a widely used simulation technique that involves generating random values within the range of a variable, and then repeating the process thousands or millions of times to create a distribution of possible

outcomes. Monte Carlo simulation can be used to analyze the distribution of data, estimate probabilities, and simulate different scenarios.

2. *Agent-based modeling*: this simulation technique involves creating virtual agents that behave according to certain rules, and then simulating their interactions with each other and with their environment. Agent-based modeling can be used to analyze complex systems, such as ecosystems or social networks, and to explore how different factors affect the behavior of the system as a whole.
3. *Discrete event simulation*: this technique involves modeling a system as a series of events that occur over time, such as arrivals, departures, or resource allocations. Discrete event simulation can be used to analyze the behavior of complex systems, such as supply chains or manufacturing processes, and to optimize resource allocation and utilization.
4. *System dynamics modeling*: this technique involves creating a model of a system that includes feedback loops, delays, and other dynamic factors. System dynamics modeling can be used to analyze the behavior of complex systems over time, and to explore how different factors, such as policy changes or external shocks, might affect the system's behavior.

Overall, simulation strategies can provide valuable insights into how data behaves and how it might change under different conditions. By using simulation techniques, you can explore different scenarios, optimize resource utilization, and make better-informed decisions based on data analysis.

6.5 Monte Carlo Simulation

Monte Carlo simulation is a widely used simulation technique that is used to model complex systems and analyze data. The technique is named after the Monte Carlo Casino in Monaco, which is known for its games of chance and randomness. Monte Carlo simulation involves generating random samples or values within the range of a variable, and then repeating the process thousands or millions of times to create a distribution of possible outcomes. These outcomes are then used to estimate probabilities, evaluate risks, and simulate different scenarios.

Monte Carlo simulation is an important tool in many fields, including finance, engineering, physics, and medicine. Its importance stems from its ability to handle complex systems and processes with many interdependent variables and high degrees of uncertainty. It can be used to estimate the likelihood of success or failure of a project, to determine the optimal level of investment, to identify key risk factors, and to make informed decisions under uncertainty.

Some of the key benefits of using Monte Carlo simulation include the following:

1. *Improved decision-making:* Monte Carlo simulation provides decision-makers with a better understanding of the risks and uncertainties associated with a particular decision, enabling them to make more informed decisions.
2. *Better resource allocation:* Monte Carlo simulation can be used to optimize resource allocation by identifying the most important input variables and the range of possible outcomes.
3. *Improved accuracy:* Monte Carlo simulation provides a more accurate estimate of the expected value, variability, and potential risks associated with a system or process, compared to other methods.
4. *Flexibility:* Monte Carlo simulation can be used to model a wide range of systems and processes, making it a versatile tool in many fields.

Overall, Monte Carlo simulation is a powerful and important tool for modeling complex systems and processes, providing decision-makers with valuable insights and information that can help them make better decisions under uncertainty.

Sample problem statement: a company wants to determine the expected revenue from a new product launch. The product is expected to have a selling price of $50 per unit, but the company is uncertain about the demand for the product. The company estimates that the demand follows a normal distribution with a mean of 500 units and a standard deviation of 100 units. The company wants to use Monte Carlo simulation to estimate the expected revenue from the product launch.

Solution:

Step 1: Define the problem

The problem is to estimate the expected revenue from the new product launch using Monte Carlo simulation.

Step 2: Generate random inputs

The random input variable in this case is the demand for the new product, which follows a normal distribution with a mean of 500 units and a standard deviation of 100 units. We will generate 10,000 random values for the demand using a random number generator in Microsoft Excel.

Step 3: Run the simulation

For each value of demand, we can calculate the revenue using the formula:

$$\text{Revenue} = \text{Demand} * \text{Price}$$

where the price is $50 per unit. We will calculate the revenue for each of the 10,000 demand values.

Step 4: Analyze the results

We can analyze the results by calculating the mean and standard deviation of the revenue values. The mean revenue is the expected revenue from the product launch. We can also create a histogram of the revenue values to visualize the distribution of possible revenue outcomes.

Step 5: Draw conclusions

Based on the Monte Carlo simulation, the expected revenue from the product launch is $25,000, with a standard deviation of $5,000. The histogram shows that there is a range of possible revenue outcomes, with some outcomes being much lower or higher than the expected value. The company can use this information to make informed decisions about the product launch and to plan for various revenue scenarios.

MATLAB code that implements the Monte Carlo simulation for the problem statement we defined earlier and plots a histogram of the revenue outcomes

```
% Define parameters
price = 50; % price per unit
mean_demand = 500; % mean demand
std_demand = 100; % standard deviation of demand
n_simulations = 10000; % number of simulations to run

% Generate random demand values
demand_values = normrnd(mean_demand, std_demand, n_
  simulations, 1);

% Calculate revenue for each demand value
revenue_values = demand_values * price;

% Calculate mean and standard deviation of revenue
mean_revenue = mean(revenue_values);
std_revenue = std(revenue_values);

% Plot histogram of revenue values
histogram(revenue_values, 'Normalization', 'pdf');
title('Revenue Histogram');
xlabel('Revenue ($)');
ylabel('Probability Density');
grid on

% Print mean and standard deviation of revenue
fprintf('Expected revenue: $%.2f\n', mean_revenue);
fprintf('Standard deviation of revenue: $%.2f\n',
  std_revenue);
```

This code generates a simulation of demand and revenue for a single product with a fixed price per unit. It assumes a normal distribution of demand with a given mean and standard deviation, and then calculates the revenue for each demand value based on the fixed price. The code then calculates the mean and standard deviation of the revenue values, and plots a histogram of the revenue distribution. Finally, it prints the expected revenue and the standard deviation of revenue. This simulation can be useful for analyzing the expected revenue and risk associated with a particular pricing strategy for a product. By adjusting the price and distribution parameters, the simulation can be used to explore different scenarios and make informed pricing decisions.

The graph shown in Figure 6.3 generated by the code is a histogram of revenue values. The *x*-axis represents the range of possible revenue values, while the *y*-axis represents the probability density of each revenue value. Since the revenue is calculated based on the randomly generated demand values, the histogram shows the distribution of revenue that could be expected given the assumptions of the simulation. The shape of the histogram will depend on the distribution of demand and the fixed price per unit. In this case, since the demand is assumed to be normally

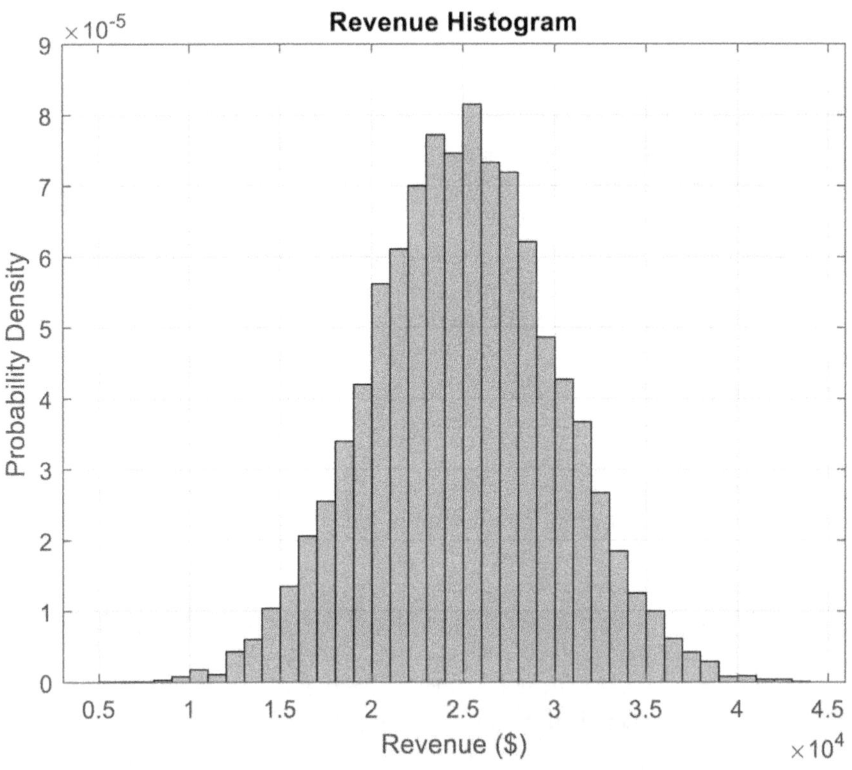

Figure 6.3 Monte Carlo simulation to calculate expected revenue of a company.

distributed and the price is fixed, the resulting revenue distribution will also be approximately normal. The histogram may also reveal information about the risk associated with the pricing strategy. For example, a wider distribution or greater variance in the revenue values may indicate higher risk, while a narrower distribution may indicate lower risk. Overall, the graph provides a visual representation of the revenue distribution that can be used to better understand the expected revenue and risk associated with a particular pricing strategy.

Another example problem related to agriculture is estimating the yield of a particular crop given uncertain weather conditions. Monte Carlo simulation can be used to estimate the yield distribution and provide insight into the range of possible outcomes under different weather scenarios. Suppose we want to estimate the yield of a particular crop based on the following information:

1. The average yield in good weather conditions is 500 bushels per acre.
2. The standard deviation of yield in good weather conditions is 50 bushels per acre.
3. The yield is expected to decrease by 10% in average in poor weather conditions.
4. The probability of poor weather conditions is estimated to be 30%.
5. To solve this problem using Monte Carlo simulation, we can follow these steps:
   I. Define the parameters:
      - mean_yield_good = 500
      - std_yield_good = 50
      - yield_decrease_poor = 0.1
      - poor_weather_prob = 0.3
      - n_simulations = 10,000
   II. Generate random weather conditions:
      - Generate n_simulations random numbers between 0 and 1.
      - If the random number is less than poor_weather_prob, set the weather condition to "poor", otherwise set it to "good".
   III. Generate random yield values:
      - For each simulation, generate a random yield value based on the weather condition.
      - If the weather condition is "good", generate a random value from a normal distribution with mean mean_yield_good and standard deviation std_yield_good.
      - If the weather condition is "poor", generate a random value from a normal distribution with mean (1-yield_decrease_poor)*mean_yield_good and standard deviation std_yield_good.
   IV. Calculate the mean and standard deviation of the yield distribution.
   V. Plot the yield distribution as a histogram.

Here's the code to implement these steps in MATLAB:

```
% Define parameters
mean_yield_good = 500; % average yield in good wea-
   ther conditions (bushels per acre)
std_yield_good = 50; % standard deviation of yield
   in good weather conditions (bushels per acre)
yield_decrease_poor = 0.1; % percentage decrease in
   yield in poor weather conditions
poor_weather_prob = 0.3; % probability of poor wea-
   ther conditions
n_simulations = 10000; % number of
   simulations to run

% Generate random weather conditions
weather_conditions = rand(n_simulations, 1) < poor_
   weather_prob;
weather_conditions = categorical(weather_
   conditions, [0 1], {'good' 'poor'});

% Generate random yield values
yield_values = zeros(n_simulations, 1);
for i = 1:n_simulations
    if weather_conditions(i) == "good"
    yield_values(i) = normrnd(mean_yield_good,
      std_yield_good);
   else
   yield_values(i) = normrnd((1-yield_decrease_
     poor)*mean_yield_good, std_yield_good);
    end
end

% Calculate mean and standard deviation of yield
mean_yield = mean(yield_values);
std_yield = std(yield_values);

% Plot histogram of yield values
histogram(yield_values, 'Normalization', 'pdf');
title('Yield Histogram');
xlabel('Yield (bushels per acre)');
ylabel('Probability Density');
grid on
```

```
% Plot probability of good and poor weather
figure;
pie([1-poor_weather_prob, poor_weather_prob],
   {'Good weather', 'Poor weather'});
title('Probability of Weather Conditions');

% Plot yield boxplot for different weather
   conditions
figure;
boxplot(yield_values, weather_conditions, 'Labels',
   {'Good', 'Poor'});
title('Yield vs. Weather Condition');
xlabel('Weather Condition');
ylabel('Yield (bushels per acre)');
grid on;

% Print mean and standard deviation of yield
fprintf('Expected yield: %.2f bushels per acre\n',
   mean_yield);
fprintf('Standard deviation of yield: %.2f bushels
   per acre\n', std_yield);
```

The code simulates a sample problem related to agriculture using Monte Carlo simulation. The problem assumes that weather conditions affect the yield of crops, and the goal is to estimate the expected yield and the associated uncertainty.

Figure 6.4 shows the histogram of the yield distribution generated by the simulation. The *x*-axis shows the yield values in bushels per acre, and the *y*-axis shows the probability density. The shape of the histogram is roughly Gaussian (bell-shaped), with the peak at the mean yield value. The spread of the distribution is determined by the standard deviation of the yield values. The histogram provides a visual representation of the distribution of the yield values, and can be used to estimate the probability of different yield values.

Figure 6.5 shows a scatter plot of the weather conditions in each simulation run. The *x*-axis shows the index of the simulation run, and the *y*-axis shows the weather condition (good or poor). The color of each point indicates the associated yield value. The scatter plot provides a visual representation of the relationship between weather conditions and yield values. The plot shows that poor weather conditions are associated with lower yield values, while good weather conditions are associated with higher yield values. The scatter plot can be used to identify any patterns or trends in the data, and to explore the relationship between different variables.

Figure 6.6 shows the pie chart has two slices, one for “Good weather” and the other for “Poor weather”. The size of each slice represents the probability of that weather condition occurring. The title of the chart is “Probability of Weather

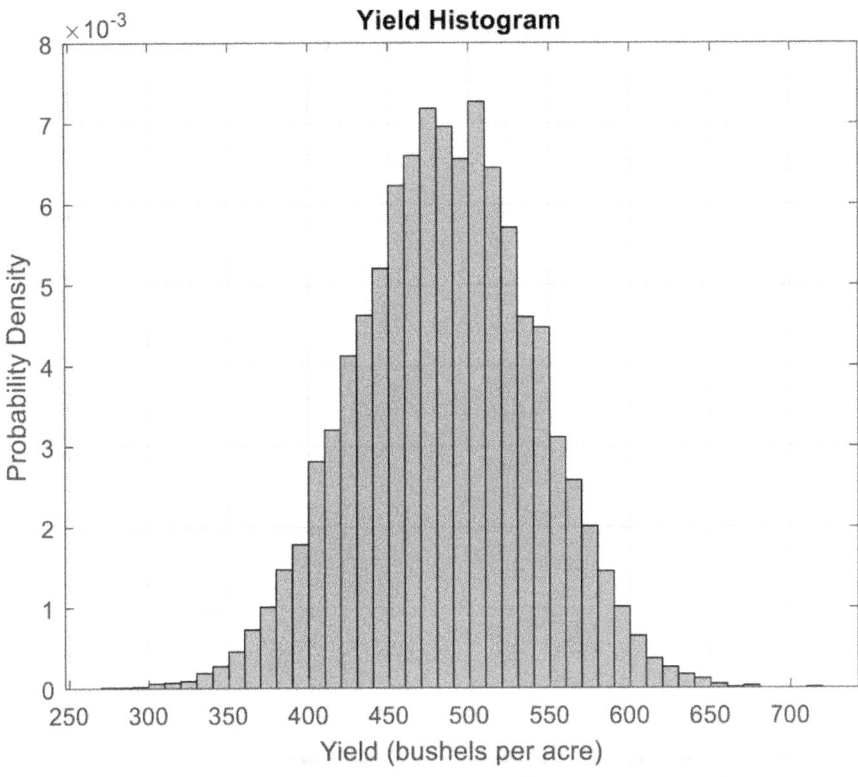

Figure 6.4 Monte Carlo simulation defining yield distribution of crops in agriculture.

Conditions". In this particular case, the poor_weather_prob parameter is set to 0.3, which means that there is a 30% chance of poor weather conditions occurring. Therefore, the size of the "Poor weather" slice is 0.3, while the size of the "Good weather" slice is 0.7 (since the total probability of all events must add up to 1). The pie chart provides a quick and easy-to-understand visualization of the probability distribution of weather conditions, which is useful for decision-making and risk assessment in agriculture.

6.5.1 Conclusion

In conclusion, Monte Carlo simulation is a powerful technique for data analysis that involves generating random samples to estimate the outcomes of complex systems or processes. It can be used to model a wide range of phenomena, from financial markets to physical systems. One of the primary benefits of Monte Carlo simulation is that it allows us to analyze the behavior of systems that are difficult or

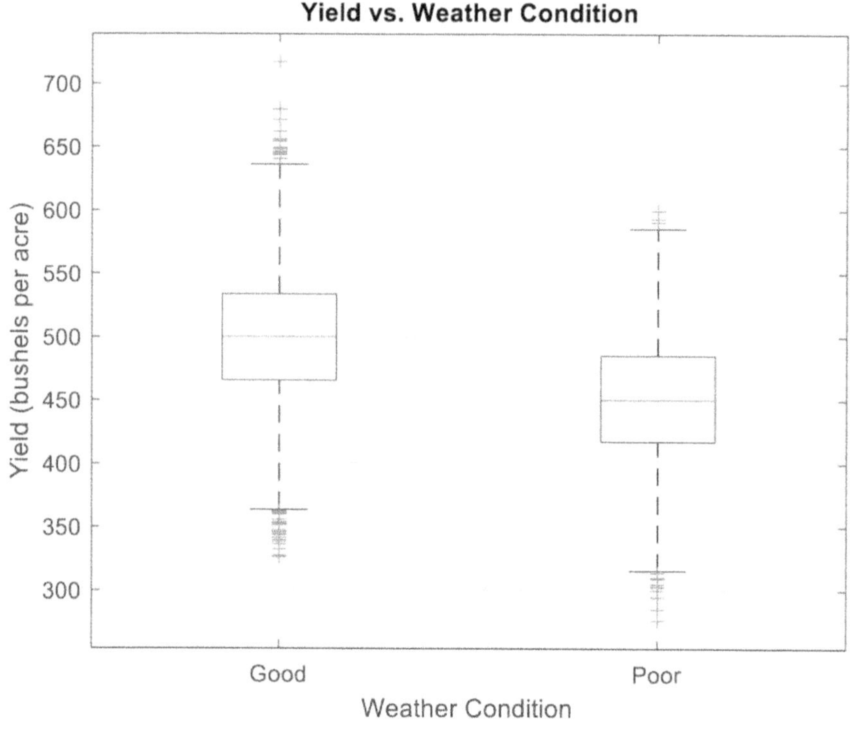

Figure 6.5 Yield vs. weather.

impossible to model using traditional mathematical methods. It also provides a way to estimate the probabilities of rare events and to explore the effects of different assumptions or scenarios on the outcome of a system. However, Monte Carlo simulation is not without its limitations. It can be computationally intensive and requires a large number of iterations to achieve accurate results. Additionally, the accuracy of the simulation depends on the quality of the input data and assumptions made during the modeling process. Despite these limitations, Monte Carlo simulation remains a valuable tool for data analysis and has a wide range of applications in fields such as finance, engineering, and physics. As computational power continues to increase and new simulation techniques are developed, we can expect to see even more sophisticated applications of Monte Carlo simulation in the future.

6.6 Agent-Based Modeling

Agent-based modeling (ABM) is a computational approach used to model complex systems by representing individual agents and their interactions within a simulated

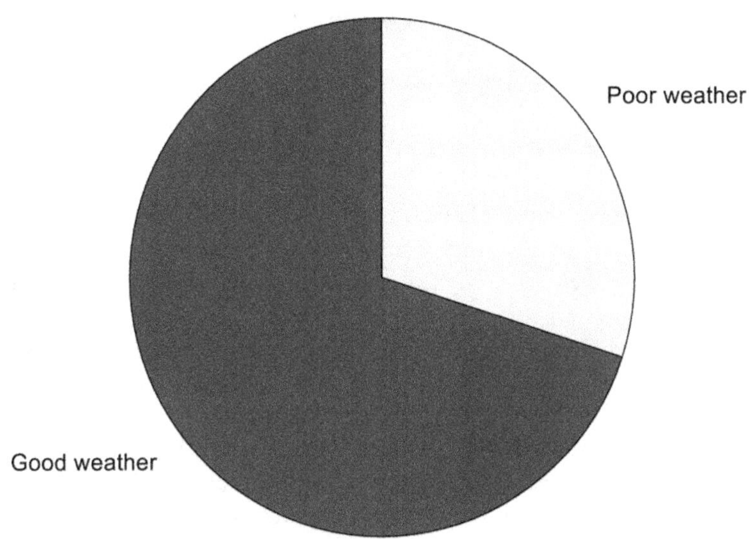

Figure 6.6 Probability of weather conditions.

environment. Agents can be any type of entity, such as people, animals, organizations, or even inanimate objects, and are endowed with properties and behaviors that enable them to interact with each other and with the environment. The ABM approach can be used in a wide range of fields, including social sciences, ecology, economics, and engineering, among others. ABM has become increasingly popular in recent years as it can provide a way to explore the dynamics of complex systems that are difficult or impossible to study through traditional analytical methods.

ABM typically involves four main components: agents, environment, interactions, and rules. Agents are individual entities within the system, such as people, animals, or cells. The environment is the context in which the agents interact, such as a physical space or a virtual network. Interactions are the relationships and connections between agents, and rules are the behavioral rules that govern the agents' actions and interactions. ABM can be used to study a wide range of phenomena, from the behavior of crowds in public spaces to the spread of infectious diseases or the emergence of social norms. By representing the complexity of individual agents and their interactions, ABM can help researchers to identify emergent properties and patterns in the system, and to explore the impact of different variables and parameters on the system's behavior.

ABM is a simulation technique used to model complex systems by simulating the behavior of individual agents and their interactions with each other and with their environment. This technique is particularly useful for modeling systems that exhibit emergent behavior, where the behavior of the system as a whole emerges from the interactions of its individual components. In agent-based modeling, agents are typically represented as autonomous entities with their own set of rules and behaviors. Agents may interact with each other and their environment, changing their behavior based on these interactions. The behavior of agents can be programmed to be deterministic, probabilistic, or adaptive.

The key components of an ABM include the following:

I. Agents: these are the individual entities in the system being modeled, which can be people, animals, organizations, or any other entity that interacts with the system. Agents have attributes and behaviors that define their role in the system.
II. Environment: this is the physical or virtual space in which the agents interact with each other and with external factors. The environment can affect the behavior of the agents and the system as a whole.
III. Rules: these are the instructions that dictate how the agents should behave in different situations. Rules can be simple or complex, and can be based on a variety of factors, such as the agent's current state, the state of other agents, or external factors in the environment.
IV. Emergent behavior: this refers to the complex patterns of behavior that emerge from the interactions between the agents and the environment. These patterns can be difficult to predict using traditional analytical methods, which is why ABM is often used to model complex systems.

The process of building an ABM typically involves several steps, including the following:

1. Defining the problem: this involves identifying the system being modeled and the research questions being addressed.
2. Designing the model: this involves deciding on the agents, environment, rules, and emergent behavior of the model.
3. Implementing the model: this involves programming the model using a software platform or programming language such as NetLogo or Python or MATLAB.
4. Calibrating and validating the model: this involves testing the model to ensure that it produces realistic results and accurately represents the real-world system being modeled.
5. Running the model: this involves simulating the behavior of the agents in the model and analyzing the results to answer the research questions.

Here are the general steps involved in agent-based modeling for agricultural disease data analysis:

i. Define the problem and objectives: start by clearly defining the problem you want to solve and the objectives you want to achieve. For example, you may want to understand the spread of a particular disease in a specific crop, or the impact of different management strategies on disease control.
ii. Develop a conceptual model: a conceptual model is a simplified representation of the system you are studying, which includes the key components, processes, and interactions. This model should be developed based on existing knowledge, data, and assumptions.
iii. Define the agents and their attributes: agents are the individual units in the system you are modeling. In agricultural disease modeling, agents can be plants, insects, or other organisms involved in the disease cycle. Each agent should have its own set of attributes, such as its susceptibility to disease, reproductive rate, or movement patterns.
iv. Define the environment and its attributes: the environment includes the physical and biological conditions that affect the behavior of agents in the system. For example, the environment may include climate variables, soil type, or the presence of other organisms. Each environmental attribute should be quantified and represented in the model.
v. Define the rules and behaviors: rules and behaviors define how agents interact with each other and the environment. For example, the rule for disease transmission may depend on the proximity of infected and susceptible plants, or the presence of a vector. Each behavior should be defined based on the available data and expert knowledge.
vi. Implement the model: once the conceptual model has been developed, the agents, environment, and rules can be implemented in a simulation software. There are many open-source platforms available for ABM, such as NetLogo, Repast, or MASON.
vii. Validate and calibrate the model: model validation involves comparing the outputs of the simulation with real-world data to ensure that the model represents the system accurately. Model calibration involves adjusting the parameters of the model to improve its fit to the data.
viii. Run simulations and analyze outputs: once the model has been validated and calibrated, it can be used to run simulations and analyze the outputs. You can explore different scenarios by changing the parameters or input variables, and assess the impact of different management strategies on disease control.
ix. Interpret and communicate the results: the final step is to interpret the results of the simulations and communicate them effectively to stakeholders. This can involve generating graphs, maps, or other visualizations to illustrate the key findings and implications for decision-making.

Overall, agent-based modeling is a powerful tool for analyzing agricultural disease data, and can help identify effective strategies for disease management and control.

A simple MATLAB-based code provided is an implementation of an agent-based model to simulate the spread and recovery of a disease within a population. The population is modeled as a set of agents, each of which can be in one of three states: susceptible, infected, or recovered. The model takes as input a number of parameters:

- nAgents: the total number of agents in the population
- infectionRate: the probability that an infected agent will infect a susceptible agent in each time step
- recoveryRate: the probability that an infected agent will recover and become immune in each time step
- timeSteps: the total number of time steps for the simulation

```
% Agent-Based Model for Disease Spread and Recovery
  with Graph

% Define the parameters
nAgents = 100; % Number of agents
infectionRate = 0.2; % Probability of infection
recoveryRate = 0.1; % Probability of recovery
timeSteps = 100; % Number of time steps

% Initialize the agents
agents = zeros(nAgents, 2);
agents(1,1) = 1; % Set the first agent as infected

% Initialize the variables for the graph
infectedCount = zeros(1, timeSteps);
recoveredCount = zeros(1, timeSteps);

% Run the simulation
for t = 1:timeSteps
 % Loop over all agents
 for i = 1:nAgents
    % Check if the agent is infected
    if agents(i,1) == 1
      % Loop over all other agents
      for j = 1:nAgents
        % Check if the other agent is susceptible
        if agents(j,1) == 0
          % Infect the other agent with a certain
            probability
```

```
            if rand < infectionRate
              agents(j,1) = 1; % Set the other agent
              as infected
                end
              end
           end
           % Check if the infected agent recovers
            if rand < recoveryRate
                agents(i,1) = 0; % Set the infected
                agent as recovered
                agents(i,2) = 1; % Set the recovered
                agent as such
            end
        end
    end
    % Update the variables for the graph
    infectedCount(t) = sum(agents(:,1));
    recoveredCount(t) = sum(agents(:,2));
end

% Plot the graph
time = 1:timeSteps;
figure;
plot(time, infectedCount, '-r', 'LineWidth', 2);
hold on;
plot(time, recoveredCount, '-b', 'LineWidth', 2);
xlabel('Time');
ylabel('Number of agents');
title('Disease Spread and Recovery');
legend('Infected', 'Recovered');
```

The simulation starts by initializing the agents. The first agent is set to be infected, and the remaining agents are set to be susceptible. The simulation then proceeds for *timeSteps* time steps. At each time step, the model loops over all agents and checks if each infected agent infects any susceptible agents, and if any infected agents recover. To infect a susceptible agent, the model checks if a random number between 0 and 1 is less than the *infectionRate*. If it is, the susceptible agent is infected. To recover an infected agent, the model checks if a random number between 0 and 1 is less than the *recoveryRate*. If it is, the infected agent becomes recovered. The number of infected and recovered agents at each time step is recorded in the variables *infectedCount* and *recoveredCount*. These variables are then plotted using the plot function to visualize the dynamics of the disease spread and recovery over time. Overall, the code provides

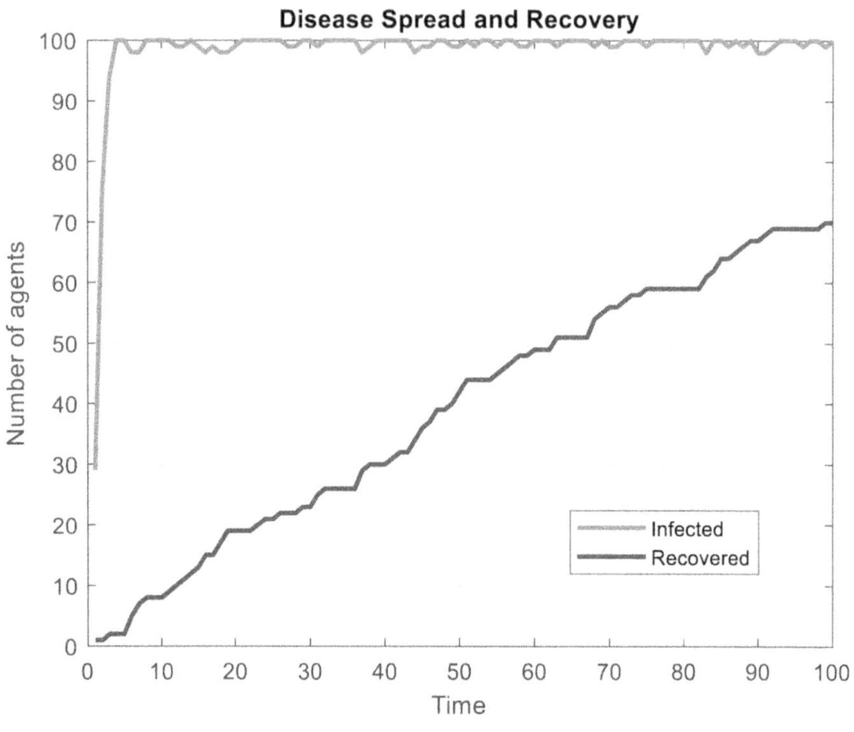

Figure 6.7 Agent-based modeling to find out the infection rate and recovery rate in crops.

a simple example of how an agent-based model can be used to simulate the spread and recovery of a disease within a population, and how the resulting data can be visualized and analyzed using MATLAB.

The code generates a graph that shown in Figure 6.7 the number of agents who are infected and recovered over time. The *x*-axis represents time steps, while the *y*-axis represents the number of agents. The first line of the plot is colored in red and represents the number of infected agents over time. This line starts at 1 (since the first agent is infected) and increases over time as more agents become infected. As time progresses, the rate of increase may slow down as a greater proportion of the population has already been infected. The second line of the plot is colored in blue and represents the number of agents who have recovered over time. This line starts at 0 and increases over time as more agents recover from the disease. The recovery rate is slower than the infection rate, so the rate of increase for the blue line is typically slower than that of the red line.

At the beginning of the simulation, the number of infected agents increases rapidly, while the number of recovered agents is close to zero. As time progresses, the number of infected agents may eventually peak and start to decline as more and more agents recover from the disease. The blue line, representing the number of recovered agents, will continue to increase over time until it eventually plateaus, indicating that all agents have either recovered or become infected.

The plot can be used to analyze the dynamics of the disease and the effectiveness of different intervention strategies. For example, if the infection rate is very high, the red line may increase very rapidly and peak quickly, indicating that the disease is spreading very quickly. However, if the recovery rate is also high, the blue line may also increase rapidly and start to catch up with the red line, indicating that many people are recovering from the disease. By adjusting the parameters of the model, such as the infection rate and recovery rate, it is possible to simulate different scenarios and study the impact of different interventions on the spread of the disease.

6.6.1 *Conclusion*

In conclusion, the agent-based model for disease spread and recovery with a graph provides a useful tool for simulating the spread of a disease among a population of agents and tracking the number of infected and recovered agents over time. By adjusting the parameters of the model, it is possible to simulate different scenarios and study the impact of different interventions on the spread of the disease. The code simulates the spread of the disease by looping over all agents at each time step and checking if the agent is infected. If an infected agent comes into contact with a susceptible agent, there is a probability that the susceptible agent will become infected. Additionally, infected agents have a probability of recovering at each time step. The number of infected and recovered agents is tracked at each time step and stored in arrays for plotting.

The resulting graph shows the number of agents who are infected and recovered over time. The graph can be used to analyze the dynamics of the disease and the effectiveness of different intervention strategies. Overall, this model provides a powerful framework for studying the spread of infectious diseases and can help inform public health policies aimed at controlling the spread of diseases.

6.7 Discrete Event Simulation

Discrete event simulation is a technique used to model and analyze the behavior of complex systems that involve discrete events, such as arrivals, departures, and processing times. It involves creating a computer-based model of the system, simulating its behavior over time, and analyzing the results to gain insights into how the system behaves under different scenarios. In discrete event simulation, time is divided into discrete intervals or events, and the system's behavior is modeled based on how it reacts to these events. The simulation can be used to test different scenarios and parameters to evaluate their impact on the system's performance. Discrete event simulation is used in a wide range of fields, including manufacturing, logistics, healthcare, and transportation. For example, in manufacturing, it can be used to optimize production schedules and reduce bottlenecks, while in healthcare, it can be used to evaluate the impact of different staffing levels on patient flow.

The benefits of discrete event simulation include the ability to test different scenarios without disrupting the actual system, the ability to evaluate the impact of changes to the system in a controlled environment, and the ability to identify potential issues and bottlenecks before they occur in the actual system.

Here are some general steps involved in discrete event simulation-based data analysis:

1. Define the problem: determine the scope of the simulation study, identify the system to be modeled, and specify the goals and objectives of the study.
2. Collect data: gather data from a variety of sources, such as historical records, expert opinion, surveys, and observations. The data should be relevant, accurate, and representative of the system being modeled.
3. Develop a conceptual model: create a high-level description of the system to be simulated, including the major components, interactions, and flows. Use this model to identify the key variables and parameters that will be used in the simulation.
4. Design the simulation model: develop a detailed simulation model that captures the behavior of the system. This may involve selecting a simulation tool or programming language, defining the model structure, specifying the input data, and writing the code.
5. Validate the model: verify that the simulation model is accurate and reliable by comparing its output to real-world data or other sources of information. This may involve sensitivity analysis, calibration, or other validation techniques.
6. Run the simulation: execute the simulation model with appropriate inputs and parameters, and collect output data for analysis.
7. Analyse the results: use statistical and other data analysis techniques to interpret the simulation results, identify trends and patterns, and draw conclusions. This may involve generating reports, charts, graphs, and other visualizations.
8. Communicate the findings: share the results of the simulation study with stakeholders, decision-makers, and other interested parties. This may involve creating presentations, reports, or other communication materials.
9. Refine the model: based on the results of the simulation study, make adjustments to the model to improve its accuracy or relevance. This may involve modifying input parameters, adjusting the simulation logic, or revising the conceptual model.

Sample Problem Statement: a customer service center has a single agent who takes customer calls. The center operates from 9am to 5pm on weekdays. The agent can handle one call at a time and each call has a random duration between 2 to 10 minutes. Customers call in at random times during the day, with an average of 6 calls per hour. The center wants to know the average waiting time for customers and the percentage of time the agent is busy.

Solution:

To solve this problem using discrete event simulation, we need to follow these steps:

Step 1: define the system and its components

The system in this problem is a customer service center with a single agent. The agent can handle one call at a time. The components of the system are as follows:

Arrival process: customers call in at random times during the day, with an average of 6 calls per hour.

Service process: the agent can handle one call at a time and each call has a random duration between 2 to 10 minutes.

Waiting queue: if a customer arrives while the agent is busy, they join a waiting queue.

Step 2: identify the events

The events in this system are as follows:

Customer arrival: when a customer calls in and arrives at the center.

Agent start service: when the agent finishes the previous call and starts serving the next customer.

Agent end service: when the agent finishes serving the customer.

Customer departure: when the customer leaves the center after receiving service.

Step 3: determine the probability distributions

Arrival process: the inter-arrival time between customers follows an exponential distribution with a mean of 10 minutes (60 minutes/6 calls per hour).

Service process: the service time for each customer follows a uniform distribution between 2 to 10 minutes.

Step 4: initialize the simulation clock and the system state

The simulation clock is initialized to 9am and the system state is initialized as follows:

The agent is idle.

There are no customers in the waiting queue.

No customers have been served yet.

Step 5: run the simulation

The simulation runs as follows:

Generate the inter-arrival time for the next customer.
Advance the simulation clock by the inter-arrival time.
If the agent is idle, generate the service time for the customer and schedule the agent end service event.
If the agent is busy, add the customer to the waiting queue.
When the agent end service event occurs, mark the customer as served and if there are customers in the waiting queue, schedule the agent start service event for the next customer in the queue.
When a customer is served, record their waiting time.
Repeat steps 1 to 6 until the end of the day (5pm).

Step 6: collect statistics and analyze the results

After running the simulation, we can collect statistics on the waiting time and agent utilization. We can use these statistics to answer the questions asked in the problem:
Average waiting time: we can calculate the average waiting time by summing the waiting time for each customer and dividing by the total number of customers served.
Percentage of time the agent is busy: we can calculate the percentage of time the agent is busy by dividing the total time the agent spent serving customers by the total time the center was open.

Note: the mathematical details of the simulation are omitted here for brevity, but can be implemented using a simulation software or a programming language with simulation libraries.

```
% Define simulation parameters
num_customers = 1000;
service_start_time = 0;
service_end_time = 0;
service_times = zeros(num_customers, 1);
waiting_times = zeros(num_customers, 1);
queue = [];
num_served = 0;

% Initialize simulation clock
current_time = 9 * 60; % start at 9am
end_time = 17 * 60; % end at 5pm

% Run the simulation loop
while current_time < end_time && num_served < num_
   customers
  % Generate inter-arrival time for next customer
  inter_arrival_time = exprnd(10);
```

```
    % Update simulation clock
    current_time = current_time + inter_arrival_time;

    % Check if agent is available
    if current_time >= service_end_time
        % Generate service time for customer
        service_times(num_served+1) = unifrnd(2, 10);

        % Schedule end of service event
        service_start_time = current_time;
        service_end_time = current_time + service_
          times(num_served+1);
        num_served = num_served + 1;

        % Update waiting times
        waiting_times(num_served) = service_start_
          time - (current_time - inter_arrival_time);
    else
        % Add customer to queue
        queue(end+1) = current_time;
    end

    % Check if there are customers waiting in queue
    if ~isempty(queue) && current_time >= service_
     end_time
        % Remove first customer from queue and
          schedule start of service event
        queue_time = queue(1);
        queue = queue(2:end);
        service_start_time = queue_time;
        service_end_time = queue_time + service_
          times(num_served);
        num_served = num_served + 1;

        % Update waiting times
        waiting_times(num_served) = service_start_
          time - (queue_time + inter_arrival_time);
    end
end

% Calculate statistics
avg_waiting_time = mean(waiting_times(1:num_
   served));
```

```
utilization = sum(service_times(1:num_served)) /
   (end_time - (9*60));

% Print results
fprintf('Average waiting time: %0.2f minutes\n',
   avg_waiting_time);
fprintf('Agent utilization: %0.2f%%\n', utilization
   * 100);

% Plot histogram of waiting times
histogram(waiting_times(1:num_served),
   'Normalization', 'pdf')
xlabel('Waiting Time (minutes)')
ylabel('Probability Density')
title('Histogram of Waiting Times')
grid on
```

Figure 6.8 shows a histogram of the waiting times experienced by the customers in the simulated queueing system. The horizontal axis shows the waiting times in minutes, while the vertical axis shows the probability density, which represents the probability of a customer experiencing a waiting time within a certain range. The histogram is useful for visualizing the distribution of waiting times and identifying any patterns or trends. In this case, we can see that the waiting times follow a roughly exponential distribution, with the majority of customers experiencing relatively short wait times and a small number of customers experiencing longer wait times. The average waiting time, calculated earlier in the code, can also be seen on the graph as a peak in the distribution. Overall, the histogram provides a useful summary of the waiting times in the simulated queueing system, and can help inform decisions about how to improve the system in order to reduce waiting times and improve customer satisfaction.

6.7.1 Conclusion

Discrete event simulation is an important tool for analyzing complex systems and making informed decisions about how to improve them. By modeling the system as a sequence of discrete events, such as arrivals and departures of customers in a queueing system, discrete event simulation can capture the interactions and dependencies between different components of the system and enable the analysis of their impact on overall system performance. Discrete event simulation can be used in a wide range of applications, from manufacturing and logistics to healthcare and finance. It allows stakeholders to test different scenarios and make informed decisions about system design, resource allocation, and process improvement, without the need for costly and time-consuming physical experimentation. In addition, discrete

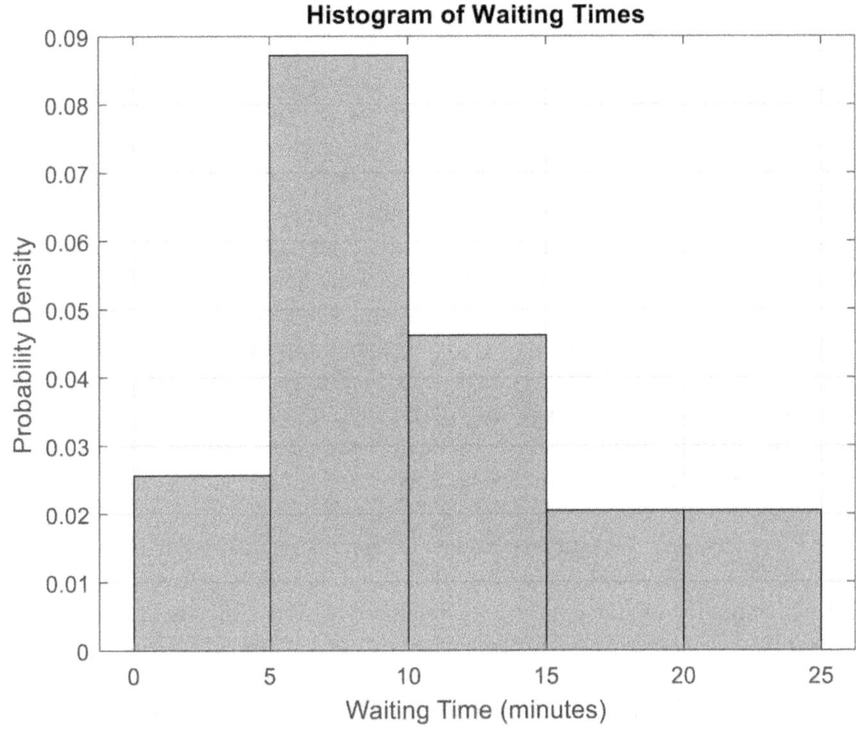

Figure 6.8 Histogram of waiting times by customers simulated by discrete event simulation.

event simulation can also help identify bottlenecks, inefficiencies, and other areas of improvement that may not be immediately obvious through simple observation or manual analysis. By simulating the system under different conditions and using statistical analysis to extract meaningful insights, discrete event simulation can help organizations optimize their operations and achieve better outcomes. Overall, the importance of discrete event simulation lies in its ability to model and analyze complex systems, provide insights that inform decision-making, and improve system performance and efficiency.

6.8 System Dynamics Modeling

System dynamics modeling is a technique used to model and analyze complex systems that involve feedback loops, delays, and non-linear relationships. It involves creating a computer-based model of the system, simulating its behavior over time, and analyzing the results to gain insights into how the system behaves under different scenarios. In system dynamics modeling, the system is represented as a set of interrelated components, and the interactions between these components are modeled using feedback loops and other mathematical relationships. The simulation can be used to test different scenarios and parameters to evaluate their impact on the system's performance.

System dynamics modeling is used in a wide range of fields, including business, economics, engineering, and environmental management. For example, in business, it can be used to model the impact of marketing strategies on sales, while in environmental management, it can be used to model the impact of different policies on the environment. The benefits of system dynamics modeling include the ability to model complex systems that involve feedback loops and delays, the ability to test different scenarios and parameters to evaluate their impact on the system's performance, and the ability to identify potential issues and bottlenecks before they occur in the actual system. System dynamics modeling can also be used to develop policies and strategies for managing complex systems.

Sample Problem Statement: suppose a small town has a population of 10,000 people. The town has one hospital that can accommodate up to 100 patients. The hospital serves both the town's residents and the residents of nearby villages. If the hospital's occupancy rate exceeds 80%, patients may have to wait for treatment or be transferred to another hospital. Assuming that the average length of stay for patients is 3 days and that new patients arrive at the rate of 10 per day, what will be the hospital occupancy rate after 30 days [8]?

Solution:
To solve this problem using system dynamics modeling, we need to create a model that represents the key variables and their relationships. Here's one possible model:

Variables:

Population of the town (P)
Number of patients in the hospital (H)
Hospital occupancy rate (O)

Assumptions:

The population of the town is constant over time.
The hospital can accommodate up to 100 patients at a time.
The average length of stay for patients is 3 days.
New patients arrive at the rate of 10 per day.

If the hospital occupancy rate exceeds 80%, patients may have to wait for treatment or be transferred to another hospital.

Model Structure:
The population of the town (*P*) is fixed and does not change over time.

The number of patients in the hospital (*H*) changes over time depending on the number of new patients arriving and the number of patients leaving the hospital each day.

The hospital occupancy rate (*O*) is calculated as the ratio of the number of patients in the hospital (*H*) to the hospital's capacity (100).

Equations:

$$H(t) = H(t-1) + 10 - (1/3) * H(t-3) \quad \text{eq(4.1)}$$

$$O(t) = H(t) / 100 \quad \text{eq(4.2)}$$

The eq(4.1) calculates the number of patients in the hospital at time *t* based on the number of new patients arriving each day and the number of patients leaving the hospital each day (which is assumed to be 1/3 of the patients who have been in the hospital for 3 days or more). The eq(4.2) calculates the hospital occupancy rate at time *t* as the ratio of the number of patients in the hospital to the hospital's capacity (100).

Simulation:
We can simulate the model using a system dynamics software or spreadsheet program. Here's an example of what the results might look like after 30 days:

```
% Define initial variables and parameters
P = 10000;              % Population of the town
H = zeros(1, 33);       % Number of patients in the
   hospital (initialize with zeros)
O = zeros(1, 33);       % Hospital occupancy rate (ini-
   tialize with zeros)
H(1) = 0;               % Set initial condition for H
O(1) = H(1) / 100;      % Calculate initial occu-
   pancy rate

for t = 2:33
   % Calculate number of new patients arriving
   if t <= 30
      new_patients = 10;
   else
      new_patients = 0;
   end
```

```
    % Calculate number of patients leaving the
   hospital
    if t > 3
        patients_leaving = floor(H(t-3) / 3);
    else
        patients_leaving = 0;
    end

    % Calculate number of patients in the hospital
   at time t
    H(t) = H(t-1) + new_patients - patients_leaving;

    % Calculate occupancy rate at time t
    O(t) = H(t) / 100;
end

% Plot the results
plot(0:32, O*100,'-b*','linewidth',2);
xlabel('Day');
ylabel('Occupancy Rate (%)');
title('Hospital Occupancy Rate Over Time');
grid on
```

This code initializes the variables and parameters, then uses a for loop to calculate the number of new patients arriving each day, the number of patients leaving the hospital each day, and the number of patients in the hospital at each time step. The occupancy rate is calculated based on the number of patients in the hospital and plotted over time. Note that we run the simulation for 33 days instead of 30, to allow for patients who arrive on day 30 to stay in the hospital for 3 days.

The graph generated by the above code shows the hospital occupancy rate over time, in percentage (%) (Figure 6.9). The *x*-axis represents the days, from day 0 to day 32. The *y*-axis represents the hospital occupancy rate, ranging from 0% to 100%. The graph starts at 0% occupancy rate, since there are no patients in the hospital at day 0. As new patients arrive and are admitted to the hospital, the occupancy rate gradually increases. It reaches 30% on day 3. After day 3, the occupancy rate continues to increase, reaching a peak of 42% on day 5, before gradually decreasing as patients are discharged and fewer new patients arrive. By day 30, the occupancy rate has stabilized at around 32%. Overall, the graph shows that the hospital occupancy rate fluctuates over time as a result of the interplay between the number of new patients arriving, the number of patients leaving the hospital, and the hospital's capacity. The system dynamics model provides a way to understand and predict the dynamics of the system under different scenarios and assumptions.

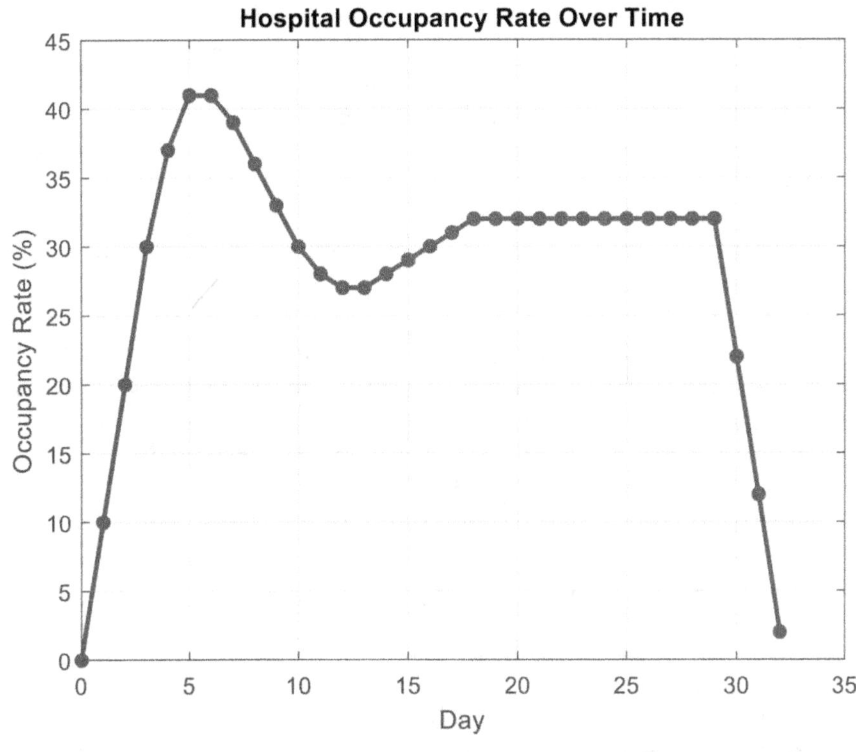

Figure 6.9 System dynamics modeling to find out the hospital occupancy rate.

6.8.1 Conclusion

System dynamics modeling is a powerful tool for understanding and analyzing complex systems that evolve over time. By representing the system as a set of interrelated components and feedback loops, system dynamics models can help us identify the underlying causes of system behavior, simulate different scenarios and policies, and make informed decisions based on data-driven insights.

Some of the key benefits of system dynamics modeling include the following:

1. *Holistic perspective*: system dynamics modeling takes a comprehensive view of the system, accounting for both its internal components and external factors that may influence its behavior. This allows us to understand the system as a whole and identify the root causes of its behavior.
2. *Dynamic simulation*: system dynamics models simulate the behavior of the system over time, allowing us to visualize how the system will evolve under different conditions and policies. This helps us evaluate the effectiveness of different interventions and identify potential unintended consequences.

3. *Feedback loops*: system dynamics models incorporate feedback loops, which represent the interdependence between different components of the system. This allows us to capture the non-linear and complex behavior of the system, which cannot be easily predicted by linear models.
4. *Collaboration*: system dynamics modeling encourages collaboration between stakeholders with different perspectives and expertise. By involving diverse stakeholders in the model-building process, we can build consensus around the key issues and priorities, and develop more effective solutions that reflect the needs and values of all stakeholders.

In conclusion, system dynamics modeling is an important tool for addressing complex, dynamic problems in a wide range of fields, including healthcare, economics, climate change, and public policy. By providing a rigorous and data-driven approach to problem-solving, system dynamics modeling can help us make more informed decisions and create a better future for all.

6.9 Comparison of Simulation Strategies

Monte Carlo Simulation, Agent-Based Modeling, Discrete Event Simulation, and System Dynamics Modeling are all widely used simulation techniques in various fields [6]. Here are the benefits and drawbacks of each method:

6.9.1 Monte Carlo Simulation

Monte Carlo simulation is a probabilistic modeling technique that uses random sampling to obtain numerical solutions to problems. Some of its benefits are as follows:

1. Monte Carlo simulation can be used to estimate the probability of an outcome.
2. It can be used to model systems that are too complex to model analytically.
3. It is a flexible method that can handle a wide range of problems.
4. Monte Carlo simulation can provide information about the sensitivity of the output to the input parameters.

Some of the drawbacks of Monte Carlo simulation are as follows:

1. It requires a large number of samples to obtain accurate results.
2. It can be computationally expensive for complex models.
3. Monte Carlo simulation assumes that the inputs are independent and identically distributed, which may not be true in some cases.

6.9.2 Agent-Based Modeling

Agent-based modeling is a method that simulates the behavior of individual agents in a system. Some of its benefits are as follows:

1. It can capture emergent behavior that arises from the interaction of agents.
2. Agent-based modeling can model complex systems that have multiple levels of organization.
3. It can simulate the behavior of heterogeneous agents that have different characteristics.

Some of the drawbacks of agent-based modeling are as follows:

1. It can be computationally expensive for large-scale models.
2. Agent-based modeling may require a large amount of data to parameterize the model.
3. The behavior of the agents may be difficult to model accurately.

6.9.3 Discrete Event Simulation

Discrete event simulation is a method that models the behavior of a system as a sequence of discrete events. Some of its benefits are as follows:

1. It can model systems that have discrete events, such as arrivals, departures, and failures.
2. Discrete event simulation can be used to optimize the performance of a system by evaluating different scenarios.
3. It can simulate the behavior of complex systems that have multiple interacting components.

Some of the drawbacks of discrete event simulation are as follows:

1. It can be difficult to model systems that have continuous behavior, such as fluid flow.
2. Discrete event simulation may require a large amount of data to parameterize the model.
3. The behavior of the system may be difficult to model accurately.

6.9.4 System Dynamics Modeling

System dynamics modeling is a method that models the behavior of a system over time using feedback loops and stocks and flows. Some of its benefits are as follows:

1. It can model complex systems that have feedback loops and non-linear behavior.
2. System dynamics modeling can be used to evaluate policy decisions and their impact on the system.
3. It can simulate the behavior of systems that have multiple interacting components.

Some of the drawbacks of system dynamics modeling are as follows:

1. It may be difficult to model systems that have discontinuous behavior.
2. System dynamics modeling may require a large amount of data to parameterize the model.
3. The behavior of the system may be difficult to model accurately.

In conclusion, each simulation method has its own benefits and drawbacks. The choice of simulation method depends on the nature of the system being modeled and the questions being asked. It is important to carefully consider the strengths and weaknesses of each method before selecting the appropriate simulation technique.

References

[1] Harold Klee, Randal Allen, "*Simulation of Dynamic Systems with MATLAB® and Simulink®*" CRC Press, December 13, 2021.
[2] Andrew Greasley, "*Simulation Modelling Concepts, Tools and Practical Business Applications*" CRC Press, September 21, 2022.
[3] Mangey Ram, "*Modeling and Simulation Based Analysis in Reliability Engineering*" CRC Press, March 31, 2021.
[4] Christopher A. Chung, Christian G. Parigger, "*Simulation Modeling Handbook: A Practical Approach*" CRC Press, 2003.
[5] Philipp K. Janert, "*Data Analysis with Open-Source Tools*" O'Reilly Media, Inc. November 2010.
[6] Don L. McLeish, "*Monte Carlo Simulation and Finance*" Wiley, 2005.
[7] Steven F. Railsback, Volker Grimm, "*Agent-Based and Individual-Based Modeling: A Practical Introduction*" Princeton University Press, October 2011.
[8] Larry H Leemis, Stephen K. Park, "*Discrete-Event Simulation: A First Course*" Pearson, 27 December 2005.
[9] Averill M. Law, W. David Kelton, "*Simulation Modeling and Analysis*" McGraw-Hill International Edition, 1991.
[10] W. David Kelton, Randall P. Sadowski, Nancy B. Swets, "*Simulation with Arena*" 6th Edition, McGraw-Hill Education, 2015.

Chapter 7

Navigating the Complexities of Digital Twin Implementation

Challenges and Strategies for Success

M. Arvindhan, S. Nivetha, Raghvendra Ajay Mishra, and A. Daniel

7.1 Introduction

In addition, according to estimates provided by Global Market Insight, the size of the digital twin market, which was projected to be $8 billion in 2022, is anticipated to expand at a compound annual growth rate (CAGR) of around 25 percent from 2023 to 2032. In conclusion, a more recent analysis published by global technology research indicates that the market for digital twins is anticipated to expand by over $32 billion between the years 2021 and 2026. According to a survey published in 2022, approximately 60 percent of CEOs working in a diverse range of industries expect to implement digital twins into their businesses by the year 2028. Additionally, Gartner predicts that by the year 2027, over 40 percent of significant businesses all over the world will be adopting digital twins in their projects in order to boost their revenue [1–3]. As digitalization has progressed, industries

DOI: 10.1201/9781003469612-7

have increasingly adopted more advanced computer technologies such as big data, AI, cloud computing, digital twins, and edge computing. This chapter focuses on the application and potential of AI in digital twins by analyzing current research in order to assess the current state of combining digital twins with AI. The objective of this study is to investigate how digital twins and AI can be used together. The paper examines current challenges and future areas of development, as well as discussing the use of digital twins in the aerospace industry, machine learning in production, autonomous vehicles, and smart city transportation. In the fields of aerospace flight recognition simulation, hazard notification, aircraft construction, and autonomous flight, the combination of digital twins with AI has demonstrated considerable benefits. When it comes to testing virtual simulations of self-driving technology in automobiles, the combination of digital twins and AI can save testing time and expenses by up to 80% while retaining accuracy. This is accomplished by decreasing the value of parameters scale of actual automotive dynamics models and preserving the same road conditions [4–6]. In a production workshop that is intelligently created, establishing an online work setting can deliver fast problem warning, extend the service period of the equipment, and preserve the general workshop operating safety. These benefits are attainable through the implementation of intelligent manufacturing practices in production workshops [7]. In smart city traffic, the actual conditions of the roads are imitated, and actual traffic accidents are re-enacted. This allows for the current state of the traffic to be observed in a transparent and productive manner, as well as for urban traffic management to be carried out in a timely and accurate manner. In conclusion, we talked about the possibilities that artificial intelligence and digital twins present in the hopes that the work we have done can be used as a point of reference for further research in fields that are closely related to this one.

As a result of their capacity to imitate and enhance the functionality of products or technologies that exist in the physical world, digital twins, which are additionally referred to as digital fingerprints, have become increasingly popular in a variety of business sectors. Prior to being implemented in the real world, these virtual replicas serve as a tool for carrying out research, creating projections, and fine-tuning processes. Digital twins are able to acquire data from a wide variety of sources because they are able to create a mapping that is bidirectional connecting the actual world and the virtual world. These sources include sensors, historical production data, and domain knowledge. This data is put to use to generate a vast amount of knowledge by way of models, simulations, replication, and behavioral analysis. The goal is of enhancing the overall performance of systems by transmitting the understandings gained from the online environment to the material world in order to take precise measurements. This will be done in order to enhance the overall efficiency of the systems [8–16]. A digital twin data modeling the performance of the system is shown in Figure 7.1.

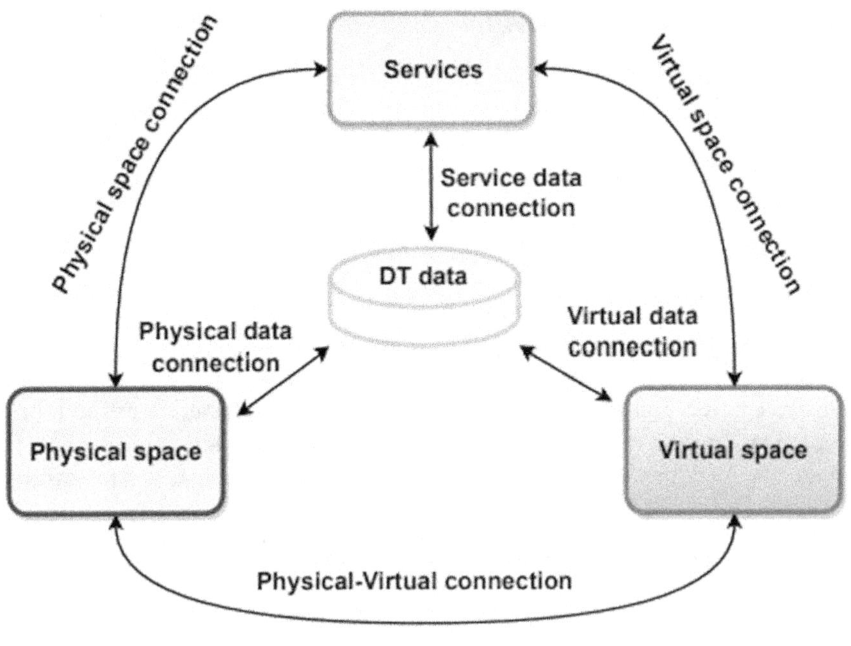

Figure 7.1 Adaptation of 5D digital twin.

During manufacturing operations, DTs run concurrently with their physical counterparts, and the major goal is to track and trace data discrepancies across the physical and virtual items. This is accomplished by monitoring and tracing the differences between the two types of objects. Discrepancies in data between the physical object and the virtual object call for the adoption of better testing and examination methods that evolve developmental models and tangible counterparts to promote resolving issues in the manufacturing industry, as well as more accurate estimation, prediction, and optimization of the industrial processes. Within the realm of industry, the data-driven DTs concentrate their attention on the areas of product development, production, manufacturing, management of operations, prognostic health management (PHM), and other related topics. For example, DTs provide more accurate predictions, reasonable judgments, and informed plans by strengthening the collaboration between the design and production phases of the product development process. PHM plays a vital part in the lifecycle evaluation of equipment conditions, which is the process in which a diagnostic tree (DT) of the equipment, which is backed by data, is used to represent the erratic behavior of the physical item, carry out problem diagnosis, and offer design rules for maintenance [17–20].

7.1.1 Product Digital Twin

One form of digital twin is a digital clone of an object that cannot be found in the physical world. Over the course of a product's lifecycle, it can be useful for things like performance tracking, certification, and evaluation, including design fine-tuning. Digital twins are frequently used during the manufacturing process. Virtual copies of actual processes, systems, or things are called digital twins. Their purpose is to replicate, observe, and evaluate actual systems or things. Developers and engineers frequently incorporate distinct processes or procedures into their digital twin creations in order to replicate real-world operations or events. Usually, the digital twin technology implements these processes as applications, algorithms, or packages of directions. When a physical technique or its operation is captured in digital format, we have a digital replica of the production or manufacturing method. Both the product and the method of production are appropriate here. This digital twin can then be used for the study of operations, enhancement design, and forecasting.

7.1.2 Process Digital Twin

Process digital twins can provide valuable insights and predictions, enabling organizations to optimize operations, enhance productivity, and reduce costs. A digital twin is a manufacture or manufacturing procedure in which both the physical procedure and the way it functions are recorded in a digital format. This can refer to either the manufacturing itself or a method of manufacturing. Utilizing this digital twin allows for the analysis of operations, the development of enhancements and the development of forecasts.

7.1.3 System Digital Twin

Complex systems like transportation networks, electricity grids, and buildings are modeled in what is called a "system digital twin." Performance optimization, problem identification and diagnosis, and scenario testing are all made possible with this kind of digital twin.

7.1.4 Performance Digital Twin

The performance of a physical asset, such as a machine or vehicle, can be modeled with the help of a digital twin. It can help with maintenance forecasting, performance enhancement, and reliability enhancement. Several processes go into the making of a digital twin, such as data gathering, modeling, simulation, and analysis. Sensors, Internet-of-Things devices, and CAD models are just some of the places where you can find the data that you need to make a digital twin. Developing a computer model of something and then using data and algorithms to replicate its behavior is what we mean when we talk about modeling and simulation. Data analysis as well

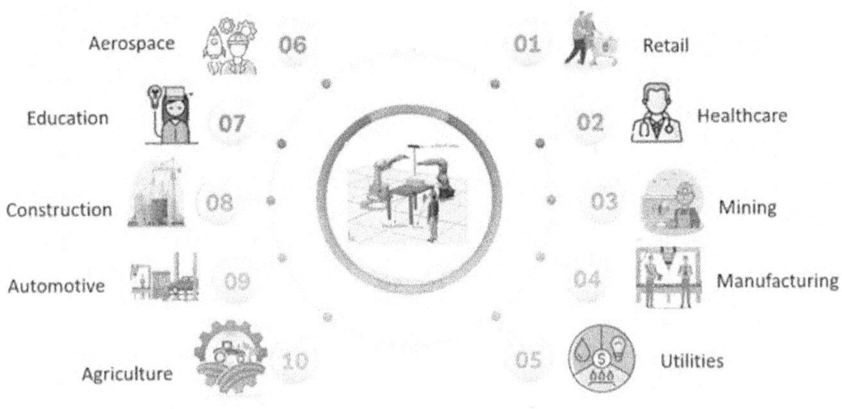

Figure 7.2 Industrial digital twin.

as machine learning methods are applied to the digital twin in order to enhance performance, foresee outcomes, and spot issues. As a real-time digital model of a physical object, the term "digital twin" was coined to describe the convergence of physical and virtual products. They provide detailed representations of physical items over an international connection. They integrate not just static product designs like CAD models, but also psychological dynamics. Through continuous information transmission, a digital representation of a physical item can be created and coexist with the latter. The intelligence and utility of digital twin technology, in particular for speedy identification and evaluation of design faults, have both been strengthened by the application of big data technology [21–25].

7.2 Digital Twin Technologies

Data modeling, data application, and data collecting are the three fundamental facets of digital twins. To gather and store real-time data, gather knowledge to offer priceless insights, and produce a digital replica of a physical thing, digital twin makes use of four technologies. The Internet of Things (IoT), artificial intelligence (AI), extended reality (XR), and cloud are some of these technologies (Figure 7.2). Additionally, depending on the type of application, digital twin makes use of a specific technology more or less [26].

7.2.1 IoT (Internet of Things)

The IoT refers to the global infrastructure of interoperable "things." This could be a connection between objects, individuals, or groups. The Internet of Things is at the heart of every digital twin's use case. By 2027, over 90 percent of the IoT systems will support digital twinning. In the IoT, sensors are used to collect data from real-world objects. Data transmission in the IoT is used to create digital models of real-world items. The digital version then lends itself to review, revision, and enhancement. With the help of the IoT, apps can create a digital twin of a material thing that is constantly updated with new information. Therefore, IoT is the backbone of every digital twin implementation.

7.2.2 Cloud Computing

The delivery of hosted services via the Internet is referred to as cloud computing. Data is effectively stored and accessed through the Internet thanks to the technology. Data computing and cloud data storage are made available to digital twins by cloud computing. With the use of cloud computing, digital twin, which has a lot of data, may store that data in a fictitious cloud and conveniently access the needed data from anywhere. Digital twins can efficiently shorten the time it takes for complicated systems to compute and get over the challenges of storing a lot of data thanks to cloud computing [27].

7.2.3 Artificial Intelligence (AI)

Artificial intelligence, or AI, is a subfield of computer science that tries to create machines with intelligence that can perform tasks normally performed by humans. Research into artificial intelligence focuses on areas including robotics, image identification, and language recognition. Using AI, digital twins gain access to a powerful analytical tool that can automatically sift through collected data, identify patterns, forecast future events, and provide solutions to problems. Expert systems, machine learning, deep learning, and neural networks are all examples of artificial intelligence.

7.2.4 Virtual Reality (VR), Augmented Reality (AR), and Mixed Reality (MR)

VR, AR, and MR are examples of immersive technologies that fall under the umbrella term extended reality (XR). These innovations have the power to broaden our perception of reality by fusing the actual and virtual worlds. Digital and physical items coexist and interact in real-time in the representations of objects made by XR. Users can engage with digital content thanks to digital twins, which utilize XR capabilities to digitally replicate real-world things [28–30]. The use case and application of a digital twin are shown in Figure 7.3.

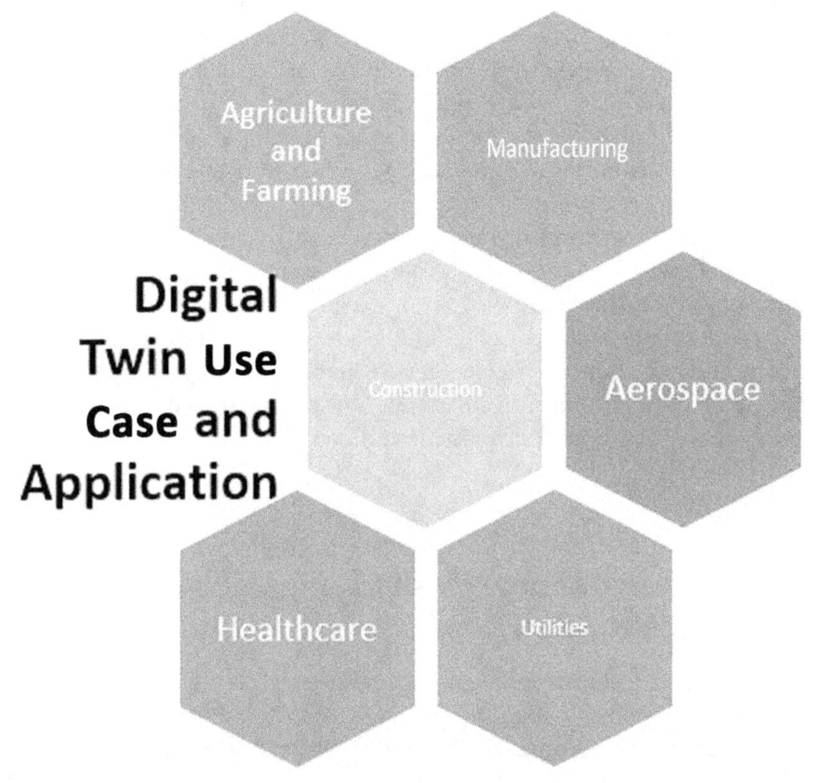

Figure 7.3 Use case and application of digital twin.

7.2.5 Digital Twins' Drivers and Challenges

The broad implementation of Industry 4.0 and the proliferation of the Internet of Things has sped up the deployment of digital twin technologies. Industry 4.0, also known as the 4th Industrial Revolution, incorporates cutting-edge methods and tools for product creation such as cloud computing, IoT, data analysis, digital twins, electronic scanning, 3D printing, additive manufacturing, and AI. Many efforts under Industry 4.0 rely heavily on the use of digital twins. Industry after industry is turning to digital doppelganger solutions for managing their assets and products over their lifetimes. Organizations can use this technology to their advantage by establishing an online representation of their products and procedures to help them plan ahead and make educated choices. Despite its many advantages, the method of digital twins covered in this chapter is nevertheless plagued by problems that plague AI and the IoT as a whole. The difficulties of building such a system and altering current ones, including data standards, information processing, and data security, are all part of the picture. The report also found that there are issues with communication, confidentiality, and protecting sensitive data, as well as the need to modernize existing IT infrastructure. High deployment costs, rising power and storage needs, worries about compatibility with current hardware or software that is proprietary, and the digital twin market's inherent complexity are just some of the obstacles that are expected to impede its growth. Digital twin-based solutions involve significant investments in technology platforms (sensors, software), infrastructure development, maintenance, quality of information authority, and security solutions. Furthermore, the capital investment required to run and maintain the digital twin architecture might be considerable. It is anticipated that the widespread adoption of digital twin technologies would be slowed by the high initial investment and complicated infrastructure needed for them [31–33].

7.4 The Future of Energy will Change the Way Work is Done

Keeping one's assets first instead of focusing on the supply chain. Previously, mining operations relied on energy supply networks to guarantee constant fuel delivery, full storage, and/or electricity contracts. Solar photovoltaic systems, turbines that generate electricity, electrolyzes, and batteries for storage are all examples of renewable energy assets that miners may be more likely to own and run. It is important for mining firms to know how their electrical infrastructure and renewable resources are doing, even whether they are owned and controlled by the company or the community at large. The mining industry consumes a lot of power. Because of this, mining companies will need to hire permanent workers to operate their equipment or work out an arrangement with the owners of the underlying assets to secure consistent electricity. [34].

- Integrating energy consumption and workflow planning into a single digital hub. The mining industry has to establish digital neural centers and work toward more dynamic and interconnected energy and operation control. This would help improve their power networks' reliability while decreasing the amount of space needed for storage. To achieve this sort of optimization, however, the energy system must operate in a predictable manner, and the control center must have a complete understanding of the energy ecological systems (including visibility of the present condition of the supply of energy and storage capacity, as well as visibility of the availability of electricity in the forthcoming days and days). Nerve centers will need both AI and skilled workers to monitor and optimize the functioning of the system as a whole. Example of energy management, workflow coordination, and integration: using hydrogen fuel vehicles to load materials into warehouses at night, then processing this material during the day, whenever solar energy is abundant. Instead of spending a lot of money on energy storage in order for keeping energy for a period of 24 hours, work might be maintained in its place using this strategy [35].

- Using digital strategies, we can have a more significant, beneficial impact on the planet. The ability to remotely availability, analyze, and manage a digital twin, for example, allows mining companies to do tasks like discovery, troubleshooting, and budgeting with minimal impact on the natural environment. Resource forecasting, distribution, monitoring, and oversight could all benefit from the use of digital twins of drill and blast designs and digital modeling of life-of-mine plans. Digital twins, virtual models, and work plans necessitate new and different skill sets, such as analytical ability, data visualization, and making intelligent choices based on predictive analytics [36].

7.5 The Influence of Technology on Work

As technologies like sensors, drones, ROS, and the cloud continue to proliferate, the mining industry is increasingly adopting robotics and automation. These cutting-edge tools enhance exploration, mining, processing, transportation, and commerce, but they also have far-reaching effects on the nature of work in the mining industry and the expertise needed to do it. This is due to the fact that extraction of natural resources is crucial to the mining sector. Companies in the mining industry now need people with expertise in data science and technology literacy; change management; sophisticated system creation and cooperation; business information technology operations as well as analysis; and, of course, superior operational and planning chops. The ability to work efficiently with AI, machine learning tools, virtual world tools, and robotics, and to successfully incorporate these innovations into regular operations, is expected to become increasingly important for mining companies in the future [37]. Digital twin influence and its technology development phases are shown in Figure 7.4.

(a) Driven by a sense of purpose and intent on achieving sustainable results: a firm that is enthusiastic about producing business results that are closely connected with the organization's purpose and the goals it has set for sustainability, and which offers appropriate incentives and rewards to its executives as well as its employees.

(b) Adaptable and flexible—demonstrates the ability to be flexible and adaptable in terms of where, how, and what work is provided. The workforce

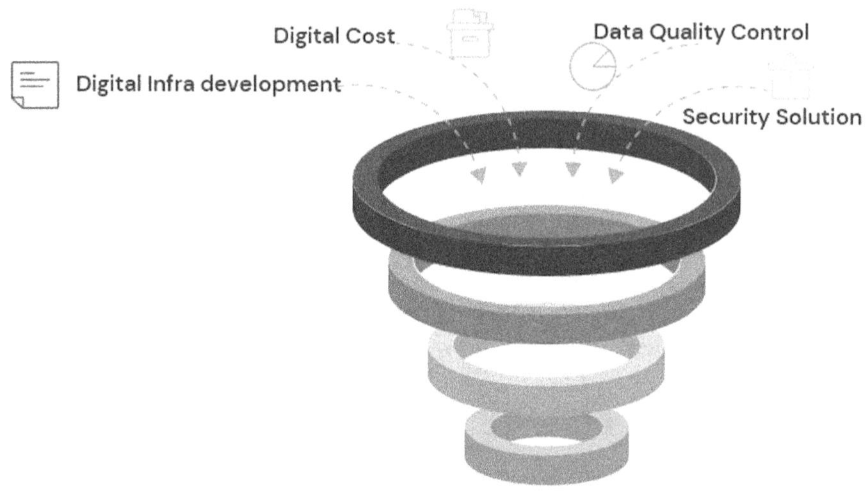

Figure 7.4 Digital twin influence and its technology development phases.

for net zero should have enhanced communication and cooperation abilities in order to draw capabilities from the larger talent ecosystem of the organization.

(c) Digitally driven, meaning that it is capable of learning through enhanced teaching methods that include simulation and virtual reality, as well as being digitally skilled and prepared to enhance what it can do through creative technological solutions [38].

(d) Making the most of human potential by concentrating on finding solutions, being inventive, and drawing on the highest levels of human creativity and analytical thinking, in addition to having strong personal connections.

(e) Equipped with transferable technical skills—assisted in the acquisition of a set of soft as well as technical abilities that enables it to operate cooperatively and nimbly with other energy sectors (Table 7.1).

7.6 Commonly Used Key Concepts Related to Digital Twin

a. Digital twin: a digital representation of a real-world component that can be tracked, analyzed, and predicted.
b. Cyber-physical system (CPS): what we are talking about here is a system where the computational and physical elements are tightly coupled and may exchange information with one another.

Table 7.1 Description of Different Terminology in Digital Twin

Term	*Description*
Physical Object	A tangible item in the real world, such as a product, system, or model.
Virtual Object	A digital representation of the physical object.
Physical Environment	The current state of any and all physical or digital item or surrounding attributes.
Virtual Environment	The action of changing the characteristics of the twin or physical thing.
State	The present situation of any and all physical or digital item or surrounding attributes.
Realization	The action of changing the characteristics of the twin or physical thing.
Metrology	The act of determining how the physical or digital duplicate is doing.
Twinning	Aligning the actual and digital manifestations of a component or twin.

c. The Internet of Things (IoT) is shorthand for a system of sensing and other devices that may communicate with one another and potentially share data.
d. Machine learning means teaching a machine to do its job without having to explicitly instruct how to do it, and it is a subfield of data science.
e. Artificial intelligence (AI) is a branch of computer science concerned with the creation of programs and devices to endow robots with the intelligence necessary to do tasks previously reserved for humans.
f. Cloud computing is a form of computing in which resources, including servers, storage, and applications, are delivered over the internet.
g. Big data is an industry term for exceptionally massive and complicated data sets that cannot be easily processed using conventional methods of data processing.
h. Digital thread: a framework for connecting and integrating data throughout the entirety of the product lifecycle, from the stage of product design all the way through to the stage of product disposal [39].
i. Simulation is the process of constructing a digital model of a system that exists in the real world and utilizing that model to test and optimize the system's performance.
j. Using a mobile smartphone or a headgear, users can experience augmented reality, which is a technology that adds to electronic data onto the physical environment.

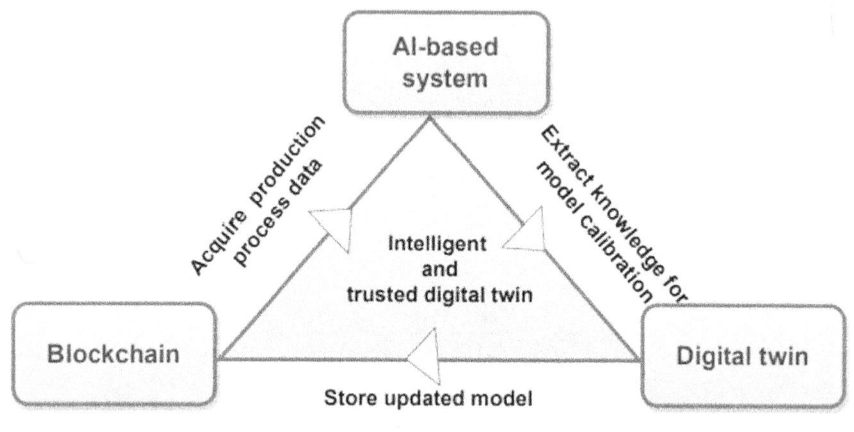

Figure 7.5 Blockchain-based digital twin with AI model.

7.7 Combining Blockchain-Based Digital Twins and Intelligence

Integrating blockchain-based digital twins with AI has the potential to revolutionize many industries by enabling more secure and efficient data sharing, as well as providing more accurate and timely insights. Figure 7.5 shows a block chain based digital twin with an AI model. Here are some possible ways forward for this integration:

7.7.1 Investigating the Possibilities of Using Artificial Intelligence in Conjunction with Digital Twins Based on Blockchain

The combination of artificial intelligence (AI) and blockchain-based digital twins has the potential to revolutionize many different industries by enabling safer and more efficient data sharing, as well as by giving more accurate and timely insights. The following are some potential next steps toward achieving this integration:

Blockchain technology provides a platform that is both more secure and decentralized, making it ideal for the storage and distribution of data. By integrating blockchain technology with digital twins, the content may be protected from being altered, from unauthorized access, and from other types of online threats. Using AI algorithms, strong security flaws can be identified and corrected.

Enhanced Productivity: digital twins that are built on blockchain have the potential to improve productivity by making it possible to share data in a more expedient and secure manner. Artificial intelligence can be put to use to rapidly and accurately evaluate and analyze massive amounts of data, which can then facilitate more effective decision-making.

More Accurate Results: artificial intelligence algorithms may be trained using the data from blockchain-based digital twins, which enables them to deliver results and suggestions that are more accurate. This can help organizations discover patterns and trends that might not be immediately visible, which can lead to improved judgments and more

New Business Models: it is possible that new types of business models, such as data marketplaces, will emerge as a result of the marriage of artificial intelligence and digital twins established on blockchains. Businesses could open up new revenue streams by selling access to their data to other companies in order to monetize their data and create revenue [40–45].

In general, the combination of artificial intelligence with digital twins that are based on blockchain technology appears to have the potential to revolutionize a wide variety of industries by enabling more exact information, the ability to communicate information in a manner that is more reliable and effective, and new business models. However, there are still challenges that need to be conquered, including those related to the scalability of blockchain and the ethical concerns raised by artificial intelligence (AI). Nevertheless, the integration of these systems presents companies with an intriguing opportunity to leverage technology to their advantage and foster innovation in their organizations.

7.8 Conclusion

Because of the presence of multiple entangled entities and continual advanced industrial processes, industries have to take into account a number of challenging components in order to maximize their revenues. These factors include things like managing costs, risk prognosis, and increasing efficiency. The creation of a digital twin that mimics every facet of the fundamental framework or process is therefore one of the potential ways that may be taken to maximize the efficiency of industrial operations. This will make it possible to conduct analysis, make predictions, and improve the performance of all processes before they are implemented in the actual world. Because data are obtained and accessible through a variety of information sources and information silos that are owned by a great number of participating firms, a strong security architecture is essential. Employing blockchain in this manner offers businesses the ability to monitor data on a distributed record that safely logs operations pertaining to the processing of product life data. DTs improve management systems by putting an emphasis on a number of different frameworks.

References

1. Dietz M., Pernul, G. Unleashing the digital twin's potential for ics security, *IEEE Secur. Privacy*, 18 (4) (2020) 20–27. 10.1109/msec.2019.2961650
2. Dietz M., Pernul, G. Digital twin: Empowering enterprises towards a system-of-systems approach, *Bus. Inform. Syst. Eng.* 62 (2) (2020) 179–184.
3. Dietz, M., Putz, B., Pernul, G. A distributed ledger approach to digital twin secure data sharing, in: *IFIP Annual Conference on Data and Applications Security and Privacy*, vol. 11559, Springer, Cham, 2019, 281–300. doi: 10.1007/978-3-030-22479-0_15
4. Dietz, M., Vielberth, M., Pernul, G. Integrating digital twin security simulations in the security operations center, in: *Proceedings of the 15th International Conference on Availability, Reliability and Security, ARES '20*, Association for Computing Machinery, New York, NY, USA, 2020. doi: 10.1145/3407023.3407039
5. Damjanovic-Behrendt. A digital twin-based privacy enhancement mechanism for the automotive industry, in: *2018 International Conference on Intelligent Systems (IS)*, 2018, 272–279. doi: 10.1109/is.2018.8710526
6. Eckhart, M., Ekelhart, A. A specification-based state replication approach for digital twins, in: *Proceedings of the 2018 Workshop on Cyber-physical Systems Security and Privacy, CPS-SPC '18*, Association for Computing Machinery, New York, NY, USA, 2018, 36–47. doi: 10.1145/3264888.3264892
7. Eckhart, M., Ekelhart, A., Weippl, E. Enhancing cyber situational awareness for cyber-physical systems through digital twins, in: *2019 24th IEEE International Conference on Emerging Technologies and Factory Automation (ETFA)*, 2019, 1222–1225. doi: 10.1109/etfa.2019.8869197
8. Eckhart, M., Ekelhart, A. Towards security-aware virtual environments for digital twins, in: *Proceedings of the 4th ACM Workshop on Cyber-physical System Security, CPSS '18*, Association for Computing Machinery, New York, NY, USA, 2018, 61–72. doi: 10.1145/3198458.3198464
9. Eckhart, M., Ekelhart A. *Digital Twins for Cyber-Physical Systems Security: State of the Art and Outlook*, Cham, Springer International Publishing, 2019, 383–412. doi: 10.1007/978-3-030-25312-7_14
10. Eckhart, M., Ekelhart, A., Eisl, R. Digital twins for cyber-physical threat detection and response, *Ercim News: Special Theme Smart And Circular Cities* 127 (2021) 12–13.
11. Frade, M.J., Pinto, J.S. Verification conditions for source-level imperative programs, *Comput. Sci. Rev.* 5 (3) (2011) 252–277.
12. Groshev, M., Guimarães, C., Martín-pérez, J., Delaoliva, A. Toward intelligent cyber-physical systems: Digital twin meets artificial intelligence, *IEEE Commun. Mag.* 59 (8) (2021) 14–20. doi: 10.1109/mcom.001.2001237
13. Hong, T., Ghobakhloo, M., Khaksar, W. 6.04 – robotic welding technology, in: Hashmi S., Batalha G.F., Van tyne C.J., Yilbas B. (Eds.), *Comprehensive Materials Processing*, Elsevier, Oxford, 77–99, 2014. doi: 10.1016/b978-0-08-096532-1.00604-x
14. Khan, R., Maynard, P., Mclaughlin, K., Laverty, D., Sezer, S. Threat analysis of blackenergy malware for synchrophasor based real-time control and monitoring in smart grid. *4th International Symposium for ICS & SCADA Cyber Security Research* 4 (2016) 53–63.

15. Kshetri, N., Voas, J. Hacking power grids: A current problem, *Computer* 50 (12) (2017) 91–95. doi: 10.1109/mc.2017.4451203
16. Lantz, B., Heller, B., Mckeown, N. A network in a laptop: Rapid prototyping for software-defined networks, in: *Proceedings of the 9th ACM Sigcomm Workshop on Hot Topics in Networks, HotNets-IX,* Association for Computing Machinery, New York, NY, USA, 2010. doi: 10.1145/1868447.1868466
17. Qi, Q., Tao, F., Zuo, Y., Zhao, D. "Digital twin service towards smart manufacturing," *Procedia CIRP* 72 (2018) 237–242.
18. Srivastava, S., Bisht, A., Narayan, N. "Safety and security in smart cities using artificial intelligence — A review," in: *2017 7th International Conference on Cloud Computing, Data Science Engineering – Confluence*, pp. 130–133, Jan. 2017.
19. Elijah, O., Rahman, T. A., Orikumhi, I., Leow, C. Y., Hindia, M. N. "An Overview of Internet of Things (IoT) and data analytics in agriculture: Benefits and challenges," *IEEE Internet of Things J.* 5 (Oct. 2018) 3758–3773.
20. Sezer, O. B., Dogdu, E., Ozbayoglu, A. M. "Context-aware computing, learning, and big data in Internet of Things: A survey," *IEEE Internet of Things J.* 5 (Feb. 2018) 1–27.
21. Moujahid, A., Tantaoui, M. E., Hina, M. D., Soukane, A., Ortalda, A., ElKhadimi, A., Ramdane-Cherif, A. "Machine learning techniques in ADAS: A Review," in: 2018 *International Conference on Advances in Computing and Communication Engineering (ICACCE)*, pp. 235–242, June 2018.
22. Shinde, P. P., Shah, S. "A review of machine learning and deep learning applications," in *2018 Fourth International Conference on Computing Communication Control and Automation (ICCUBEA)*, pp. 1–6, Aug. 2018.
23. Castellani, A., Schmitt, S., Squartini, S. Real-world anomaly detection by using digital twin systems and weakly supervised learning. *IEEE Trans. Ind. Inform.* 17 (2021) 4733–4742.
24. Murillo, A., Taormina, R., Tippenhauer, N., Galelli, S. "Co-simulating physical processes and network data for high-fidelity cyber-security experiments," in *Proceedings of the Sixth Annual Industrial Control System Security (ICSS) Workshop*, 2020, ICSS 2020, Austin, TX, USA, 8 December 2020, 13–20.
25. Saad, A., Faddel, S., Mohammed, O. Iot-based digital twin for energy cyber-physical Systems: Design and implementation. *Energies* 13 (2020) 4762. [Google Scholar] [CrossRef]
26. Suhail, S., Jurdak, R., Matulevicius, R., Seon Hong, C. Securing Cyber-physical Systems through Blockchain-based digital twins and threat intelligence. *arXiv* 2021. arXiv:2105.08886
27. Chukkapalli, S. S. L., Pillai, N., Mittal, S., Joshi, A. "Cyber-physical System security surveillance using knowledge graph based digital twins—a smart farming usecase," in *Proceedings of the 2021 IEEE International Conference on Intelligence and Security Informatics (ISI)*, Antonio, TX, USA, 2–3 November 2021, pp. 1–6.
28. Zhao, G., Liu, S., Lopez, C., Lu, H., Elgueta, S., Chen, H., Mileva Boshkoska, B. Blockchain technology in agri-food value chain management: A synthesis of applications, challenges and future research directions. *Computers in Industry*, 109 (2019) 83–99. https://doi.org/10.1016/j.compind.2019.04.002

29. Zheng, Z., Xie, S., Hong-Ning Dai, Chen, W., Chen, X., Weng, J., Imran, M. An overview on smart contracts: Challenges, advances and platforms. *Future Generation Computer Systems*, 105 (2020) 475–491.
30. Li, M., Li, Z., Huang, X., Qu, T. Blockchain-based digital twin sharing platform for reconfigurable socialized manufacturing resource integration. *Int. J. Prod. Econ.* 240 (2021) 108223. https://doi.org/10.1016/j.ijpe. 2021.108223
31. Saini, P., Dwivedi, Y. K., Rana, N. P. Digital twin technology: A systematic literature review and future research directions. *J. Clean. Prod.* 286 (2021) 125429.
32. Tao, F., Zhang, M., Liu, A., & Nee, A. Y. C. Digital twin in industry: State-of-the-art. *IEEE Trans. Ind. Inform.* 15(4) (2019) 2405–2415.
33. Jiang, P., Liu, B., Yang, L. T. A survey on digital twin: From methodology to industry applications. *J. Intell. Manuf.* 30(2) (2019) 487–500.
34. Zhang, Y., Zeng, B., Zhang, D., Luo, J. Digital twin: Technologies, challenges, and future directions. *IEEE Trans. Ind. Inform.* 17(2) (2021) 1162–1171.
35. Liu, W., Guo, S., Qian, J., Huang, G. Q. Digital twin-enabled sustainable manufacturing: State of the art, challenges and opportunities. *J. Clean. Prod.* 318 (2021) 128504.
36. Fan, J., Xiang, Q., Li, X., Liu, Z. Challenges and opportunities: Digital twin for sustainable manufacturing. *J. Clean. Prod.* 261 (2020) 121045.
37. Song, Y., Liu, S., Wang, Y., Chen, X., Wang, Y. A survey of digital twin: Platforms, technologies, and applications. *J. Manuf. Syst.* 60 (2021) 41–52.
38. Chen, J., Shen, W., Zhu, W. From digital twin to industry 4.0: An overview. *J. Ind. Inf. Integra.* 13 (2019) 1–6.
39. Reymus, D., Niggehoff, F., Kuhlen, T. W. Digital twin technology: A comprehensive analysis of the state of the art. *Procedia CIRP* 88 (2020) 28–33.
40. Wang, L., Wang, X. Digital twin: A survey from data perspectives. *IEEE Access* 8 (2020) 131547–131563.
41. Luo, J., Zhang, Y., Zhang, D. Digital twin: The backbone of cyber-physical systems. *IEEE Access* 7 (2019) 133237–133246.
42. Kusiak, A. The digital twin: The future of the smart factory. *Int. J. Prod. Res.* 59(4) (2021) 959–964.
43. Raza, M. A., Noor, N. A systematic literature review on digital twin for sustainable manufacturing. *J. Clean Prod.* 303 (2021) 126937.
44. Zhan, T., Deng, C., Wang, W. From digital twin to digital world: A survey of cyber-physical systems. *IEEE Access* 8 (2020) 202195–202216.
45. Eynard, B., Gomes, S., Fontana, M. Digital twin and model-based systems engineering: A promising synergy for future industry. *Computers in Industry*, 123 (2020) 103313.

Chapter 8

Using Improved Finite Element Modeling to Combat Cardiovascular Disease

A Review of a Developing Area at the Intersection of Several Disciplines

V. Sheeja Kumari, T. Manikandan, A. Selva Kumar, S. Ponmaniraj, and K. Prabu

8.1 Introduction

In the real world, a digital twin can be used to simulate how things work and study how they change over time. This will improve operations, and big changes can be evaluated more accurately. Moreover, the concept of a digital twin has been on every single one of Garner's lists of the "Top 10 Strategic Technology Trends" since 2017. Some industries have shown a lot of interest in this technology in recent years. Among these industries are smart cities, retail, automotive, oil and gas, aerospace, mining, and healthcare [1].

It is evident that all of the fields above have a large amount of untapped potential that still remains to be exploited. Particularly in the health and medical field,

DOI: 10.1201/9781003469612-8

where data from the Internet of Things (IoT) and digital twin technology will be a key part of the reforming of the health and medical sector. A person who looks healthy might choose to ignore the small signs of what are probably normal symptoms. The digital twin keeps an eye on a person's medical records, compares them to patterns that have already been recorded, and does an analysis of the symptoms of the disease.

As a result of the data, a digital model of a typical healthy patient can be created, which allows for the definition of new criteria for "healthy." A digital twin is a digital representation of a physical object or a device that is generated through simulation using finite element analysis [2]. Also, digital twins can be used to test new designs before they are put into production in order to see how well they work. Using this technology, NASA was able to explore space for the first time in 2002. Eventually, the technology was applied to many different fields, such as manufacturing, oil and gas, telecommunications, aerospace, as well as other fields. A number of technologies such as Industry 4.0, machine learning, artificial intelligence (AI), data analytics software, reinforcement algorithms are making it much easier to collect and analyze data [1]. A virtual copy of a physical object is made using software programs such as MATLAB, Simio Simulation, Abaqus, and many others in order to make a virtual copy of the physical object. There are three important steps to the process: first of all, in the preprocessor stage, an image is created of the real object constructed on the input constraints; then, in the second step, a virtual image is split into different parts, and the virtual copy is analyzed using an agile-based method; and, finally, to learn from the analysis, the image is put back together. An example of simulation is the process of simulating a real thing, such as a machine or a human organ, in order to be able to carry out experiments on the model to gain a deeper understanding of how the system operates or to be able to compare different methods of running it. These methods are the most commonly used in the fields of management science and operations research. As a general rule, simulations are based on the finite element method (FEM) and in most cases it is called as finite element analysis as well [1, 2].

By using this method, one is able to predict how a product will react to the forces it will meet in the real world and how it will react to those forces. With the help of the virtual image, FEM is able to predict what the real object will behave like in the future based on what the virtual image shows. In the field of finite element analysis, Abaqus is a great piece of software that is often used. In recent years, a number of technologies have been developed to support the demand for demand analytics, such as AI, IoT, and other sensor-based devices. This technology is used in a variety of industries, such as manufacturing, healthcare, aerospace, defense, and many others. As a result of the capabilities of a digital twin, the medical field is a good place for new ideas, thanks to what it can do. There is no doubt that the future of healthcare will be personalized since each person is given the appropriate care at the right time based on their specific needs. This chapter explores various ways in which the technology of digital twins can be functional to the medicine arena in order to facilitate the development of more accurate and personalized diagnosis methods for patients [3]. In order to categorise the various types of heart diseases, a machine learning algorithm is used that employs an algorithm that utilises the principles of machine learning.

8.2 Monitoring of the Patient

With virtual patients, doctors can keep an eye on the real patient they are based on. Adaptive analytics and algorithms are used to get accurate results, and the data collection and curation processes are kept up to date at all times. Modern technologies combined with virtual twins of patients can help doctors monitor patients from a distance (RPM) [4]. With easier access to medical care, patients will be less worried and their families will have more peace of mind every day.

8.2.1 Cost Savings

By using digital twin technology to get rid of slow spots in the flow of patients, one medical facility was able to reduce its operating costs by 900%. Also, it is possible that the new technology will cost more in the beginning. But because its services lower costs, the hospitals and organizations that use them make more money over their lifetimes [4].

- With this information, the hospital would be better able to put its staff and resources, like new gadgets, beds, and places, where they are needed most.
- With the help of digital twin techniques, in a hospital easily one can predict cardiopulmonary and respiratory arrest alike medical emergencies. This, in turn, helps the hospital organization do better preventative maintenance and gives personalized costs for healthcare.
- It helps make treatment plans that are right for each patient.

Healthcare Tailored to the Patient

An administrator, nurse, and doctor can get a detailed view of how patients are coming and going by creating a "virtual twin" of the hospital. As a result, workflows can be planned more efficiently. A well-organized workflow reduces the amount of time patients spend waiting in line. Furthermore, it makes it easier to schedule appointments and keep track of inventory levels. Doctors, like cardiologists, can benefit from using technology by treating each patient individually, which results in more work being done [5].

Using artificial intelligence, for example, the "digital heart twin" made by the healthcare company Siemens Healthineers helps doctors figure out how different medicines affect each patient's body more accurately. By analyzing epidemiology data, the technology could also be used to find risk zones in order to keep track of a specific infection. Technology would be used to do this. Here are some examples of other fundamental forces that, thanks to digital twin technology, help the healthcare industry grow:

(i) The Internet of Things (IoT)
(ii) Software tools for analyzing software
(iii) Machines that teach themselves
(iii) Spatial element graphs, spatial element network graphs

Also, the new model for the virtual simulation can automatically adjust and improve itself with the help of big data. GE Healthcare, Philips, Siemens, Alacris, IBM, and Microsoft (under the Azure brand) are just some of the companies that are using digital twin technology to great effect in the healthcare industry [6]. Even though twin technology has failed in the past and is still struggling to become widespread, its future looks very bright. And thanks to the ideas that digital twins came up with, the healthcare organizations were finally able to see a pattern of inefficiency and figure out why it happened in the first place. Because of this, forward-thinking practises and ways to stop problems from happening again were created.

8.4 Literature Review

In precision medicine, CVD patients can be phenotyped more accurately if they share the same symptoms or conditions. The diagnosis and treatment are guided by clinical, imaging, molecular, and other factors. Digital twins can help us achieve this potential. Virtual patient versions are created and real-time apprises are made of a number of data entities [7]. As a result, we can predict disease and choose the best treatment for our patients. It was explored what the term "digital twin" means, the ideas that make it up, the challenges associated with a new field, and the possible CVD applications. A mapping review was conducted based on a thorough review of peer-reviewed literature. In order to locate people in the industry and patent applications, the Internet was used. We found 88 papers in the databases Compendex, EMBASE, Medline, ProQuest, and Scopus. A total of 28% of these (n = 25) were about cardiovascular conditions, 41% (n = 36) were about other conditions, and 31% (n = 27) were about the health digital twin in general. A total of 15 companies wanted to make money from digital twins or simulation models in health, and all of them focused on cardiovascular diseases. It was found that 18 patent applications were filed by 11 different individuals. 27% came from universities, while 73% came from companies. Heart health was a theme in three of the proposals. Digital twins in CVD have only been researched for a short time in both industry and academia. Globalization and interdisciplinary research are also becoming increasingly important. The majority of the applications were numerical simulation models, but there were also models similar to real-time cyber-physical systems. AI decision-making tools are difficult to use because of ethical limitations and experimental barriers. The right way to use these tools is difficult [8].

Around the world, cardiovascular disease is responsible for about one-third of all expiries. Also, this disease is the foremost root of disability-adjusted life spans, which are the years of life lost to early demises and the years survived with disabilities. A major cause of early death is ischaemic heart disease (IHD), which kills more people than any other form of cardiovascular disease. It depends on the setting, however, whether treatments that are proven to be effective are available and used. The most important thing is to make sure that risk assessments are accurate and tailored to each individual, as well as to make sure that prevention treatments are as well. The risk of cardiovascular disease and coronary heart disease is currently calculated using risk algorithms that only consider a few traditional risk factors. However, as a lot of events happened to people who were thought to have low risk under traditional algorithms and some of those who had risk factors responded well to therapy, it is clear that there are still a lot of questions that need to be answered in order to make the maximum active use of anticipatory medicines, instruments, and other remedies from a wellbeing and economic perception [7, 8].

The field of precision medicine, also known as personalized medicine, is growing rapidly. The aim is to phenotype patients with the same condition or symptoms more accurately. By doing so, screening, diagnosis, and treatment can be tailored to each individual. By using biological databases (like the genome sequence) and a variety of markers, both biological and not, it is now possible to divide patients into groups so that they can receive more targeted treatment. A mix of cells has been measured using omic technologies for a long time to measure the activity of numerous genes, molecular features and other proteins instantaneously. As a result, we now have high-volumed, multifaceted data that we call omics data, which helps us understand the relationship between genotype and phenotype [9]. Patients can be classified according to their metabolic, genomic, bacteriological, proteomic, scientific, behavioral, and health pathways. It may be possible to overcome this natural difference between people by using advanced computational techniques. As a result, clinical decisions will be more accurate and interventions will be selected more accurately. Utilizing the concept of a digital twin is one way to make precision medicine a reality. It is proposed that patient-specific therapy should be based on digital twins, which can be used to predict the outcome of a treatment and tailor a patient's prognosis (the real-life twin).

Digital twins are used in healthcare to help with strategic planning, risk analysis, quality assurance testing, and preventative maintenance. Cloud and advanced level of network generations, and forecast analysis software's are among the technologies used in these feedback systems. As a result, virtual models, physical assets, and instantaneous data gathering and conversation between them can all come together in three ways. Each twin is connected to the other via devices or sensors. Algorithms gather, store, and combine data from different roots to find updates, modifications, and outlines, predicts, and analyzes the letdowns, test different decisions, and optimize the real-world quality. Compared to modeling or simulation, pairing a process or an entity saves money and time, and it is often used in different transportations such as roadways, sea and trains, supply-chain operations, and civil engineering. Due to what was said above, this is the case. Manufacturing industries, for example, often use digital copies of machines and other factory tools. As well as smart cars, this category also includes modern consumer goods. For a plane to remain in good mechanical and structural shape throughout its lifespan, its digital twin is very important. By using testable scenarios that are based on real entity data fed to an echoed system, it is proposed that healthcare facilities and departments can improve staff sharing, sightseer/patient movement, waiting times, provision of instruments and other inner resources, spare vehicle contacts, and other service-interrelated operations. In a similar manner, and using the human genome project4 as a starting point, an individual health digital twin can be used to help make decisions and predictions about a real-world patient. These roots can be found at the beginning of the project. The population-based databank holds both types of data, and it is the focus of the overall design. Data from electronic health records as well as imaging, biological, clinical, genetic, and molecular information will be used to complete a whole phenotyping. Additionally, real-world data from a person's surroundings can be phenotyped using wearable electronics and mobile data sensors. When these continuously collected data are transformed into clinically usable knowledge, an automated, iterative process of data pretreatment, data mining, and data integration provides you with more useful information than any single data source. There are a lot of different sources of these numbers. A digital twin's phenotypic data is analyzed in cardiology using a predictive framework that combines statistics and mechanistics, allowing the twin to make its own decisions [10, 11]. In order to create a digital twin for a real person, a population-based databank will be used. Patients' closest cluster group's average traits will be shown in the digital twin. Databanks store the results of virtual interventions that were later used on real patients. Using these results, the twin is changed and new information is added to the pool of population data. For the databank to grow and maintain its physiological and demographic diversity, this dynamic loop must be in place. To combine and analyze data from large groups of patients, this paradigm makes use of bioengineering and computer science. Given that the digital twin principle has a lot of potential to help research on CVD, it would be helpful to have a better understanding of its full scope. Its goals were to describe "digital twin" research in relation to cardiovascular disease with a focus on how it is used; to summarize the key ideas; to define the obstacles for digital twin research in the context of precision medicine generally, as well as the fields that contribute to the subject [12].

This is a review of mapping that has been changed to answer more general questions. An exploratory mapping review has two purposes: to give a brief overview of a topic or to determine how much and what kind of literature is available in a field. Systematic review methodology included parts that made the search process clear and repeatable. It was necessary to make changes for the reason that the concentration was on a wide range of information, which included both research and non-research materials. To evaluate the available literature, a methodical approach was used. A mapping review, however, does not contain a synthesis of the available data or a critical evaluation of the literature's methodological quality.

Through the application of simulation software, a digital twin of the human heart is created, and the study's ultimate goal is to determine the likelihood of an individual developing heart disease. This model demonstrates how to create a diagnosis approach and how data is transferred across implantable devices. In order to strengthen the model's analytical capabilities, a machine learning algorithm was incorporated into the Philips Healthcare Model of digital twin [13]. Using the decision tree technique used in machine learning (target variable explains parameters causing heart disease), predictions can be made regarding the maximum likelihood that will affect the target variable. Based on their observations, the doctors will formulate a diagnosis based on the target variable in order to determine how accurate the model is. In the field of digital twin technology, several papers have been published, such as [2], which integrates digital twin technology into MBSE techniques and experimentation test beds as a core component of MBSE methodology. The following paper is also in this field: [2]. Additionally, it covers the advantages of combining digital twins with system simulations and online gadgets. An article [12] analyzes the architecture, implications, and trends in digital twins. An architecture for the digital twin of the product is presented in the paper. Based on a systematic analysis of the product's connotation, this design was created. A recent development with digital twin products is also discussed in the article. In order to derive more efficient approaches to the design, production, and delivery of products, the authors of [6] have used big data analytics. As a result of this study, the limitations of the virtual replica will be brought to light as well as challenges encountered during the analysis itself. This study aims to design a smart production management framework based on a digital twin [11].

Additionally, it makes a recommendation for a control measure that is applicable to complicated product production methods. Assembling components requires a number of fundamental strategies, the most important of which are real-time management, organization, and data capture. A shop floor worker uses these methods [14]. Boschert and Rosen [16] explain how to develop a model for either evaluating systems or supporting design tasks using digital twin technology. They propose a comprehensive reference model using skin model. In 2018, Tao *et al.* created a framework [9]. Forecasts on the remaining useful life of an offshore wind turbine power converter are needed. Sivalingam *et al.* [4] used a digital twin framework. Schroeder *et al.* [8] used automated ML model attributes linked to digital twins in their study. Brosinsky *et al.* [19] have improved power control system centers using digital twins, and Uhlemann *et al.* [14] have developed automated data acquisition and selection using digital twins. Using digital twins, Brosinsky *et al.* [19] have improved the application in power control system centers. Brosinsky *et al.* [19] have used digital twins to improve the applicability of power control systems. In this article, the concept of microprocessor-controlled pacemakers is dismantled, which are capable of capturing, storing, and transmitting information about the patient [3].

8.5 Concepts Fundamental to the Digital Twin in Health

It is possible to apply research conducted on digital twins to a wide range of diseases and conditions; nonetheless, there are a few core concepts that apply to all of these disorders. The most important subfield of computer science is artificial intelligence (AI). By combining breakthroughs in four major areas: computing power, processing of "big data," machine learning, and pattern recognition, AI systems replicate human reasoning [15]. The development of AI depends on these four fields. Alternatively, AI systems attempt to mimic human reasoning. The Internet of Things, or IoT, refers to the process of facilitating the flow of data between physical sources connected to a network (including the so-called "big data").

In this context, the term "Internet of Things" (IoT) is used. With the help of IoT-enabled artificial intelligence systems and cloud computing, it is possible to construct a digital representation of an existing physical system. Furthermore, a cyber-physical system cannot have a digital twin unless there is a constant or, at the very least, periodic data exchange between its physical and digital equivalents. A cyber-physical system (CPS) has these distinguishing characteristics, as depicted in Figure 8.1. Twin subtypes are mostly active or passive depending on how frequently the data is fed, in the end. A "what if" situation can be recreated using

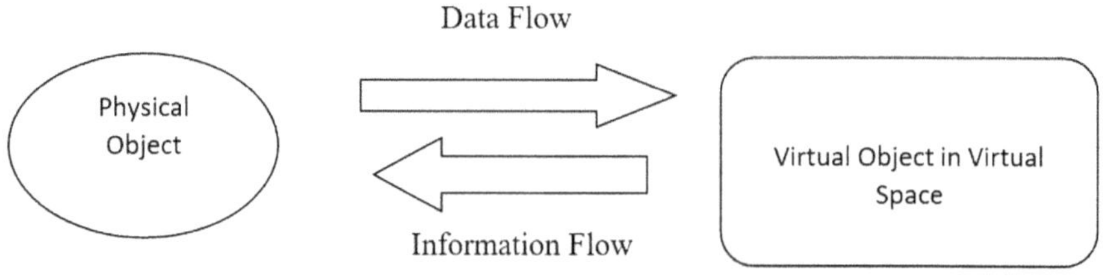

Figure 8.1 The concept of a cyber-physical system.

a digital twin. The twin can also track, identify, and foresee issues because of the cumulative, real-time, real-world data flow within the CPS. This is made possible by a method called closed-loop optimization, wherein the constant synchronization between the twins enables the virtual twin to quickly reconfigure itself as it adopts the traits of its physical twin, foresees potential problems, and tests potential solutions before implementation. In addition to fast adjusting to its physical twin, closed-loop optimization enables the virtual twin to swiftly restructure itself. These operational and intelligent aspects distinguish a digital twin from a model that is just a simulation since the latter is a digital representation of a physical object.

Data-driven predictive and prognostic AI models are more therapeutically valuable than conventional data processing methods as a result of the combination of the aforementioned characteristics—AI systems, IoT approaches, and bidirectional data sharing. This was accomplished using IoT devices, AI systems, and bidirectional data interchange. Associative data processing is a widespread practise. This technique applies epidemiological and statistical models based on structured, descriptive, and retrospective data pertaining to a previously formed population. A data pool like this is less dynamic and more static in a localized centralized storage system. Furthermore, information from other sources, such as digital models, wearable technology, and health records, is frequently merged with data that is already fragmented and stored independently [14].

The information obtained from other sources is not combined with this information. A digital twin paradigm, however, involves processing data that is multidimensional, unstructured, and distributed across multiple locations. They include (almost continual) data flow from the surrounding environment, gathered in a prospective and exponential manner from a variety of sources. In data processing, this aspect is called context awareness, and that is the name given to it. It is possible to analyze big data using data fusion both prescriptively and predictively. Combining different types of big data can be used for these analyses. Supervised and unsupervised machine-learning processes are included in these studies.

Exploratory pattern identification and the application of physiologic, clinical, and social causal pathways can differentiate these processes. Convolutional neural networks and deep learning are two subfields of machine learning that could revolutionize the way illness outcomes are predicted. These developments are one of the main reasons why researchers are so excited about them. It is obvious that when these data-driven technologies are considered collectively, the level of reasoning that can be accomplished with them is higher than the processing that is typically done by statistical methods like logistic regression and decision trees, as well as numerical, physics-based analyses, like those used in hemodynamic models of coronary vessels and heart valve mechanics. Reasoning power outperforms both of these approaches with modern data-driven technologies.

8.6 Applications of Digital Twin Technology in Cardiovascular Disease

Further detail was provided on the research that coined the term "digital twin" to address specific CVD-related clinical problems. Computer, bioengineering, and mathematical sciences techniques were used to simulate and model the data. Despite this, seven of the articles were described as antecedents to either active or semi-active digital twins. In spite of the fact that none of the original research papers were fully incorporated into ordinary clinical practice, seven of them were described as such [16]. Research publications can be viewed as both proof of concept and model validation, which is the most accurate description. Nine papers explored various aspects of precision cardiology in the form of discussions. Combining statistical and mechanistic modelling approaches, machine learning applications, and combined statistical and mechanistic modelling techniques were all taken into consideration in addition to clinical acceptability, translation, possible benefits, and limits. Most digital twin components are utilised at the organ level in terms of scale. Modeling of these components was based on a structural, biomechanical, or electrical feature of the heart.

In addition to echocardiography, tomography, and magnetic resonance imaging, these characteristics were derived from databases of electrocardiograms (ECG) and mathematical models. Descriptive information from medical records as well as prospective physiological and behavioral data from smart wearables were combined to create a twin model with "human" features at the scale of a virtual patient. Health records were used to create these models. Data were collected in real time as they occurred. In one of the investigations, electronic health records were used to create a profile database. This study aimed to identify a number of clinical and demographic characteristics that can be used to select an antihypertensive medication based on the patients' individual needs. Another study examined novel applications of edge computing in the context of ischemic heart disease. The ECG dataset was used to train and test the AI dataset, and the cellphones were coupled with external sensor equipment to create a body area network based on the ECG data. There is no doubt that body area networks are becoming increasingly important in the medical field. In order to achieve the desired results, these two approaches were combined with one another in order to achieve the desired outcome. In order to share biometric information between the smartphone that belonged to the actual twin and the smartphone that belonged to the digital twin, Bluetooth connectivity and services provided by the 5G network were used, as depicted in Figure 8.2. Data fusion and analysis of the data were carried out at the digital twin, where the data was combined and analyzed.

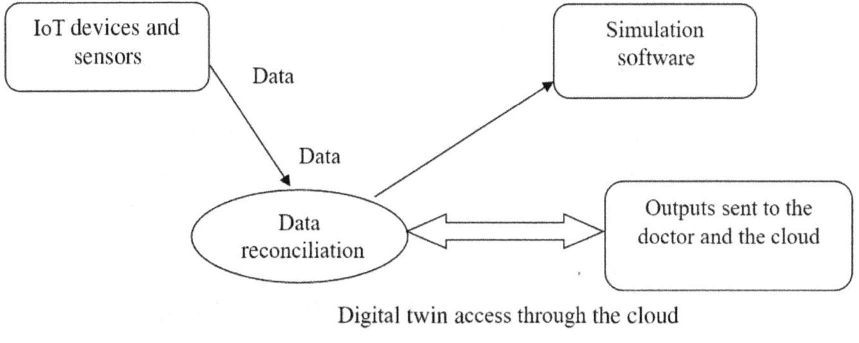

Figure 8.2 Digital twin model of the heart.

8.7 The Viability of the Health Digital Twin and its Implementation

Seven interconnected topics were discussed throughout the included papers, and each one provided an illustration of an important translational challenge for health digital twin research. They are as follows:

1. Big data hazards
2. Computational power needs
3. Data sharing and intellectual property issues
4. Cyber security
5. Professional barriers
6. Ethical
7. Governance and regulatory

8.7.1 Big Data Hazards

AI for inductive reasoning presents methodological hazards, which is one of the issues that have been recognized. To determine whether or not the findings can be generalized, an independent validation is required using fresh patient cohorts, cohorts from different centres, cohorts from different geographical regions, and cohorts spanning time. Furthermore, there is the possibility of confounding bias, which occurs when one variable out of a large body of data gives the misleading impression that the two are connected. Selection biases can affect conclusions or lead to models that further societal biases related to race, gender, or other categories when weighting is used to combat them. Selection biases can be combated by weighting [17,18]. There is a risk that weighting can lead to models that further societal biases. Data that is readily available generally requires transparency, the plausibility of computer predictions, and independent validation.

8.7.2 Computational Power Needs

The Internet of Things presents a number of challenges for medical technology, particularly for devices with limited processing power, memory, and power. Computing power and costs associated with supporting infrastructure that support the capture, storage, and processing of data are important components of large data sets [18]. To ensure interoperability with shared networks, it is more difficult to meet the standards for scalability when cybersecurity concerns exist.

8.7.3 Intellectual Property and Data Sharing Problems

The establishment of databanks and the exchange of information within and between countries requires compliance with data protection requirements concerning the obtaining of consent, the anonymization of data, the notification of data breaches, and the secure transfer of data internationally. While national and international data privacy regulations protect the health information of individuals, compliance requires ongoing vigilance against data breaches, whether they are caused intentionally or accidentally [18]. Using large amounts of data and machine learning raises questions about who owns the intellectual property of the results. These results raise questions about who owns the results produced by individuals, institutions, and businesses.

8.7.4 Cybersecurity

A medical CPS or digital twin programme must meet certain requirements when applying transformative technologies. These specifications include reduced patching needs, secrecy, dependability, safety, and secure coding. Just like more traditional physical media are prone to storage and infrastructure defects, so is the cloud computing system to processing flaws. Businesses are exposed to a number of risks from third parties, including eavesdropping, malicious software, and distributed denial-of-service attacks. In order for commercial entities to continue having an interest in gathering, keeping, and analyzing personal and medical data, research that involves data harvesting will cause public aversion to the use of big data, abuse, breaches, and theft.

8.7.5 Professional Barriers

The credibility of virtual patient models and the trust required of the computational processes that generate them represent potential barriers to incorporating these models into routine workflows. Neither the models themselves nor the processes that deliver them can be trusted without trust. In order to guarantee that the models produce correct results, this level of trust is necessary. Software and other forms of technology have some people concerned that one day they will be replaced or become less skilled in their work, despite the fact that physicians are responsible for providing care to patients. Furthermore, providing care for patients is one of a physician's primary responsibilities [20]. In light of enhanced productivity and personalization, this worry constitutes a barrier to translation. As well as concerns about the technology itself, concerns have been raised about the possibility that AI applications could undermine social interactions between clinicians and patients. Concerns about the technology itself have been raised along with these concerns. Patients and professionals alike would be affected by this change.

8.7.6 Ethical Barriers

When it comes to matters of ethics, developing a patient's digital twin for the purposes of precision medicine presents a number of problems regarding legacy, privacy, and identity, as well as what happens to the digital twin after the real twin has passed away. For example, to develop a databank representative of that community and make it accessible for research, access to the electronic medical records of a large population requires bringing the practises around informed consent up to date. The result is the creation of a research databank. Who can profit from other people's biological data is a pertinent issue in terms of equity [16]. It is unknown whether the development of health digital twins exacerbates racial or other sorts of social prejudices, discriminates against people with the least wealth, or discriminates against communities underrepresented in model cohorts with no demographic or ethnic variability. A health digital twin may also discriminate against those who are wealthy in society, although that is unknown. The group will exclude who? Could the quantification of these differences and their dissemination to the community as a whole give rise to new concerns about equality? Currently, people differ in terms of their physical capability, health state, and expected life expectancy.

8.7.7 Governance and Regulatory

The potential regulatory and legal issues relating to a health digital twin have not yet been identified, but they could be particularly difficult for medical cyber-physical system-related devices that contain a lot of embedded software for sensing and monitoring human activity. There are currently no recognized regulations or laws pertaining to health digital twins. The verification, scalability, and proof documentation and submission processes may need to be modified due to the interconnectedness of these devices and the intricacy of their design [17]. It is possible that models of a more passive type that employ twinning concepts will be the only ones that can be submitted to regulators due to the complexity of the databases that need to be compiled, curated, and expanded in order to enable an active digital twin for routine clinical decision-making. Products that plan procedures by employing individualized computational modeling and simulation are currently on the market and can be purchased. These products are currently available. For instance, there are goods that are currently available on the market that are advertised for use in structural heart disease, cardiac catheterization, and aneurysm repair. These products are currently available. Although approval procedures of regulatory authorities differ by country or region in terms of purpose, cost, timeframe, perceived level of rigor, and perceived level of rigor, these kinds of items are already available for purchase on the market [18].

When this science becomes more visible in the marketplace, the requirements for certification and approval may force national regulatory bodies to change their traditional methods. It is possible that these changes will be driven by certification and approval requirements. In the United States, the Food and Drug Administration just started a program to pre-qualify appropriate evaluation tools (like a digital tool or a computer model) so that they can be used later in actual regulatory submissions. As a result, the application will be processed more quickly. Medical device manufacturers can speed up and simplify the product development process by doing this. This program is intended to assist the Food and Drug Administration (FDA) in achieving these objectives. This is consistent with the company's acknowledgment of the expanding market for simulation software as a medical device as well as the potential for computational modeling and artificial intelligence tools to support device evaluation and reduce costs throughout the entire regulatory pipeline [19]. The business is aware of the potential of simulation software as a medical device as well as the growing market for such products.

Digital twin governance mechanisms could, for example, be adapted from the standard procedures that are currently in place for designing, regulating, and inspecting medical databases and biobanks. It is also possible that the governance mechanisms necessary to protect the rights of individuals with digital twins will be developed separately. When genome data are expanded to include biological and behavioral data, existing privacy safeguards may need to be enhanced. The reason for this is that genomics data are becoming increasingly complex [18, 19]. The result is that appropriate law—legislation that is timely and harmonized across jurisdictions—must evolve and adapt with the adjustments made by other agencies in response to technological advancement. As well, this law needs to be completed in a timely manner. Academic institutions, companies in the private sector, and organizations in the public sector often collaborate to develop standard methodologies and ensure software and protocol interoperability.

8.8 Research Methodology

The basic components of digital twins are presented in this study and discussed. In order to evaluate variables like chest discomfort and resting blood pressure, information is gathered via equipment implanted in the patient's heart, such as a pacemaker and an implantable cardioverter-defibrillator (ICD). A cardioverter-defibrillator is one of these devices [20]. Through these many devices, information is gathered. A digital model of the human heart will be generated using the parameter by the simulation software. Through the use of simulation, a biophysical model of the patient will be created. Using this model, you can simulate the behavior of a genuine patient in a way that is comparable to what the patient would do. Accessing real-time data necessitates the utilization of data analytics methods such as big data analytics and cloud computing; this connects the digital twin of the patient with other patients who have had cases that are comparable, and then chooses the one that has the most accurate computed result possible [1].

To categorize the conditions of patients according to the degree of precision possessed by the target variable, additional machine learning methods can be used, such as decision trees. Real-time intervention advice is based on the ideal scenarios selected for consideration. Any actions taken or unplanned events that occur during the operation are updated in real-time in the digital twin of the patient. This information is gathered in order to determine which course of action will be most effective in relation to the actual object. Using the technologies and processes described above, the model can be developed with the help of the digital twin of the heart and connected to the physical creation, allowing for real-time monitoring and the detection of future heart diseases using predictive and prescriptive analytics. It is possible to achieve this. In the entire process, an accuracy of 79% was maintained, and the decision tree splits the parameters based on the precision. Due to its relatively low complexity and ease of application, we chose to use a decision tree approach as opposed to other algorithms. Consequently, we decided it was the best option [19,20].

The progression of the process includes a number of distinct steps, including the following:

1. The patient will have a pacemaker and an implanted cardioverter defibrillator surgically placed in their heart during this stage. These devices will be used to monitor the patient's heart problems and collect data.
2. The next step is to upload the data tò the cloud where it will be stored. The data that was collected takes into account all of the characteristics that were assessed. These parameters include blood pressure, chest discomfort, fasting blood sugar, and others.
3. In the third stage, the data are first uploaded to the cloud, where they are then analyzed by simulation software in order to build a digital duplicate. The simulation software known as SimScale can be used to assist in the production of a virtual instance. This programme is hosted in the cloud.
4. The information is transferred from the cloud to the simulation programme during the fourth stage of the process.
5. After the obtained parameter has been entered into the simulation software, a finite element analysis will be performed in order to generate a digital twin of the physical object.

6. The sixth stage is when the digital twin begins to demonstrate behaviors and responses that are comparable to those of the physical heart.
7. The seventh stage is when digital twins are able to get real-time data from the cloud, and any modifications that are made to the parameters will be reflected in the virtual instance as soon as they are made.
8. In the eighth stage, a machine learning algorithm that is commonly referred to as a decision tree is utilized to develop a strategy for identifying the possibility of a disease based on the risk parameters and the degree to which those parameters deviate from the standards. This strategy is developed on the basis of the risk parameters and the degree to which those parameters deviate from the standards.
9. The algorithm is also able to access the patients' past medical records and can adjust its diagnosis method accordingly.
10. The algorithm has successfully diagnosed the patient. In order to lower the number of mistakes made during this step of the process, it is initially carried out in the virtual heart.
11. The diagnostic process will then be carried out on the actual patient by the algorithm.

8.9 Discussion

An examination of the application of the term "digital twin" and the core ideas underlying this topic can be found in the following article [21]. The idea of a continuous real-time multi-source data feed into the virtual twin holds a prominent position among these fundamental concepts. This idea, along with the artificial intelligence technologies that optimize these data into useful information about the physical twin, is also known as the "digital twin." Intelligent systems that are able to accurately characterize, comprehend, cluster, and categorize extensive amounts of data are known as "health digital twins." When it comes to making decisions regarding the diagnosis and prognosis of disease, it has been suggested that these systems should complement rather than take the place of human intellect. Estimating and stratifying risks, forecasting progression, choosing an intervention, and predicting the outcome of that intervention using data that is streamed and integrated from a variety of sources are all examples of the roles that digital twins play in making the possibilities of precision medicine a reality. It is critical to recognize that the combination of inductive and deductive thinking is the driving force behind these activities within the cardiovascular digital twin model. This is one of the most important things to keep in mind [22].

When it comes to cardiovascular disease (CVD) therapy, the potential of AI systems to more accurately phenotype patients with the same symptoms or conditions and overcome the limitations of existing risk stratification algorithms, resulting in less response-dependent treatment selection may become possible. These are based on the average person and the responses predicted by the individualized model. This could potentially allow treatment selection to be based on the response predicted by an individualized model, rather than the response of the average person. Predictions about which treatment will be most beneficial to an individual are no longer based on the person's current or past conditions. Instead, they are evaluated using forward-looking simulations. This replaces the current practice of making such predictions based on an individual's current or past condition. Collecting and combining multiple changing molecular, physiological, behavioral, and other characteristics of real humans in real-time to extract accurate data for intelligent digital representation is a challenging task. Translating technology requires overcoming many obstacles. There are various ethical constraints and hurdles that prevent physicians from adopting decision support systems based on artificial intelligence systems for disease management [23]. Processing power requirements, cybersecurity issues, data sharing issues, and myriad ethical barriers are examples of these challenges.

It is significant to emphasize that, in contrast to the well-established complicated modeling science that is employed to make treatment predictions and inferences in disease, including CVD, the characterization of an active health digital twin as described in this review is very new. It is necessary to highlight this since it is significant. The lack of a clear bidirectional data flow with a real patient may restrict the use of the phrase "digital twin" in mechanistic models. This is because a model must be extremely complicated in order for it to be able to assert such a claim in a way that is true to reality. Take, for example, the iterative process that integrates imaging data with engineering sciences in order to arrive at a prototypical digital twin for clinical translation. This method aims to arrive at a digital twin that can be used in clinical settings. This emphasizes the possibility that a clear distinction between the uses of numerical modeling and the idea of a data-driven twin, as described in this review, may not be a reasonable expectation. Even more appealing is the idea that combining mathematical (deductive) and data-driven (inductive) modeling could get over limitations specific to each approach and build connections between the knowledge gleaned from them. For instance, within the digital twin design and validation, which is centered on a population-based databank, complementary deductive and inductive data modeling methods are at work [24]. As a result, forecasts about the population as a whole can be made with greater accuracy. The deductive, mechanistic model includes both clinical and experimental data to discover causes and predict results based on anatomical and mechanical knowledge of a physical system as well as hypothesised linkages. This is carried out to ascertain how a physical system functions. This thorough understanding of the system also provides a descriptive advantage over inductive or empirical models and avoids the need to retrain a predictive mechanistic model to take into account new information or conditions. This is because such knowledge eliminates the need to account for data that has not been seen before. It is possible that the assumptions that are utilized will restrict the benefits that such simulations may have to provide in terms of clinical interpretability. The choice of boundary conditions for measurements that are necessary to solve equations can be one of these assumptions, as can the impact those circumstances have on the situation.

On the other hand, the inductive and statistical pathway makes advantage of procedures associated with machine learning in order to train, test, and edit complex data. It finds predicted linkages, patterns, and correlations when mechanisms are either poorly known, too complex to represent mechanistically, or when missing facts must be inferred. One of the many appealing aspects of empirically derived models is their capacity to process a large amount of multivariable data from a variety of external sources, such as biological databases, wearable sensors, and others, in an efficient manner. This capacity is one of the many appealing features of empirically derived models. When compared to mechanistic or physics-based models, machine learning systems have a lesser level of generalization, which, combined with the available volume and heterogeneity of data variables that are used to train machine learning systems, can be considered to be two different types of constraints. As a consequence of this, making progress in the field of digital twins will probably need maintaining a focus on data-driven as well as knowledge-based mathematical modeling.

On the scale of the heart, for instance, a retrospective analysis of aortic valve recipients has revealed the viability of inductive-deductive model synergy for predicting conduction problems. The irregularities in the electrical activity of the heart can be identified using this synergy. This combination was used to predict the presence of anomalies in the electrical impulses that the heart sends out [22, 23]. A development above more conventional device-anatomy simulations was the application of machine learning to the classification of anatomical, procedural, and mechanistic data. This made it possible to increase and incorporate the characteristics that may be used to accurately forecast the recurrence of bundle branch block or the need for a permanent pacemaker. The secret to this endeavor's success was the inclusion of machine learning to more traditional device-anatomy simulations. The highly recognised international project known as the virtual physiological human describes both the accomplishments that have been done thus far and the significant future hurdles that need to be addressed in order to combine diverse physiological body systems. All around the world, this project is being carried out.

This chapter agrees with earlier studies' findings that research applications for digital twins only include some of the components of the prototypical synchronized cyber-physical system that is provided by Internet of Things connectivity. The fact that this review was written helped to facilitate the discovery of this information. So, in order to make it easier to make comparisons between study and implementation, there has to be more unanimity over the definition of the digital twin. A further vital objective is to direct research efforts towards the development of simple, user-friendly interfaces that enable non-technical end users to interact with digital twin systems in their respective fields. This is an important step that needs to be taken as soon as possible. In the future, advances will include refining the ways of visualization that allow non-expert users of a mirrored, networked system to engage with the AI-derived information about the physical twin. This will be possible through the use of a system that will be improved. The most up-to-date technology will be utilized in order to facilitate this engagement. There are now available techniques like 3D avatars, multi-dimensional holographic projections, and augmented reality [23]. However, standardization amongst several prospective digital twins in a system would enable both their seamless communication within the network tying the virtual and physical twins together as well as the involvement of people who are not data science professionals. This would be beneficial for a number of reasons. It is anticipated that the commercial sector will be the principal driver of the key improvements in the technologies that enable digital twin systems. This expectation is based on previous experience. One example of this would be platforms for big data analytics that prioritize decentralized storage over centralized storage and parallel processing of structured and unstructured data over serial processing. In other words, these platforms favor distributed over serial processing. As a result of ongoing developments in communication infrastructure, the Internet of Things (IoT) will continue to see growth and development. These developments will include a greater availability of 5G mobile and internet networks, cheaper sensor devices, and flexible cloud and edge computing environments to satisfy the requirements of networked services for the storing and processing of data. Other difficulties include ensuring the timely update and correct reproduction of data derived from intricate physical twin components (such as human organs) onto the digital twin to prevent postponing crucial physical twin intervention. This can be accomplished by ensuring that the data is replicated in a timely manner. The application of artificial intelligence is the means by which this can be accomplished. It is necessary for the various digital twins that are participating in a shared interaction, such as the human body, to communicate with one another and synchronize their data. Because of this, advances in application programming interfaces are required. These interfaces need to be able to optimize consistent and predictable software-to-software interoperability [24]. A significant step forward in the field of physics-based modeling is the reduction in price of computational hardware while simultaneously increasing its level of accuracy. This not only speeds up the process of solving the equations that govern a physical system (like how vessels deform or interact with turbulent blood flow), but it also improves the quality of 3D visualizations and accelerates the availability of training data sets for machine-learning models, both of which are required to create a digital twin.

In the evolution of model-based personalization for decision support, it is essential for the clinical end-user that the data be accurate, reproducible, and consistent, and that measurement uncertainty be incorporated into the data assimilation process within the model. First, the accuracy of the data, and second, that the measurement uncertainty be incorporated into the data assimilation process. It is necessary for the clinical end-user to take into mind the socio-technical aspects of how such advances flow into practise. Clinicians may get the false impression that artificial intelligence operates in a mysterious and unrelatable manner due to the so-called "black-box" nature of model-based applications that do not provide details of the effects of uncertainties within the clinical data that underpin the model. These applications are based on models and do not provide details of the effects of uncertainties within the clinical data that underpin the model [25]. As a consequence of this, one of the challenges associated with adoption is to increase the level of transparency surrounding the evidence that supports the development and validation of a digital twin solution to either assist in making therapeutic decisions or draw prognostic conclusions. This is one of the challenges that comes as a result of the fact that one of the benefits associated with adoption is to assist in making therapeutic decisions. It is highly likely that the gradual incorporation of AI-enabled platforms into conventional methods of patient care will call for the collaborative expertise of professionals in the fields of technology, biomedicine, and behavioral sciences in order to cultivate the conditions necessary for widespread adoption and use. This is because the gradual incorporation of AI-enabled platforms into conventional methods of patient care is expected to take place over the course of several years. When models are validated in terms of an original notion, one of the most crucial steps to take is to expand the model to a more generic patient cohort. This patient cohort should have physiological and demographic features that are not controlled as strictly as the initial cohort. Due to the difficulty in accessing large populations and the information they contain, the development of so-called mega cohorts, which are prospective digital health "data lakes," as well as big data infrastructure (including the techniques characterized as data fusion), is necessary in order to phenotype the volunteers who have signed up for these initiatives and to optimize the use, re-use, and sharing of these data. Due to the availability of several trustworthy databanks, researchers from all over the world have access to data that has been anonymized (National Institutes of Health).

8.10 Conclusions

This chapter possesses a lot of strengths, one of which is that its mapping intent permitted the inclusion of material sourced from the internet in addition to at least six different forms of scholarly publications. This is only one of the chapter's many strengths. As a consequence of this, the reader is presented with a picture that is both more accurate and comprehensive of the various disease conditions in which digital twin science is being applied, as well as of the academic and non-academic stakeholders operating within this multidisciplinary field. In the framework of a study into the idea of the digital twin, various limits should be brought to the forefront of the discussion [21]. Because the researchers took a pragmatic approach to isolating work that was referred to as a digital twin, it is possible that relevant papers that described predictive models but did not explicitly use the term were excluded from the search process. This is because the researchers took a pragmatic approach. As a result, it is probable that the research that has been published on digital twins for CVD drastically underestimates the scope of their application. Future study in the subject might instead conduct a systematic review of a more focused question and a smaller collection of publications with a more uniform research design. We used a systematic search approach inside a mapping style of review. We conducted a review using a mapping technique with a methodical search strategy. Under the framework of a mapping-style review, we utilized a methodical search strategy. The methodology evaluations that are included in that form of study may lead to the production of a variety of viewpoints on the subject after they are completed [22]. Material obtained from electronic databases and websites represents information that was available on the date that the search was conducted; however, there is a possibility that the review does not adequately represent all of the information that is currently available on a subject that is continuously undergoing development. If only materials written in English are used, it is conceivable that not all of the relevant literature in the topic will be taken into consideration. Because just one author tracked the information that was collected from the various sources, there is a greater possibility that the screening judgements would have been different if there had been more than one reviewer. In addition, it has been hypothesized that AI and digital twin technology might be able to derive "intelligence" from vast quantities of patient data in order to enhance the quality of the patient experience, care coordination, scheduling, and other service-oriented operations at the level of healthcare facilities [7, 8], but this possibility has not been investigated. The included publications that made no attempt to test any of the hypothesized economic benefits of digital twins applied to CVD care. Nonetheless, this is an essential area that future research should investigate, as it is a growing field of enquiry [23].

In conclusion, the research on digital twins for cardiovascular illnesses as a whole comprises proof-of-concept studies. These studies highlight the use of data-driven methodologies, which are representative of the objectives of precision medicine. The promise of a fully functional digital twin of either the human heart or the human organism in order to improve clinical decision-making is still quite a ways off. However, this does not mean that this technology will never be developed. Even though it is based on AI, its success is contingent on a wide range of human factors. These include citizen populations that are willing to prospectively contribute bio-data towards mega cohorts of patients; an adaptable, agile ethical and regulatory landscape; and multi-disciplinary scientific expertise. Its progression is hindered by difficulties in practicability and implementation that are not specific to CVD alone. To be more specific, these difficulties originate from the fact that although AI, at its foundation, depends on numerous human elements for its effectiveness, despite being AI itself. The University of Sydney Cardiovascular Initiative recently hosted a collaborative workshop of clinicians, (bio)engineers, data scientists, and others required to advance research pathways for cardiovascular disease. This workshop provided important contextual insights to the field in general and for cardiovascular disease in particular. Clinicians, (bio)engineers, data scientists, and other necessary participants attended the session. For this session, clinicians, (bio)engineers, data scientists, and others who are required to increase research paths for C gathered [24].

The application of an algorithm that is able to learn from the data produced by machines makes it possible for the digital twin technology to more effectively personalize medical care for each individual patient (decision tree) [25]. The data relating to the patients, both current and historical, are intended to be captured by the model, which was developed with this capability in mind. The findings will serve to direct medical professionals, healthcare organizations, nurses, and patients in the application of simulation technologies to the prediction and management of heart disease, as well as the construction of a model that could be used for further diagnosis. This will be accomplished through the utilization of simulation technologies. Individuals will have access to medical treatments that are more specifically designed to meet their unique requirements as a result of these technologies, which will allow for the development of medical therapies that are both more efficient and specialized.

References

[1] Cunbo, Z., Jianhua, L., Hui, X., Xiaoyu, D., Shaoli, and Gang, W. (2017). Connotation, architecture and trends of product digital twin. *Computer Integrated Manufacturing Systems*, 23(4), 753–768.

[2] Madni, A., Madni, C., and Lucero, S. (2019). *Leveraging digital twin technology in model based systems engineering. Systems*, 7(1), 7.

[3] Sanders, R., Martin, R., Frumin, H., and Goldberg, M. (1984). Data storage and retrieval by implantable pacemakers for diagnostic purposes. *Pacing and Clinical Electrophysiology*, 1228–33.

[4] Sivalingam, K., Sepulveda, M., Spring, M., and Davis, P. (2018). *A review and methodology development for remaining useful life prediction of offshore fixed and floating wind turbine power converter with digital twin technology perspective. Renewable and Sustainable Energy Reviews*, 97, 1–16

[5] Joseph, P. et al. (2017). Reducing the global burden of cardiovascular disease, part 1. *Circulation Research*, 121, 677–694.

[6] Tao, F., Cheng, J., Qi, Q., Zhang, M., Zhang, H., and Sui, F. (2017). Digital twin-driven product design, manufacturing and service with big data. *The International Journal of Advanced Manufacturing Technology*, 94(9–12), 3563–3576.

[7] Vernon, S. T. et al. (2017). Increasing proportion of ST elevation myocardial infarction patients with coronary atherosclerosis poorly explained by standard modifiable risk factors. *European Journal of Preventive Cardiology*, 24, 1824–1830.

[8] Schroeder, G., Steinmetz, C., Pereira, C., and Espindola D. (2016). Digital twin data modeling with AutomationML and a communication methodology for data exchange. *IFAC-PapersOnLine*, 12–17.

[9] Tao, F., Sui, F., Liu, A., Qi, Q., Zhang, M., Song, B., Guo, Z., Lu, C., and Nee, A. (2018). Digital twin-driven product design framework. *International Journal of Production Research*, 56(24), 3935–3953.

[10] Collins, F. S., and Varmus, H. (2015). A new initiative on precision medicine. *New England Journal of Medicine*, 372, 793–795.

[11] Zhuang, C., Liu, J., and Xiong, H. (2018). Digital twin-based smart production management and control framework for the complex product assembly shop-floor. *International Journal of Advanced Manufacturing Technology*, 94(1–4), 1149–1163.

[12] Petrova-Antonova, D., Spasov, I., Krasteva, I., Manova, I., and Ilieva, S. A digital twin platform for diagnostics and rehabilitation of multiple sclerosis. In *Computational Science and Its Applications—ICCSA 2020* (eds Gervasi, O. et al.) 503–518 (Springer International Publishing, Cham, 2020).

[13] Thuemmler, C., and Bai, C. Health 4.0: application of industry 4.0 design principles in future asthma management. In *Health 4.0: How Virtualization and Big Data are Revolutionizing Healthcare* (eds Thuemmler, C. and Bai, C.) 23–37 (Springer Int Publishing, Cham, 2017).

[14] Uhlemann, T., Lehmann, C., and Steinhilper, R. (2017). The digital twin: Realizing the cyber-physical production system for industry 4.0. *Procedia CIRP*, 61, 335–340.

[15] Figtree, G. A. et al. (2021). Mortality in STEMI patients without standard modifiable risk factors: A sex-disaggregated analysis of SWEDEHEART registry data. *Lancet*, 397, 1085–1094 .

[16] Boschert, S., and Rosen, R. (2016). Digital twin—The simulation aspect. *Mechatronic Futures*, 1(1), 59–74.

[17] Lal, A., Pinevich, Y., Gajic, O., Herasevich, V., and Pickering, B. (2020). Artificial intelligence and computer simulation models in critical illness. *World Journal of Critical Care Medicine*, 9, 13–19.

[18] Bhattad, P. B., and Jain, V. (2020). Artificial intelligence in modern medicine—the evolving necessity of the present and role in transforming the future of medical care. *Cureus*, 12, e8041–e8041.

[19] Brosinsky, C., Westermann, D., and Krebs, R. (2018). Recent and prospective developments in power system control centers: Adapting the digital twin technology for application in power system control centers. *Electric Power Systems Research*, 166, 138–147.

[20] Ravì, D. et al. (2017). Deep learning for health informatics. *IEEE Journal of Biomedical and Health Informatics*, 21, 4–21.

[21] Naplekov, I. et al. (2018). Methods of computational modeling of coronary heart vessels for its digital twin. *MATEC Web of Conferences*, 172, 01009.

[22] Vukicevic, M., Vekilov, D. P., Grande-Allen, J. K., and Little, S. H. (2017). Patient-specific 3D valve modeling for structural intervention. *Structural Heart* 1, 236–248.

[23] Semakova, A., and Zvartau, N. (2018). Data-driven identification of hypertensive patient profiles for patient population simulation. *Procedia Computer Science*, 136, 433–442.

[24] Martinez-Velazquez, R., Gamez, R., and El Saddik, A. Cardio twin: A digital twin of the human heart running on the edge. In *2019 IEEE International Symposium on Medical Measurements and Applications (MeMeA)* 1–6 (IEEE, 2019).

[25] Jimenez, J. I., Jahankhani, H., & Kendzierskyj, S. Health care in the cyber-space: Medical cyber-physical system and digital twin challenges. In *Digital Twin Technologies And Smart Cities. Internet of Things (Technology, Communications And Computing)* (eds Daneshkhah, A. et al.) 79–92 (Springer Nature Switzerland, 2020).

Chapter 9

Comprehensive Study of Digital Twin in Smart and Customized Healthcare

Salna Joy, R. Baby Chithra, and Ajay Sudhir Bale

9.1 Introduction

The world is currently experiencing a significant shift brought about by digital transformation. With the continuous emergence of new technological innovations, the DT has resulted in one most of the talked-about technologies in recent years. The medical field, which has come into close scrutiny because of the COVID-19 epidemic, has come within reach of this innovation [1]. The word "DT" refers to a simulation method that uses testable theories, instruments, and statistical data to combine different physical and transdisciplinary components, representing the full lifetime of a physical product or piece of equipment by mapping it to a digital domain. Digital twins, digital reflections, and electronic maps are other names for it.

The digital twin has gained popularity as a building block for the metaverse, a rapidly emerging platform that offers immersive digital experiences and interactions similar to real life. Decades of technological advancements in cognitive and artificial intelligence, coupled with the accelerating pace of digitalization across various sectors and the consequent increase in data processing and analysis capabilities, have all contributed to the growth of DT technology [2].

DOI: 10.1201/9781003469612-9

The rapid evolution of the digital landscape, including advancements in the Internet of Things (IoT) and cloud computing, has led to a significant upgrade in the general digital infrastructure. These elements work together to produce a more open environment for the use of DT technologies now than they did when the idea was initially proposed. Currently, a digital twin is regarded as a crucial element of the Fourth Industrial Revolution in both academic and industrial sectors. To realize the promise of tailored, bright, and preventive medical care, DTs are crucial to the industry's transformation. As personalized healthcare continues to evolve to deliver the proper care at the proper time and in the proper way, there is an increasing need to build virtual representations of persons [3].

Healthcare dynamic technologies can dynamically reflect sources of data including digital medical records, illness databases, and representation, demographics, and behavioral information of a person as time passes. Physiological, behavioral, hormonal, intellectual, and psychiatric data have become more diverse, accurate, and accessible as a result of technological advancements like IoT and AI. As a consequence, there has been a significant rise in curiosity in DT study and possible therapeutic uses.

If properly employed, DTs have had the ability to allow linked care's untapped talent and alter the ongoing treatment of long-term illnesses, lifestyle-related illnesses, and health issues. In addition to what is now accessible, wristbands and mhealth apps may offer a variety of real-time data points to "feed" DTs. Care professionals might use a person's DT to influence therapy choices or offer suggestions that are unique to that person [4]. The possible uses of DT in the field of healthcare are depicted in Figure 9.1.

A reliable data architecture must be developed to generate a DT model. On the basis of solid computer-aided design (CAD) models, the designer's look, shape, movement organization, and connection structure are defined. Moreover, to support new device kinds and functionalities, including such robotic mobility, human touch among machine tips and components, and product specifications representations, virtualized activities must also be built using scripting. This method makes it possible to quickly construct projects and build a sizable model library for significant manufacturing machinery.

The 5D model of a digital twin is portrayed in [5], which includes

- The patient as the real-time entity
- The patient's virtual replica as DT
- DT data
- DT services
- Data exchange networks.

DT data consists of real-time patient data as well as machine learning or mathematical modeling-based simulated data. The DT healthcare systems provide continuous body metric monitoring, well-timed diagnosis, and treatment personalization. Data

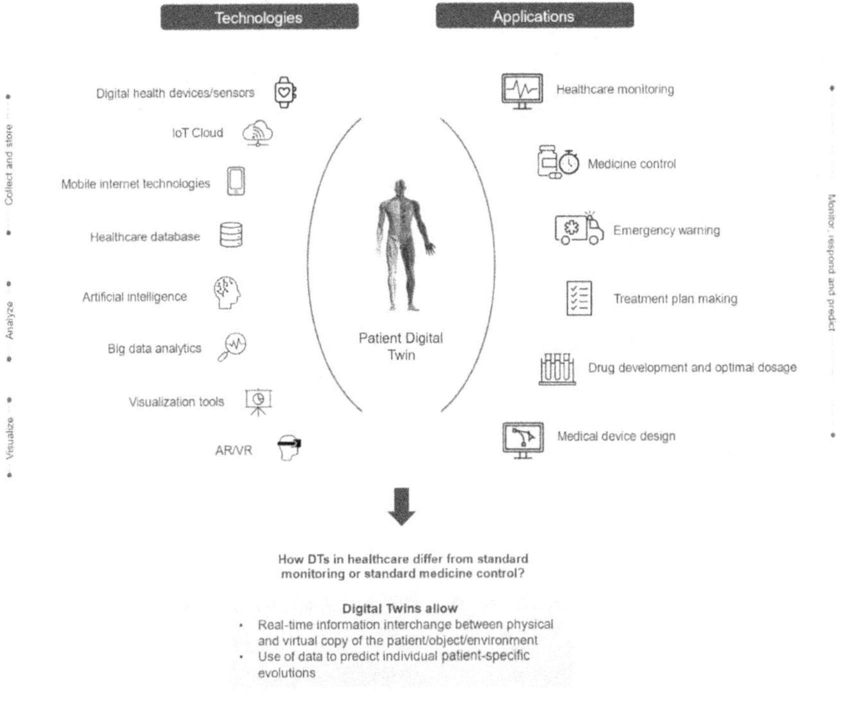

Figure 9.1 Digital twin in healthcare [4].

Source: Open Access Article.

exchange networks take care of the data transfer between the patient and DT. The implementation of the 5D model in a healthcare context is challenging as the said domain is extremely complex and sensitive, demanding ethical as well as privacy concerns that need to be thought about.

An extended life cycle DT model consistent with the fundamental 5D model supports human life at three different phases: the preconception phase, lifespan, and post-life phase [6]. The preconception phase focuses on genetic risk factors and prebirth care. Early intervention of genetic risk factors can be analyzed through genetic screening and testing to decide on an early treatment plan. Genetic models and data analytics developed via applying different genetic algorithms to digital twin help enhance the future generation's health perspectives. Efficient genetic research programs with the help of DT lead to the discovery of new early diagnostic techniques and treatment plans and the concept of a "Lifetime Genomic Record" of individuals.

The second phase of the healthcare model encompasses the lifespan of the individual. It incorporates the continuous real-time monitoring of body metrics via wearable health sensors, which lend a hand in slowing down or staying away from chronic disease developments with necessary lifestyle and nutritional intake changes, and the prevention of abrupt life-threatening conditions. Surgery and treatment simulations performed on digital twins, support treatment optimization and result in high-precision surgical procedures. Personalization of treatment plans and medicines facilitates smart customized healthcare systems. DT opens new doors to efficient clinical trials and research with no ethical, social, or health-consequence issues. DT framework has been employed in early cancer detection and prediction and advanced oncology research domains [7]. Post-life phase pacts with organ donation and transplantation sector. Better prediction for acceptance and cost-effectiveness is various attributes of transplantations. The "Lifetime Genomic Record" of the donor considerably improves the success rate of organ transplantation through precise simulations and a personalized matching process [8].

Industry 4.0 puts forward the perception of digital twinning for every entity in the real world. Molecular-level cells, DNA, organs, and the entire human body create digital counterparts, and the derived database and created DT model facilitate real-time health monitoring, early detection and health disorder prevention, treatment optimization, personalized medicine, and enhanced clinical trials for novel outcomes. Digital twins offer new means of healthcare education and medical knowledge sharing. DTs of healthcare facilities such as hospitals and surgical procedures contribute to efficient hospital management and resource allocation [9]. The HospiT'Win framework projected in [10] optimizes patient channels and hospital resource allocation and foresees the impact of unpredicted occurrences in hospital operations.

Nevertheless, there are a number of basic issues with the DT concept, including accessibility to devices made by various makers, standardizing quality and file formats, handling enormous volumes of data, and analyzing the information in real-time. Reliable, quick, and safe channels of communication are necessary to facilitate in-field, actual contact between both the twin model and the field. Multisensor character data parser systems like OPC UA, Modbus, and CAN bus could be utilized to create both downstream and upstream communication in cases involving several sensors and concurrent and integration of data. The connection central controller must choose the proper data transfer forms and protocols (e.g., XML or JSON). The effective implementation of a DT system and the multisensor standardized data computerized systems depend on this link among downstream real-time data and upstream evaluation results [11].

9.2 Personalized Digital Twin

The healthcare industry is undergoing a revolution, and DTs are playing a crucial role in enabling more personalized, preventive, and sophisticated care. The rise of personalized healthcare has created a pressing need for a virtual replica of individuals that can provide appropriate care when and how it is needed. A more sophisticated form of the DT with possibilities for meaningful information is the idea of a personal DT (PDT). By simplifying precise decision-making and assuring appropriate therapy choice and optimization, PDTs can provide individuals with considerable advantages [12, 13].

The PDT is a digital copy of an individual that encompasses various aspects of their life, including physical, biological, mental, and social factors [14]. Its purpose is to provide individuals with support in self-care, personal growth, and self-reflection. By offering real-time private health information that may be used by health services and educational environments to spot potential health hazards, especially among elderly persons and those who have chronic diseases, PDT revolutionizes customized care. By providing customized advice and choices for its clientele, this draws health professionals and partners to maximize their businesses. The significant concentration of PDT includes both the real world and the internet. Physical sensors used in phones, watches, Fitbits, and other fitness trackers acquire private information about users by monitoring their body's temperature, pulse rate, hypertension, glucose and insulin sensitivity, and number of walks completed.

The proposed DT framework based on neural networks by Haya Elayan *et al.* [15] to detect abnormal health conditions is depicted in Figure 9.2, which is inspired from [15]. The framework aims at constant monitoring of the patient for possible anomalies in body metrics. The attached IoT-enabled wearable sensors

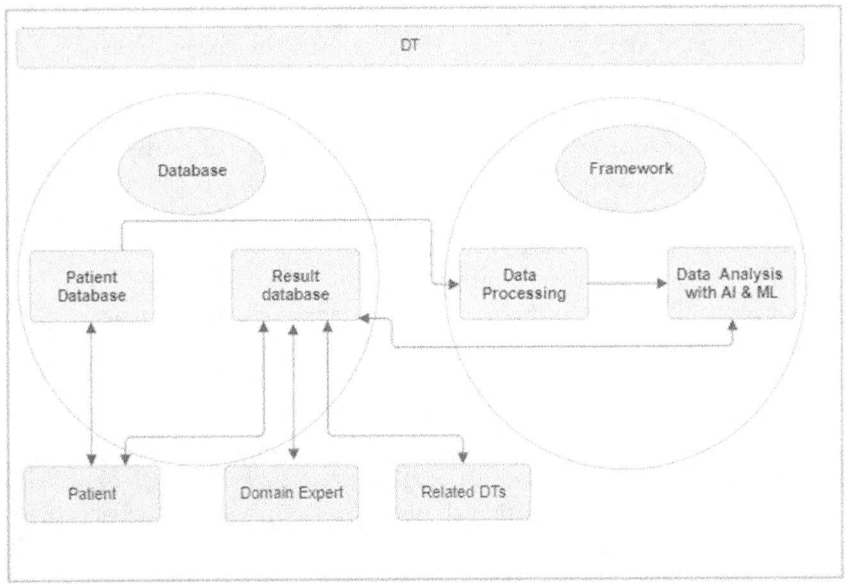

Figure 9.2 ML-based DT framework.

on the patient's body collect vital body parameters and transfer them to the digital twin. After preprocessing and analysis, this data will be applied to different machine learning and neural network-based algorithms capable of detecting abnormalities early and aiding in enhanced preventive measures. DT enables the testing and tracking of different treatment plans, thereby improving life expectancy. The three different segments of the framework are as follows:

1. *Data collection and predictive model creation:* the real-time data collected from wearable IoT sensors is temporarily stored in a cloud database and applied to machine learning (ML) algorithms to create a predictive model based on classifiers. The end result from ML predictive model is stored in the result database, which is accessible to the patient and healthcare professional.
2. *Consistent monitoring and model optimization:* the real-time inputs from different body metric sensors are constantly fed to the ML predictive model, which identifies anomalies in the parameters and proactively prevents critical health scenarios and reduces lifestyle diseases. The added descriptive feedback in the result database optimizes the model and serves as an informative decision-making factor for patients with similar health issues or future cases.
3. *Analysis of similar cases for DT enhancement:* patient DT is enhanced by comparison with other patients with similar health issues. The data result and data from similar patients' DT result databases are compared with current patients' parameter values and predictive model results. This improves the accuracy of the model by expanding its domain with actual results. Healthcare professionals can utilize the results for more accurate treatment plans.

A cardiovascular classifier was implemented on a patient's DT to diagnose heart problems. The wearable IoT sensor captures ECG signals and converts them into digital data and then applied them to the ML algorithms to derive a predictive model for the detection of heart conditions. For better accuracy SVC algorithm, convolutional neural network, LSTM, logic regression, and MLP algorithm were employed and the results were used for DT optimization. MIT-BIH arrhythmia-based dataset [16] is used as the training and testing data in the model training and evaluation phase for the five ML models. After performance analysis and optimization predictive DT model is derived. The accuracies obtained were 0.9709, 0.9667, 0.956, 0.756, and 0.676, respectively, for LSTM, CNN, MLP, SVC, and regression models, respectively. Confusion matrix reveals more false positives and false negatives for SVC, MLP, and LR models against LSTM and CNN, which in turn brings down positive prediction accuracy (precision), ratio of actually identified positive samples (recall), and the harmonic mean between them (F1 score). The evaluation metrics point to the conclusion that the neural network-based CNN and LSTM perform superior over conventional ML algorithms (SVC, MLP, and logistic regression) in terms of accuracy, precision, F1 score, area under the curve (AUC), and recall in deriving the DT model.

The operational data intelligence, which comprises three levels and is a feature of the reference framework for creating a personalized medical program based on PDTs, is represented in Figure 9.3. The physical devices make up the first layer, which also includes wearables, cellphones, diagnostic supplies, and other similar gadgets. The second layer consists of industrial technologies, such as cooperating twins for various body parts and data collecting tools for gathering information from real-world sensors and simulating a similar world. The third layer is made up of the application fields, which include blockchain technology (BT) based interface, explainable AI, knowledge base management, synchronization, streaming data analysis, and data analysis [17].

The physical devices layer has a significant role in the smart personalized medical care system, where biosensors are used to collect personal health data from different body parts [18]. The industrial technologies layer employs data acquisition technologies to collect data from the sensors then introduce it towards the modeled setting, where real-time data analysis takes place [19]. Effective and efficient analysis requires collaboration among a group of DTs, as a single DT cannot perform it alone [20]. As a result, PDTs rely on collaborative twins for various applications, including cell phones, wearable technology, bioelectronics, and hospital instruments. These DT sets, they include several bodily organs, aid in comprehending the overall PDT status of personalized healthcare systems.

The content is constantly retrieved from the current database via the standard query element. And the data modeling technologies semantically represent the personal health data for digital objects [21]. To detect trends and make forecasts for tailored medical care, the data management module takes forecasting analytics built on DTs. By combining with DT-based individualized trial information of specific patients, the created predictions may be tailored for each patient, which aids in the generation of personalized good degree of accuracy prognosis, diagnosis, as well as the effective optimization of new treatments [22]. Under the suggested standard framework, the understandable AI element is employed to analyze the assumptions made by the ML models [23].

To maintain synchronization constancy inside a customized public health system, the synchronization element must be built as a PDT-based synchronization gateway. Medical professionals and engineers may try out situations and make evaluations on a virtual version of surgical supplies using modeling. DT tech is used in streaming data analysis to obtain real-time data for assessing the condition of tangible assets. There are two scenarios for blockchain and DTs in which the necessary personal information around PDTs must be unchangeable and the blockchain is crucial to their security [24]. The PDT should also communicate with physicians, medical facilities, medical providers, etc., necessitating the use of blockchain technology.

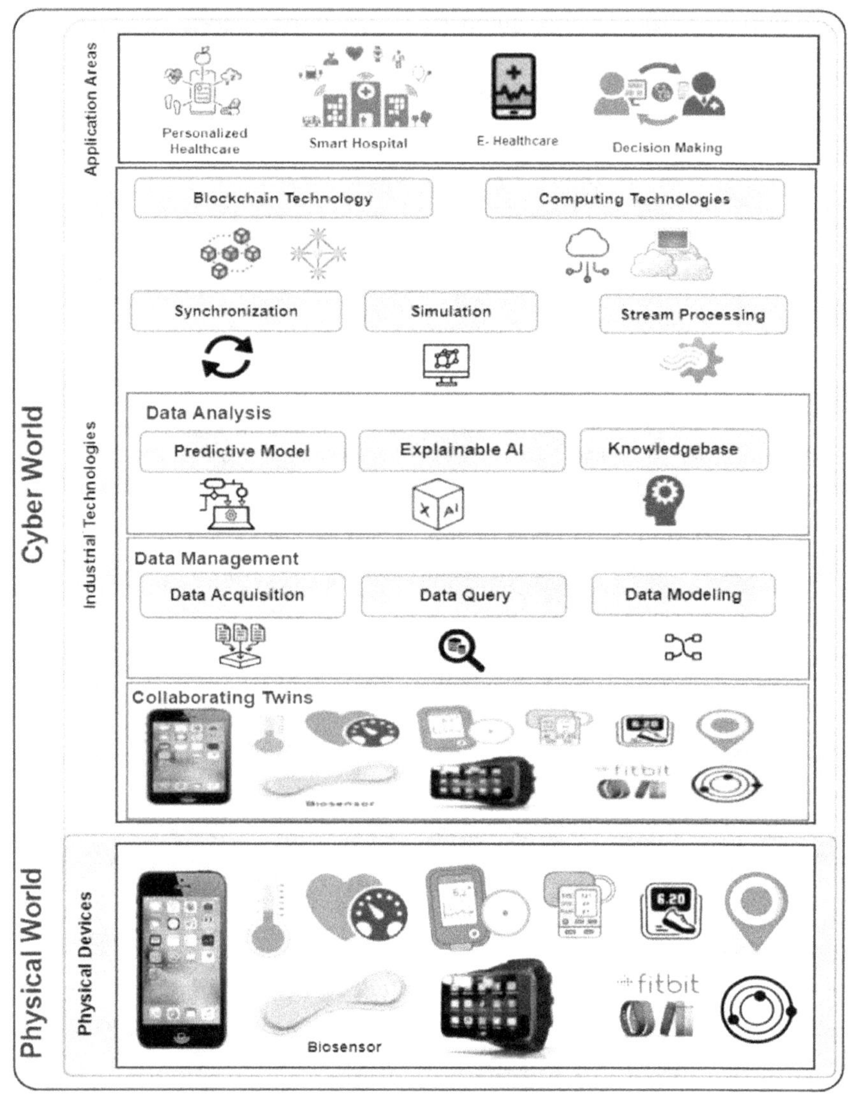

Figure 9.3 PDT-based personalized healthcare framework [17].
Source: Open Access Article.

A DT-based customized medicine architecture is shown in Figure 9.4. The model first concentrates on a single patient exhibiting local illness symptoms. Next, by integrating millions of illness data using high-performance computing, virtualized replicas of the patient's DT are produced. Every virtual twin is provided with a distinct medicine to test as a possible illness therapy, leading to one computerized

Figure 9.4 Conceptual personalized medicine [4].
Source: Open Access Article.

treatment program that benefits the individual. For the patient's professional treatment, the optimal medication therapy is chosen in the end [4].

9.2.1 PDT-Based Pandemic Alerting Framework

The PDT has been implemented to combat the spread of COVID-19 by collecting, analyzing, modeling, and disseminating epidemic relevant information on people's PDTs [25]. The smart personalized healthcare framework, based on PDTs, can be utilized to report individuals with COVID-19. Once identified, those in close proximity to the infected individual will be alerted via their PDTs, prompting them to practice utilizing antibacterial wipes, keep their distance from others, and refrain from handling anything you interact with.

Users can share and exchange current COVID-19 information with the blockchain network, as shown in Figure 9.5, including possibly infected people, diseased individuals, physicians, medics, and chemists, in addition to federal agencies like healthcare organizations. The purple arrows in Figure 9.5 indicate how individuals in quarantine zones can use their PDTs to communicate their condition to blockchain networks. As shown by the blue arrows, this will return the information to the user. Physicians can examine the condition and advise the contract to take the required medicine. The black arrows in Figure 9.5 show how the input on the medicine from pharmacists and physicians is subsequently transmitted to the public blockchain.

The epidemic warning can send out prompt distant notifications to stop COVID-19 epidemics by tracking the illness rate and locating places where the virus is growing. Figure 9.5 shows alert and warning signals as orange and red arrows, etc. Alerts will be issued to all who are on the network if the PDT shows that a person has COVID-19 infection. PDTs will notify those in close vicinity to the sick individual

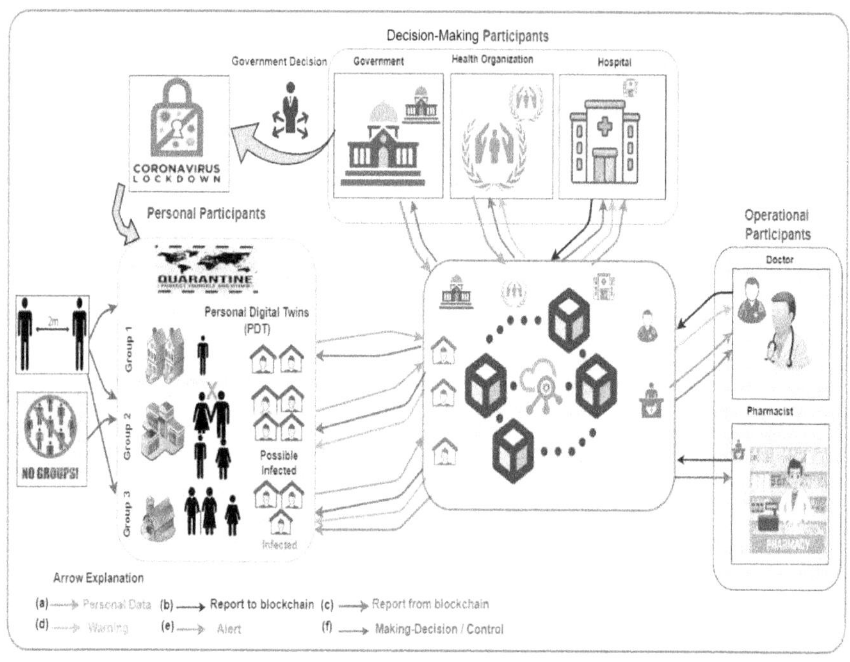

Figure 9.5 PDT-based COVID-19 alerting framework [25].
Source: Open Access Article.

to use sanitizer, keep their distance from application components, and practice interpersonal separation. PDTs can also assess a person's local health condition and give the pandemic warning the most recent data on verified confirmed samples [17, 25].

9.3 The Role of AI in Digital Twin

DTT is a digital image of a real-world product, procedure, or system. It permits the development of a real-time digital representation of a physical item or system for the purposes of monitoring, simulation, and operational optimization. Gathering information from sensors as well as additional resources embedded in a tangible product or system is the first stage in developing a digital twin (DT). After collecting this information, a digital replica of the system is developed. The DT may be used to keep an eye on how well the physical thing or system is doing, analyze data to spot problems, and run what-if scenarios to find the best way to improve [26]. Figure 9.6 depicts the use of AI in DT.

DTT has several applications outside of the realms of industry and academia. It enables continuous monitoring of machinery and systems, as well as predictive

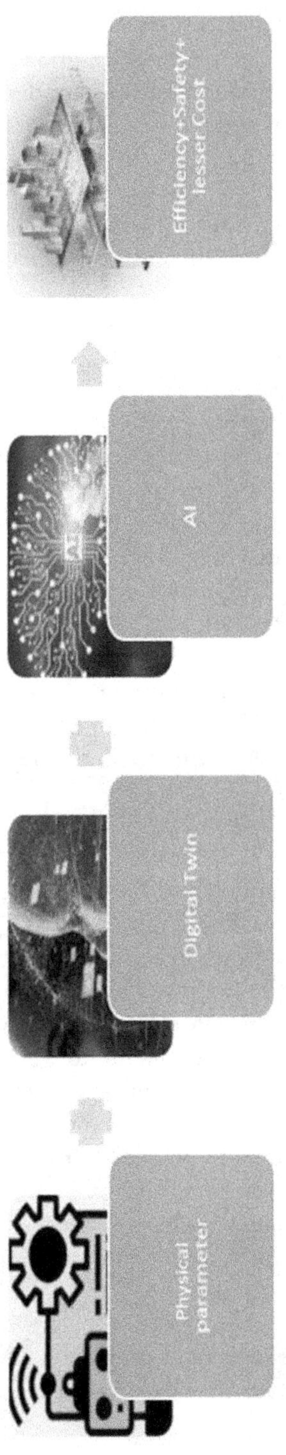

Figure 9.6 DT with AI.

servicing and efficiency enhancements. DTs may also be used to validate changes to designs and plans before they are implemented in the real world. Overall, DTT's real-time insights and optimization possibilities have the ability to completely alter how we plan, construct, and run physical systems. From the building and industrial industries to healthcare and transportation, DTT is finding widespread use. In the industrial industry, for instance, DT may be used to foresee future bottlenecks or inefficiencies by simulating the production process. Modeling the human body and simulating its response to various treatments or interventions are two examples of how DT might be utilized in healthcare [27].

Digital twin (DT) is the process of using computer models to generate a virtual counterpart of a physical thing or system, such as a machine, building, or even a whole city, and then using that replica to mimic, monitor, and optimize the performance of the original object or system in real time. Sensor data, data from IoT devices, and other information are all used to construct DTs. The DT is then trained to act in a manner similar to the real-world counterpart, revealing how effective the item or system is and where adjustments need to be made [28].

Overall, DTT has the potential to improve efficiency, reduce costs, and enhance the performance of a wide range of physical systems and objects. The combination of physical parameters, DT technology, and AI has the potential to revolutionize many industries as depicted in Figure 9.6. The creation of virtual reproductions or simulations of actual things, procedures, or systems is a function of DT technology. DT technology relies heavily on AI algorithms to expand its functionality and offer insightful data. Here are a few AI techniques that are frequently applied in DT implementations:

Machine Learning (ML): ML algorithms are used to train models on historical and real-time data from the physical system to learn patterns, correlations, and anomalies. These models can then be used to predict behavior, optimize performance, and identify potential issues in the physical twin. Incorporating DT into the healthcare industry would enhance the procedures by fetching the patients and the health providers in a single gateway. The human body parameters are made to be the inputs to the AI enables ECG classifier models that follow machine learning algorithms and used for continuous monitoring of heart functions and to find out the anomaly [29].

Deep Learning: Deep learning techniques, in particular neural networks, are a subset of ML that are used for challenging pattern recognition tasks. They are helpful when working with high-dimensional data and can be applied to time-series analysis, picture identification, and natural language processing in DTs [30].

Predictive Analytics: this entails predicting future behavior based on historical data using statistical and machine learning approaches. Predictive analytics can be used to foresee the need for maintenance, performance decline, and other crucial physical system events. Through its use in precision medicine, cancer care and treatment modeling, predictive analytics and machine learning, and in combining different spectrum of clinician viewpoints, the DT could detect neurological problems. The distinct areas make up the field of pediatric cancers: diagnostics, a range of therapeutic choices, and extensive follow-up. This industry is the ideal case study for the confluence of DT and machine learning technologies because to the numerous data points that are thus generated across time and location.

Predictive analytics is performed with the combination of statistics and computer science converge. It is important to carefully address the ethics of artificial intelligence with regard to access to continuously created patient health information and privacy. Although the idea of a DT is still relatively new, continuing discussions about internet security and privacy exist in all sectors. However, by establishing access requirements for primary care doctors and verified patient circles of care, institutions that have widely implemented EHR systems have already established a precedent. Second, integrating or expanding the usage of electronic health records would increase the cost of a cancer patient's total healthcare, as would the price of new wearable technology and the networking required to complete the DT trifecta. Last but not least, certain DT models, such as black box algorithms, largely rely on deep learning computations. [31].

Natural Language Processing (NLP): NLP algorithms are employed when the DT interacts with humans through natural language interfaces, enabling seamless communication and data input. Bi-LSTM and with Deep VR models can easily identify the presence of cancerous cells in the lungs by using Bi-LSTM as a middle layer in the RNN network. A self-attention layer weights are visualized in our DeepVR Model to make it simpler to understand and for better demonstration as far as the attention mechanism learns. In particular, a mapping between the input words and the attended weights are created and the parameters are extracted from the trained model's self-attention layer. The input code words are in line with their appropriate weights. Various weights indicate varying degrees of influence on the model's decision-making. The more impact a code word has on the detection model, the darker its coloring. In particular, the detection model will focus more on the darker word.

It is suggested that the traditional approach of developing IoT systems in isolation from the physical world often leads to suboptimal performance and reliability. By creating a DT of the physical system, it is possible to simulate and test IoT systems in a virtual environment, thereby improving their integration with the physical system. An overview is provided which consists of the DT concept and its applications in industrial settings, highlighting the benefits of using DTs for IoT system design, simulation, and optimization. The authors also discuss various challenges in implementing DT technology, such as data integration and model calibration, and propose potential solutions.

A case study of a DT implementation in a smart manufacturing system, demonstrating how DT technology can improve the performance and reliability of IoT systems. Overall, the work provides valuable insights into the potential of DT technology to enhance the virtual-real integration of industrial IoT systems [32, 33].

Reinforcement Learning: in some cases, DTs may incorporate reinforcement learning algorithms to optimize control strategies. By simulating and testing different actions and their outcomes in the virtual twin, the physical system's control algorithms can be improved. Background: techniques for categorizing patients into two or more diagnostic groups or predicting the likelihood of a specific event have been created using machine learning. Clinical trials and patient management could benefit from the use of DT models, which predict the full trajectories of patient health data. In [34], a variational autoencoder-based DT system is applied to a cohort of individuals who suffered an ischemic stroke later in life. The patient symptoms were modeled using ICD codes for ischemic stroke and laboratory standards, and the DT's capacity to predict medical evaluation paths before and after the sudden healthcare occasion was assessed [34].

Anomaly Detection: AI algorithms for anomaly detection help in identifying deviations from normal behavior in the physical system. They can be used to raise alarms or trigger corrective actions when unexpected events occur. DT technology can be used to create a virtual model of a patient, which can be used to monitor their health in real-time. This can be done by collecting data from wearables, sensors, and other sources, and using this data to create a DT that mimics the behavior of the patient's body.

This allows healthcare providers to monitor a patient's health status and identify potential issues before they become serious. When compared to actual measurement data, the development of DT technologies makes it possible to simulate complicated machinery with realism. As a result, it is well suited to produce synthetic datasets for the use in anomaly detection methods. Weakly supervised innovative techniques to anomaly detection for industrial contexts are presented in this study. The methods create a training dataset using a DT that replicates the usual functioning of the equipment and a limited number of labeled aberrant measurements from the actual machinery. We specifically offer a neural architecture based on the Siamese Autoencoders (SAE) and a clustering-based method dubbed cluster centers (CC), which are designed for poorly supervised scenarios with a limited amount of labeled data samples [35].

Optimization Algorithms: these algorithms are used to find the best set of parameters or control strategies that optimize the performance of the physical system. They are often used in conjunction with simulation models in the DT, which can also be used to simulate medical procedures and treatments. This can be done by creating a virtual model of a patient's body and simulating the effects of different treatments or interventions. This can help healthcare providers to identify the most effective treatment options for a patient, and to develop new treatments and procedures. DTs can be used to optimize medical equipment and devices. For example, by creating a DT of a medical device, engineers can simulate how the device will perform under different conditions and identify areas for improvement. This can lead to more effective and efficient medical devices that improve patient outcomes. DTs can also be used to develop personalized treatment plans for patients. By creating a DT of a patient's body, healthcare providers can simulate the effects of different treatments and interventions on the patient.

This can help to identify the most effective treatment options for each individual patient, based on their unique characteristics. Emerging technologies have recently made a significant contribution to the healthcare system's ability to treat a variety of ailments, to the point that the development of such techniques is now seen as a viable means of curing many diseases. Therefore, it makes no sense to underestimate the significance of such cyber environments in the medical and healthcare system, especially in light of the recent pandemic, which has demonstrated the importance of such technologies. Combining such systems with the metaverse can also result in some truly amazing applications.

This is the reason why the DT of a medical microrobot has been created in this research. It is controlled by a stochastic model predictive controller (MPC) that is supported by a system identification based on machine learning (ML). The use of magnetic technology in this robot is advantageous. These micro-magnetic systems can be crucial in monitoring, medication distribution, or even some delicate inside procedures, given their size, control system, and requirements. Thus, when designing and simulating this magnetic mechanism, precision, robustness, and dependability were all taken into account. Finally, a Kalman filter updates the controller to account for this environment while a second-order statistical noise is added to the plant. The outcomes demonstrate the effectiveness of the suggested controller [36].

9.3.1 DT-Based System for Smart Aging

Inside the subject of providing eldercare, there's a clear disparity seen between the growing need for an enhancement to the standard of life for the elderly as they age and the unbalanced and subpar growth of the senior care business. Household, neighborhood, and hospital senior care are all included in the conventional aged concierge. Yet, the traditional senior care system fails to satisfy the demands of the elderly in the present period because of the "4-2-2" family structure and a sizable aging society. It is vital to take into account not only the patient data of the elderly but also the society, industry, health organizations, and governments according to the real demands of the elderly when developing older-oriented goods [37].

All tiers of elder care activities are reliant just on gathering and analysis of big data, making big data a core innovation of intelligent senior care, as indicated by the closed-loop diagram of the intelligent aging framework shown in Figure 9.7 [11]. The senior care service sector loop is created by integrating several sectors, such as

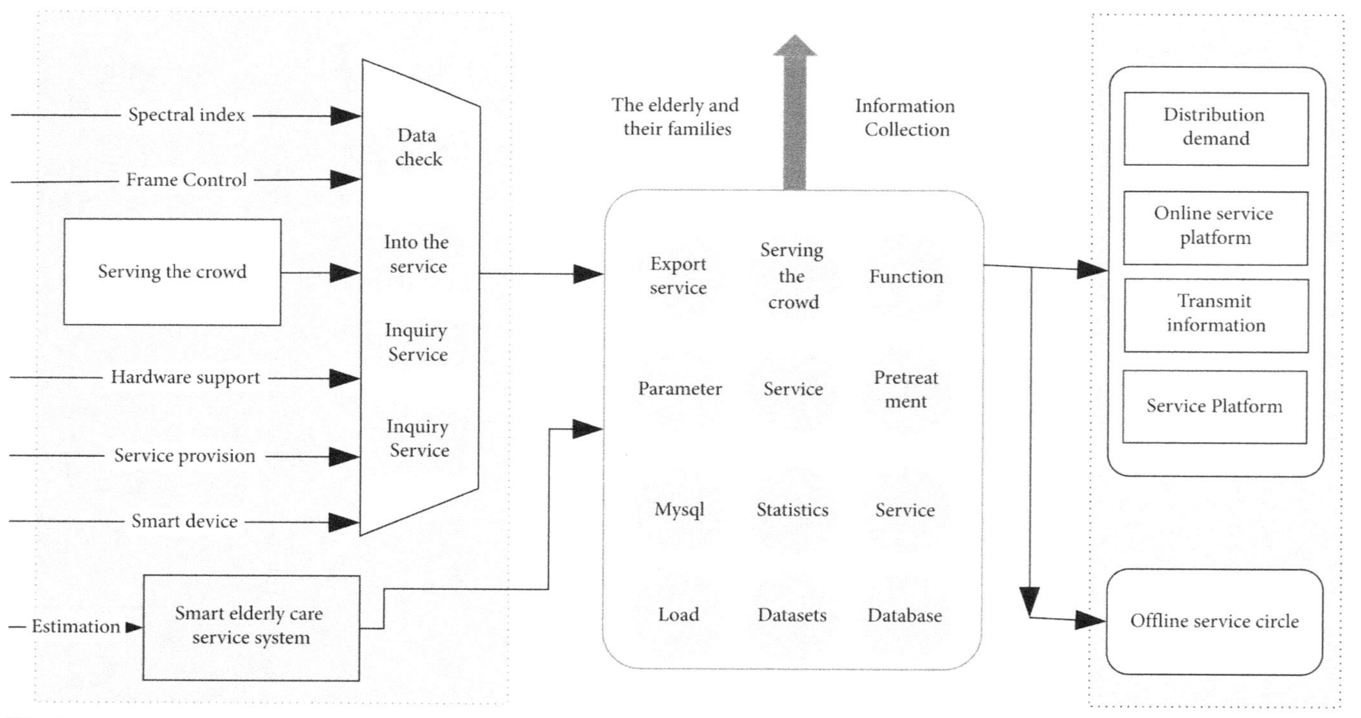

Figure 9.7 DT schematic for smart aging system [11].

Source: Open Access Article.

the Web, the finance sector, property investment, and banking, with an emphasis on activities and the intelligent senior care sector. This is the company's unavoidable growth tendency for the modern era of senior care. Intelligent gadgets, a digital service platform, and a physical services loop make up the smart elder care service. Senior citizens use technological devices that record and transfer files about their daily activities, illness, activity, rising consumer, and other topics to a cable internet network. The smart gadget, the internet service network, and the physical business loop create a complete loop for the elderly and their family. A clever senior care service platform is created with a focus on obtaining and incorporating other senior care provider knowledge assets from the local area, communicate the smart elder care provision of the services between many societies, document the basic information and provider requirements that seniors require, offer additional life leadership and amenities for them in accordance with their needs, and satisfy their "socialized" necessities. To address the issue of a lack of human resources and achieve superior elder services more simply and effectively, the system makes use of rapid integration, the supply and need relationship, sensors, big data, cloud services, and internet as linkages [38].

Using the World Wide Web, the DT design for object-oriented analytical modeling is finished. The online system structure and features of the system are first shown with the design of the DT system. Secondly, the database table appropriate for this system is developed with the performance and requires the presence of mind. The system's API code is supplied lastly.

9.4 DT in Dentistry

Orthodontic institutes using DT for distance learning raises questions about their advantages over existing remote education practices, how they differ from them, and their design and implementation. Currently, DT applications are benefitting from the convergence of artificial intelligence technology, big data, and the Internet of Things. Areas of study which have already deployed DT applications may clearly see this unification.

A portable teaching and learning DT platform is designed as a tool for remote dental education [39, 40]. It utilizes DT technology to provide a combined physical and virtual learning experience to improve students' dental skills. The system has two separate workplace kinds: the instructor workstation and the student workstation, and it enables many learning modalities, like KPIs, video streams, and visual statistics. The student workstation has all the required components and comes with the DT-student program for filming and viewing training films with physical activity [41] as shown in Figure 9.8. The instructor workstation is combined with a typical teacher work space and dentistry equipment.

Additionally, the DT framework offers PrepScanner, an automated measurement system for assessing dental performance, and the DT framework library, which acts

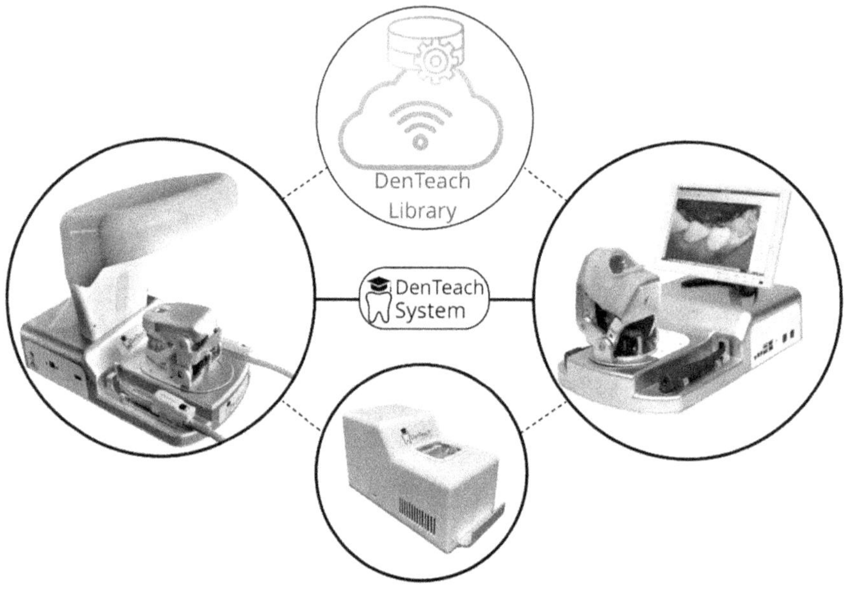

Figure 9.8 Digital-twin based framework in dentistry [41].
Source: Open Access Article.

as the central platform. The system gives monitors for teacher and pupil desktops, business intelligence, and a testbed in the mirroring paradigm of the student software in addition to the new framework DTs API being built for the collection. The DT platform's novel use of DT technology and small, transportable form make it an efficient remote teaching tool overall.

The past few decades have seen a dramatic increase in the importance of sophisticated mathematics and deep learning (DL) algorithms in the diagnosis of health variables and disorders. Dentistry is one field that needs greater attention. To take advantage of the immersion properties of this equipment as well as convert the real-life practice of dentistry to the virtual world, it is practicable and beneficial to create DTs of dental disorders in the metaverse. With the help of these innovations, medical professionals, students, and the general public will soon be able to access a wide range of healthcare options from the comfort of their own homes. Another significant benefit of these kinds of devices is the potential for immersive interactions between physicians and patients, which may greatly enhance the effectiveness of the medical system. Offering these conveniences via a blockchain system improves trustworthiness, security, transparency, and auditability of data transfers. In addition to reducing expenses, increased productivity is another benefit. Cervical vertebral maturation (CVM) is an important consideration in many types of dentistry, and work in [42] presents the conceptualization and implementation of a

DT of CVM inside a blockchain-based metaverse system. For the suggested system, a DL technique was applied to develop a computerized diagnostic procedure for future CVM pictures. MobileNetV2, a key component of this strategy, is a mobile architecture that boosts cellular model efficiency across a wide range of metrics. The suggested method of digital twinning has very little latency and computational expenses, making it accessible to general practitioners and experts alike and ideal for adjusting to the Internet of Medical Things (IoMT). The application of DL-based computer vision as a real-time measuring technique means that the suggested DT is devoid of any extra sensors. Additionally, a complete theoretical structure for developing MobileNetV2-based CVM DTs inside a blockchain environment was established and executed, demonstrating the method's scalability and compatibility. Low-cost DL has numerous potentials uses in the future digitally, as shown by the outstanding results of the suggested model on a gathered tiny dataset. The study also demonstrates how DTs may be produced and implemented for dental problems using little equipment, hence lowering patients' out-of-pocket expenses for evaluation and therapy.

The "Calculate_hash" algorithm is employed in BT to create a cryptographic hash value that serves as a fingerprint for each individual block. The information contained in a block is represented by a fixed-length sequence of letters called a hash value, which is calculated using complicated mathematical methods. The security and permanence of the blockchain are guaranteed by the fact that every block has a distinctive hash value, which also updates if any data in the block is changed. The approach uses a hash algorithm, such as SHA-256, to compute a block's hash value from the data recorded in the block. This includes the interactions, date and time, and hash value of the prior block. The resultant hash value is recorded in the current block as a means of identification and serves as a connection to the prior block in the chain. With the "Calculate_hash" function, any efforts to alter or interfere with the data on the blockchain may be quickly identified, ensuring its integrity. The flow is shown in Figure 9.9.

In BT, the "Create a Blockchain" module is a computer notion that represents the template used to generate a new blockchain. Fresh blocks may be added, the chain can be validated, and disputes can be resolved because this document provides the attributes and techniques required for constructing and operating a blockchain. Every single block in the chain is connected to the one before it by its own hash value, creating the class. Sophisticated cryptography procedures are used to protect the integrity of the whole chain, which consists of blocks that each include an assortment of transactions and a time stamp. Techniques for verifying a chain by verifying the validity of every block and its connections to earlier blocks are also included in the "Create a Blockchain" class. The blockchain is discarded and the testing process is repeated if any discrepancies or signs of manipulation are found.

The "Create a Blockchain" course also includes a mechanism to settle disagreements, which is crucial. A block creation conflict occurs when two or more

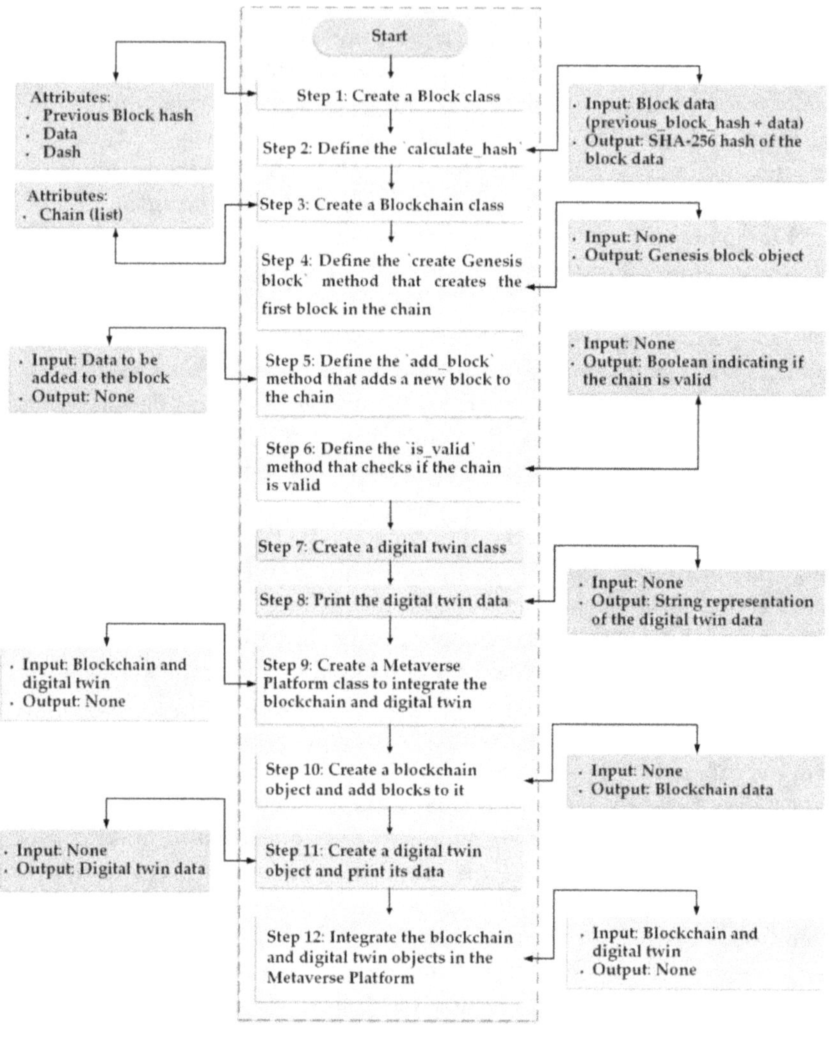

Figure 9.9 Flow of the work using blockchain technology in dentistry [42].
Source: Open Access Article.

nodes attempt to produce the identical block at the same moment, necessitating an update to the chain to reflect the right block. This procedure specifies how inconsistencies in the chain are fixed, guaranteeing uniformity throughout every network node. Using the "Create a Blockchain" class, programmers may easily launch a new blockchain, monitor its operations, and protect the data it contains.

In BT, the initial block of a new blockchain is generated using the "Create_genesis_block" procedure. You may think of this block as the "genesis block" since it kicks off the whole chain. Whenever a fresh blockchain is generated, the "Create_genesis_block" function is normally used to generate the first blocks in the chain of blocks, which has a fixed amount of data and a predetermined hash value. This is the first block in a blockchain and contains essential data about the network, including the moment the block was created, the first group of participants, and other metadata. Most implementations of blockchains use predetermined hash functions or hard coded values for the genesis block's hash. This is the initial block on the blockchain and is linked to future blocks by their hash values after it is produced.

The genesis block is the first block in the blockchain and the beginning point from which all subsequent blocks and transactions build. The "Create_genesis_block" function allows developers to quickly generate the first block of a blockchain, complete with a predetermined hash value and data, so that future blocks may be appropriately connected to it.

Blockchain technology relies on the "Add_block" procedure for adding new blocks to the chain. The newly created block's activities and any associated information are read in as input, and an individual hash value is then generated. The hash value is derived from the data included in the block and its hash value of the preceding block in the chain utilizing a hash algorithm, such as SHA-256. When an additional block is created, its hash value is computed and used to connect it to the prior block in the chain. Any effort to tamper with the data in a block would result in a distinct hash value that would be promptly discovered by the network, making the blockchain secure and immutable via this association mechanism. To guarantee that the information contained in the latest block is genuine and in accordance with the rules and regulations of the blockchain, the "Add_block" function additionally incorporates error-checking and verification algorithms. If a problem is found during validation, the block is thrown out and the procedure is resumed.

It is common for developers to be given instructions that need them to add an additional technique to the class named "str" to display the DT information. Data from the DT item is printed in a human-readable manner using the "str" method, which defines the numerical depicting an object. It is up to the coder to decide what details about the DT should be printed by defining the "str" function for the class that represents the DT. To output the text version of the DT object in a readable by human manner, the developer may utilize the "str" method after it has been defined for the DT class. When testing and troubleshooting, this may save time by letting the developer study the DT's attributes and connections in detail before diving into the code. Creating the "str" function for the DTclass is a frequent programming effort that may improve the readability and understandability of the code, as well as facilitate the programmer's ability to interact with and oversee the DT in their computer system.

The purpose of this amalgamation is to provide a system where DT may be safely kept and controlled on the blockchain, where they will be immutable and resistant to tampering. In addition to making, it simpler to examine and comprehend the interdependencies between various DT, the metaverse platform class may assist to assure the integrity and quality of the data recorded on the blockchain. The development of a metaverse, a virtual environment composed of both digital and physical components, hinges on the creation of the metaverse platform class. In a metaverse, people may interact with one another and digital items in an environment together that mixes parts of the real and digital worlds.

The tenth step is to generate a blockchain object, the data structure used in distributed ledgers to record transactions in a chronological order over a chain of blocks. The transfer of bitcoin, a change of control of an electronic asset, and the carrying out of an intelligent contract are just some of the many types of events that may be recorded on a blockchain, which is effectively a distributed ledger. A crypto hash, a timestamp, and a log of events are included in each "block" in the chain. Other nodes in the network check the validity of each new block that is added to the chain.

In the eleventh stage, a DT object is made, that is an electronic replica of a real-world system or procedure. A DT may be utilized for real-time monitoring and control, change simulation and efficiency optimization, and distant repair and upkeep. A DT is made up of three main parts: a digital representation of the physical system, a system for gathering and analyzing data from sensors, and a suite of analytical tools for drawing conclusions and making suggestions based on that data. This stage requires the use of computational tools or an infrastructure that facilitates the creation of DT, like Siemens MindSphere, Microsoft Azure DTs, or IBM Watson IoT.

The last step of the procedure is to include blockchain technology and DT items into the metaverse platform. For new applications and use cases to be made possible, integration must be achieved between the blockchain and DT objects. Its property information, revision history, and simulated outputs might all be recorded on a blockchain. It is possible to utilize a DT to gather data from sensors through an analog system, provide alarms based on rules, and initiate blockchain transactions.

This stage requires the creation of APIs and protocols that allow for two-way communication between the blockchain and the DT objects. A uniform data format, RESTful APIs, or blockchain-based smart contracts might all be part of the solution. When the two systems are fully integrated, the metaverse platform can be launched, and developers may begin creating blockchain- and digital-powered apps and activities.

Cone-beam computed tomography (CBCT) images can blur the teeth owing to artifacts, therefore more scanning is sometimes necessary to create virtual duplicates. Plaster models are widely used, yet they come with a few limitations that might be problematic. The purpose of the work in [43] is to compare the accuracy of various digital dentition simulations to that of traditional casts made of plaster. Twenty participants' imprints were cast in alginate, and their intraoral scan (IOS) and cone beam computed tomography (CBCT) pictures were captured. The alginate imprint was scanned twice, five minutes and two hours later, using a computer modeling printer. The whole arch was imaged using CS 3600 sections and i700 wireless scans in parallel on an IOS. Plaster cast DTs were layered with alginate imprint and IOS DTs. Differences and lengths between each metric were calculated. The largest deviations occurred in two-hour-old alginate imprint checks, however all of them were smaller than the 0.39 mm voxel size used in CBCT. In comparison to the plaster method, alginate imprint scans and IOS serve as beneficial additions to CBCT. Scanning the material's imprint within five minutes or scanning the complete arch intraorally with division both increase reliability [44–46].

Therefore, five pictures of each patient's teeth were made. The R2GATETM (MegaGen Implant Co., Ltd.) software fed the CBCT pictures (DICOM format) and individual dental scan images (STL format). When integrating the left and right maxillary central incisors, as well as the cusps of the mesiobuccal of the left and right maxillary first molars, they used a semi-automatic technique centered on these landmarks. Each patient had five final DTs created as shown in Figure 9.10.

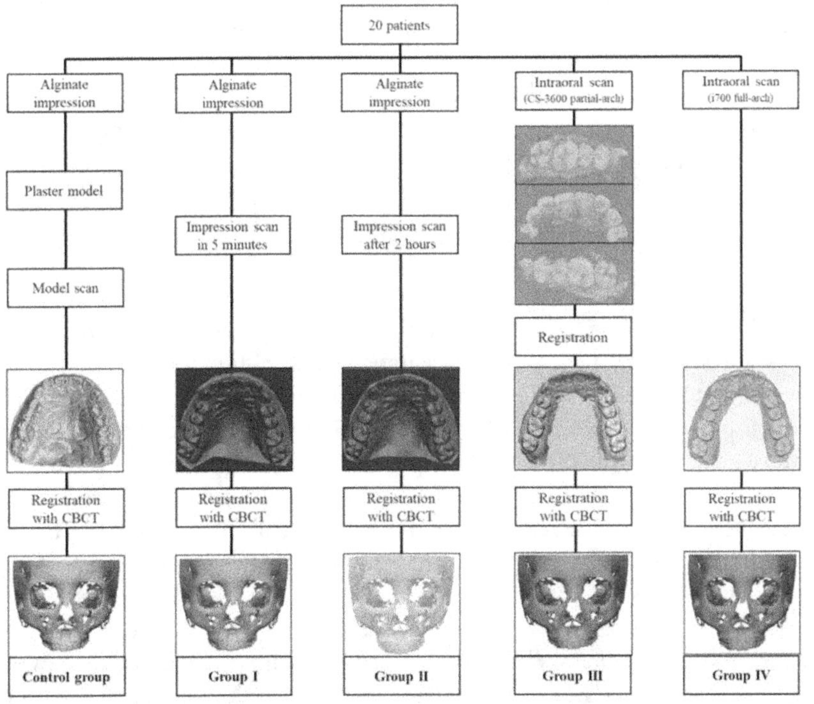

Figure 9.10 Group per group DT preparation flowchart [43].
Source: Open Access Article.

9.5 Challenges of DT in Healthcare

DTs have the possibility of making a substantial contribution to patient care, but there are design problems and worries that might prevent them from developing to their fullest capacity. As DTs incorporate a number of cutting-edge technologies, every one of these elements introduces unique socio-ethical concerns throughout the deployment phase. Also, due to technological restrictions that might range between information collecting to application development, the DT idea in medicine and the confluence of such innovations remain in their early phases [47]. Although DTs may deliver great results quickly, choosing a treatment and providing preventative medicine could prove to be possible using their forecasting accuracy.

Although DTs have long been prevalent in those other sectors, the incorporation of human life and dignity makes the implementation of DTs in medicine extremely complicated [48]. Economic, ethnic, legal, and societal issues, among others, stand in the way of DTs achieving their maximum promise in the medical industry. Most DT systems have technological challenges, and cultural-technical challenges necessitate a cultural change towards information sharing. Standardizing and disseminating the data gathered might be assisted by a central organization or network that is in charge of supporting cultural endeavors at the national level. As DT-related ethical and social issues are still in their infancy, it is crucial that academics, businesses, regulators, and professionals thoroughly explore the signaling pathways, organizational consequences, ethical considerations, and social consequences of DTs.

9.6 Conclusion

The DT is now widely used as a central component of the revolutionary the virtual world structure, which offers accurate digital leisure activities of real-world activities and interpersonal connections. Decades of study into behavioral and AI, the high pace of digitalization throughout various sectors, and the accompanying improvements in the analysis and interpretation of data have all contributed to the growth of the area of DT. To meet the evolving needs of individualized care, it is essential to create a digital representation of each patient that can provide the appropriate dose at the appropriate time. The concept of a PDT, which is an improved version of a DT is examined in the chapter. This PDT is based on the patient's physical characteristics, medical history, present condition, and anticipated future needs and can offer actionable insights for personalized diagnosis, therapy selection, and procedure planning. By giving healthcare professionals and the industry tailored information about their patients, the PDT can enhance decision-making (i.e., patients). It is also considered how to combine DTs, blockchain, and AI technology to create smart healthcare services that enhance people's lives. Nevertheless, challenges with access, privacy, security, and applicability for a range of requirements, as well as computational, ethical, and cultural issues, are now limiting the usage of DTs in healthcare.

References

1. Hassani, H., Huang, X., MacFeely, S. Impactful digital twin in the healthcare revolution. *Big Data Cogn. Comput.* 2022, 6, 83. https://doi.org/10.3390/bdcc6030083
2. Wang, Q., Jiao, W., Wang, P., Zhang, Y. "Digital twin for human-robot interactive welding and welder behavior analysis," *IEEE/CAA Journal of Automatica Sinica,* 2020, 8, no. 2, 334–343.
3. Hasselgren, A., Rensaa, J.A.H., Kralevska, K., Gligoroski, D., Faxvaag, A. Blockchain for increased trust in virtual health care: Proof-of-concept study. *J. Med Internet Res.* 2021, 23, e28496.
4. Armeni, P., Polat, I., De Rossi, L.M., Diaferia, L., Meregalli, S., Gatti, A. Digital twins in healthcare: Is it the beginning of a new era of evidence-based medicine? A critical review. *J. Pers. Med.* 2022, 12, 1255. https://doi.org/10.3390/jpm12081255
5. Tao, F., Liu, W., Zhang, M., Hu, T., Qi, Q., Zhang, H., Sui, F., Wang, T., Xu, H., Huang, Z. Five-dimension digital twin model and its ten applications. *Comput. Integr. Manuf. Syst.* 2019, 25, 1–18.
6. Boulos, K., Maged N., Zhang, P. "Digital twins: From personalised medicine to precision public health." *J. Pers. Med.* 2021, 11, no. 8, 745.
7. Mourtzis, D., Angelopoulos, J., Panopoulos, N., Kardamakis, D. A smart IoT platform for oncology patient diagnosis based on AI: Towards the human digital twin. *Procedia CIRP* 2021, 104, 1686–1691.
8. Briceno, J., Calleja, R., Hervás, C. Artificial intelligence and liver transplantation: Looking for the best donor-recipient pairing. *Hepatobiliary Pancreat. Dis. Int.* 2022, 21, 347–353.
9. Aheleroff, S., Xu, X., Zhong, R.Y., Lu, Y. Digital Twin as a Service (DTaaS) in Industry 4.0: An architecture reference model. *Adv. Eng. Inform.* 2020, 47, 101225.
10. Karakra, A., Fontanili, F., Lamine, E., Lamothe, J. HospiT'Win: A predictive simulation-based digital twin for patients pathways in hospital. In *Proceedings of the 2019 IEEE EMBS International Conference on Biomedical Health Informatics (BHI),* Chicago, IL, USA, 19–22 May 2019; IEEE: Piscataway, NJ, USA, 2019; pp. 1–4.
11. Deng, Y. "Digital Twin-based modeling of complex systems for smart aging", *Discrete Dyn. Nature Soc.* 2022, 2022, 11. Article ID 7365223, https://doi.org/10.1155/2022/7365223
12. Angulo, C., Gonzalez-Abril, L., Raya, C., Ortega, J.A. A proposal to evolving towards digital twins in healthcare. In *Proceedings of the International Work-Conference on Bioinformatics and Biomedical Engineering,* Granada, Spain, 6–8 May 2020; pp. 418–426.
13. Sundaravadivel, P., Kougianos, E., Mohanty, S.P., Ganapathiraju, M.K. Everything you wanted to know about smart health care: Evaluating the different technologies

and components of the internet of things for better health. *IEEE Consum. Electron. Mag.* 2017, 7, 18–28.
14. Shengli, W. Is human digital twin possible? *Comput. Methods Programs Biomed. Update* 2021, 1, 100014.
15. Elayan, Haya, et al. “Digital twin for intelligent context-aware IoT healthcare systems.” *IEEE Internet of Things J.* 2021, 8, 16749–16757.
16. Moody, G.B., Mark, R.G. “The impact of the mit-bih arrhythmia database,” *IEEE Eng. Med. Biol. Mag.* 2001, 20, no. 3, 45–50.
17. Sahal, R., Alsamhi, S.H., Brown, K.N. Personal digital twin: A close look into the present and a step towards the future of personalised healthcare industry. *Sensors* 2022, 22, 5918. https://doi.org/10.3390/s22155918
18. Roda, A., Michelini, E., Zangheri, M., Di Fusco, M., Calabria, D., Simoni, P. Smartphone-based biosensors: A critical review and perspectives. *TrAC Trends Anal. Chem.* 2016, 79, 317–325.
19. Kamel Boulos, M.N., Zhang, P. Digital twins: From personalised medicine to precision public health. *J. Pers. Med.* 2021, 11, 745.
20. Li, Q., Gravina, R., Li, Y., Alsamhi, S.H., Sun, F., Fortino, G. Multi-user activity recognition: Challenges and opportunities. *Inf. Fusion* 2020, 63, 121–135.
21. Oakes, B.J., Meyers, B., Janssens, D., Vangheluwe, H. Structuring and accessing knowledge for historical and streaming digital twins. In *Proceedings of the SEMANTICS Co-Located Events*, Amsterdam, The Netherlands, 6–9 September 2021; pp. 1–13.
22. Alharbi, A., Alosaimi, W., Sahal, R., Saleh, H. Real-time system prediction for heart rate using deep learning and stream processing platforms. *Complexity* 2021, 2021, 5535734.
23. Amann, J., Blasimme, A., Vayena, E., Frey, D., Madai, V.I. Explainability for artificial intelligence in healthcare: A multidisciplinary perspective. *BMC Med Inform. Decis. Mak.* 2020, 20, 310.
24. Hasan, H.R., Salah, K., Jayaraman, R., Omar, M., Yaqoob, I., Pesic, S., Taylor, T., Boscovic, D. A blockchain-based approach for the creation of digital twins. *IEEE Access* 2020, 8, 34113–34126.
25. Sahal, R., Alsamhi, S.H., Brown, K.N., O’Shea, D., Alouffi, B. Blockchain-based digital twins collaboration for smart pandemic alerting: Decentralized COVID-19 pandemic alerting use case. *Comput. Intell. Neurosci.* 2022, 2022, 7786441.
26. Dahir, H ., Luna, J ., Khattab, A ., Abrougui, K., Kumar, R. Chapter 4 – challenges of digital twin in healthcare, Editor(s): Abdulmotaleb El Saddik, *Digital Twin for Healthcare*, Academic Press, 2023, 73–95. ISBN 9780323991636, https://doi.org/10.1016/B978-0-32-399163-6.00009-3
27. Botín-Sanabria, D.M., et al. “Digital twin technology challenges and applications: A comprehensive review.” *Remote Sens.* 2022, 14, no.6, 1335.
28. Al-Ali, Abdul-Rahman, et al. “Digital twin conceptual model within the context of internet of things.” *Future Internet* 2020, 12, no.10, 163.
29. Elayan, H., Aloqaily, M., Guizani, M. “Digital twin for intelligent context-aware IoT healthcare systems.” *IEEE Internet of Things J.* 2021, 8, 16749–16757.
30. Ding, H., Yang, L., Yang, Z., Wang, Y. “Health prediction of shearers driven by digital twin and deep learning.” *China Mech. Eng.* 2020, 31, no. 07, 815.

31. Thiong'o, G.M., Rutka, J.T. "Digital twin technology: The future of predicting neurological complications of pediatric cancers and their treatment." *Front. Oncol.* 2022, 11, 781499.
32. Zhang, J., Li, L., Lin, G., Fang, D., Tai, Y., Huang, J. "Cyber resilience in healthcare digital twin on lung cancer," in *IEEE Access* 2020, 8, 201900–201913. doi: 10.1109/ACCESS.2020.3034324
33. Banerjee, I., Sofela, M., Yang, J., Chen, J.H., Shah, N.H., Ball, R. et al., "Development and performance of the pulmonary embolism result forecast model (perform) for computed tomography clinical decision support", *JAMA Netw. Open* 2019, 2, no. 8, 1760–1768.
34. Allen, A., Siefkas, A., Pellegrini, E., Burdick, H., Barnes, G., Calvert, J., Mao, Q., Das, R. "A digital twins machine learning model for forecasting disease progression in stroke patients" *Applied Sciences* 2021, 11, no. 12, 5576. https://doi.org/10.3390/app11125576
35. Castellani, A., Schmitt, S., Squartini, S. "Real-world anomaly detection by using digital twin systems and weakly supervised learning," in *IEEE Trans. Industr. Inform.* July 2021, 17, no. 7, 4733–4742. doi: 10.1109/TII.2020.3019788
36. Keshmiri Neghab, H., Jamshidi, M. (Behdad), Neghab, H.K. "Digital twin of a magnetic medical microrobot with stochastic model predictive controller boosted by machine learning in cyber-physical healthcare systems" *Information* 2022, 13, no. 7, 321. https://doi.org/10.3390/info13070321
37. Kwok, P.K., Yan, M., Qu, T., Lau, H. "User acceptance of virtual reality technology for practicing digital twin-based crisis management," *Int. J. Comp. Integr. Manufac.* 2021, 34, no. 7–8, 874–887.
38. Tao, F., Qi, Q., Wang, L., Nee, A.Y.C. "Digital twins and cyber-physical systems toward smart manufacturing and industry 4.0: Correlation and comparison," *Engineering* 2019, 5, no. 4, 653–661.
39. Maddahi, Y., Kalvandi, M., Langman, S., Capicotto, N., Zareinia, K. RoboEthics in COVID-19: A case study in dentistry. *Front. Robot. AI* 2021, 8, 612740.
40. Maddahi, Y., Kalvandi, M., Maddahi, A., Dhannapuneni, P. Automated apparatus and method for quantifying dimensions of dental preparations. *U.S. Patent* 63, 164, 759, 23 March 2021.
41. Maddahi, Y., Chen, S. Applications of digital twins in the healthcare industry: Case review of an IoT-Enabled remote technology in dentistry. *Virtual Worlds* 2022, 1, 20–41. https://doi.org/10.3390/virtualworlds1010003
42. Moztarzadeh, O., Jamshidi, M., Sargolzaei, S., Keikhaee, F., Jamshidi, A., Shadroo, S., Hauer, L. Metaverse and medical diagnosis: A blockchain-based digital twinning approach based on MobileNetV2 algorithm for cervical vertebral maturation. *Diagnostics* 2023, 13, 1485. https://doi.org/10.3390/diagnostics13081485
43. Lee, J.H., Lee, H.L., Park, I.Y. et al. Effectiveness of creating digital twins with different digital dentition models and cone-beam computed tomography. *Sci Rep.* 2023, 13, 10603. https://doi.org/10.1038/s41598-023-37774-x
44. Lin, H.-H. et al. Artifact-resistant superimposition of digital dental models and cone-beam computed tomography images. *J. Oral Maxillofac. Surg.* 2013, 71, 1933–1947.

45. Jiang, T., Lee, S.-M., Hou, Y., Chang, X., Hwang, H.-S. Evaluation of digital dental models obtained from dental cone-beam computed tomography scan of alginate impressions. *Korean J Orthod.* 2016, 46, 129–136.
46. Burzynski, J.A., Firestone, A.R., Beck, F.M., Fields, H.W.Jr., Deguchi, T. Comparison of digital intraoral scanners and alginate impressions: Time and patient satisfaction. *Am. J. Orthod. Dentofac. Orthop*. 2018, 153, 534–541.
47. Erol, T., Mendi, A.F., Dogan, D. The digital twin revolution in healthcare. In *Proceedings of the 2020 4th International Symposium on Multidisciplinary Studies and Innovative Technologies (ISMSIT)*, Istanbul, Turkey, 22–24 October 2020; pp. 1–7. Available online: https://doi.org/10.1109/ISMSIT50672.2020.9255249
48. Reis, L., Maier, C., Mattke, J., Creutzenberg, M.,Weitzel, T. Addressing user resistance would have prevented a healthcare AI project failure. *MIS Q. Exec.* 2020, 19, 279–296.

Chapter 10

Sustainable Organic Farming in Indian Rural Areas with the Aid of the Internet of Things

R. Rajesh, S. Sivakumar, and S. Premkumar

10.1 Introduction

Smart organic farming follows traditional farming systems relying on vermicomposting, green manure, crop rotation, biological pest control, poultry farming, and livestock production. The key objective of organic farming is to develop ecological biodiversity in the agricultural field and also enhance soil fertility in the plantation farm. It specifies the procedure to use animal or plant waste for pest control and as biological fertilizers. Organic farming was a solution for environmental sufferings caused by synthetic fertilizers and chemical pesticides. In other words, smart organic farming recovers and enhances the ecological balance in agricultural farms [1].

According to the Food and Agriculture Organization (FAO) "Organic agriculture improves various argi-ecosystem like a biological cycle, biodiversity and soil fertility with mechanical, biological and agronomic techniques, and this is achieved without using off-farm input materials to the farming area."

Smart organic farming is related to the second sustainable development goal of the World Health Orgnization (WHO) to "end hunger, achieve food security and improve nutrition and promote sustainable agriculture" as organic farming is a

DOI: 10.1201/9781003469612-10

unification of different elements, such as water, soil, waste materials, microbes, forestry, and agriculture to achieve sustainability and combats the high requirements for quality food in the agricultural industry. The critical and shocking issue is providing healthy and nutritious food to the next-generation population.

According to Bill & Melinda Gates Foundation research, every year around 11 million people are ending up in an early grave due to a shortage of healthy food. The research, which is published in *The Lancet*, was conducted from 1990 to 2017 in 195 countries and it concludes that with a better diet one out of five deaths can be prevented. The research summarizes that the common key factor for death is a lower consumption of food grains and the per capita income of every country in 2050 needs to be higher when compared to today's target. Due to the need for high income, people expect quality diets with fiber-rich food and minerals. With higher population growth, there is a need for high-quality food to feed the upcoming population and trends depict the rapid growth of food demand. The solution to this critical problem can be resolved with smart organic farming [2].

10.1.1 Organic Farming in India

According to the FiBL survey 2020, India stands ninth with 1.94 million hectares of land used as organic agricultural land, Australia takes the first position with 35.6 million hectares of organic agricultural land, as shown in Figure 10.1.

In India, organic farming is in a preliminary stage. The statistical data from the Union Ministry of Agriculture and Farmers' Welfare as of March 2020 reports that India has 2.78 million hectares of farmlandused for organic farming, which is 2% of the net sown farmland in India. To encourage chemical-free organic farming and increase the minimum support price to farmers, the Government of India launched two schemes: Paramparagat Krishi Vikas Yojana (PKVY) and Mission Organic Value Chain Development for North East Region (MOVCD) in 2015.

In this, 1.94 million hectares is registered with the National Programme for Organic Production (NPOP), 0.59 million hectares is with the Paramparagat Krishi Vikas Yojna (PKVY), 0.17 million hectares is under specific state schemes, and 0.07 is with the Mission Organic Value Chain Development for North Eastern Regions (MOVCDNER). In India, 96% of organic food production was under the certification control of NPOP and 4% is under the certification control of the Participatory Guarantee System (PGS). The organic food products in India are authenticated with logos like Jaivik Bharat, FSSAI, and PGS Organic India and organic farmers need 3 years to get a PGS Green certificate to sell their organic chemical production.

According to the FiBL survey 2020, with small organic farmland under cultivation (Figure 10.2), India ranked first with a greater number of organic farmers in the world. Every farmer in the Sikkim state of India practice fully organic farming.

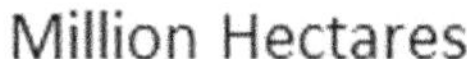

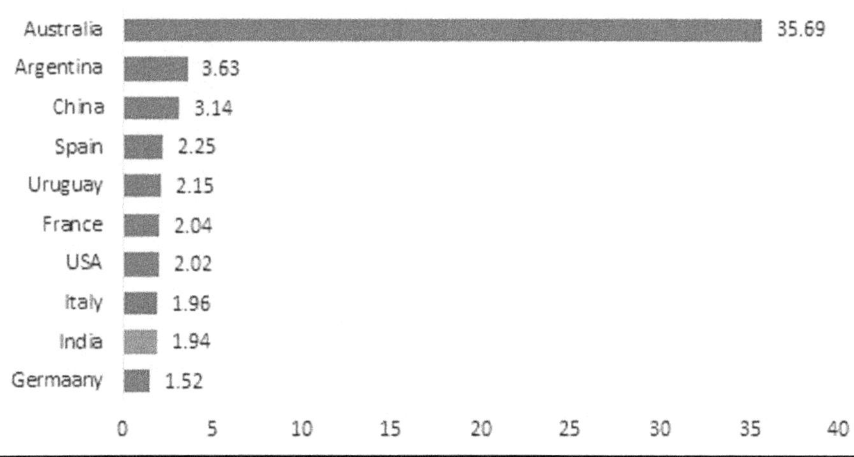

Figure 10.1 Organic agricultural land (million hectares).
Source: FiBL Survey 2020.

India has around 1.9 million organic farmers, which are about 1.3% of 146 million agricultural farmers as of March 2020. Organic farmers in tribal, hilly regions and farmers who were not registered are not counted, which may increase the total organic farmers in India.

From the Agri-export Policy 2018, India plays a major role in world organic markets, touching 5151 Crore. India exports various organic products like tea, sesame, soybean, flax seeds, rice, pulses, and medical plants, which drives around 50% of the global organic export in 2018–2019. As the need for healthy food increases in countries like UK, USA, Switzerland, and Italy, the Indian organic market has proved its potential and reached its organic products in different countries.

The Ministry of Agriculture, has launched a website, vedkrishi.com (farm-to-fork service), to boost the sale of organic food products in India. To attain our goal in organic farming, India needs to focus on improving profits and production of organic products. On this website, consumers can buy vegetables, grains, dairy products, and pulses as farm-fresh products from organic farmers. It also helps organic farmers in providing farm inputs like biofertilizers, vermicompost, and disease control products. Figure 10.3 represents the leading organic food producers in the world.

With the high demands on safe and quality foods and sustainable agricultural farming, smart organic farming act as a solution for quality food and sustainability. Indian agricultural land cultivated under organic farming has been improving from

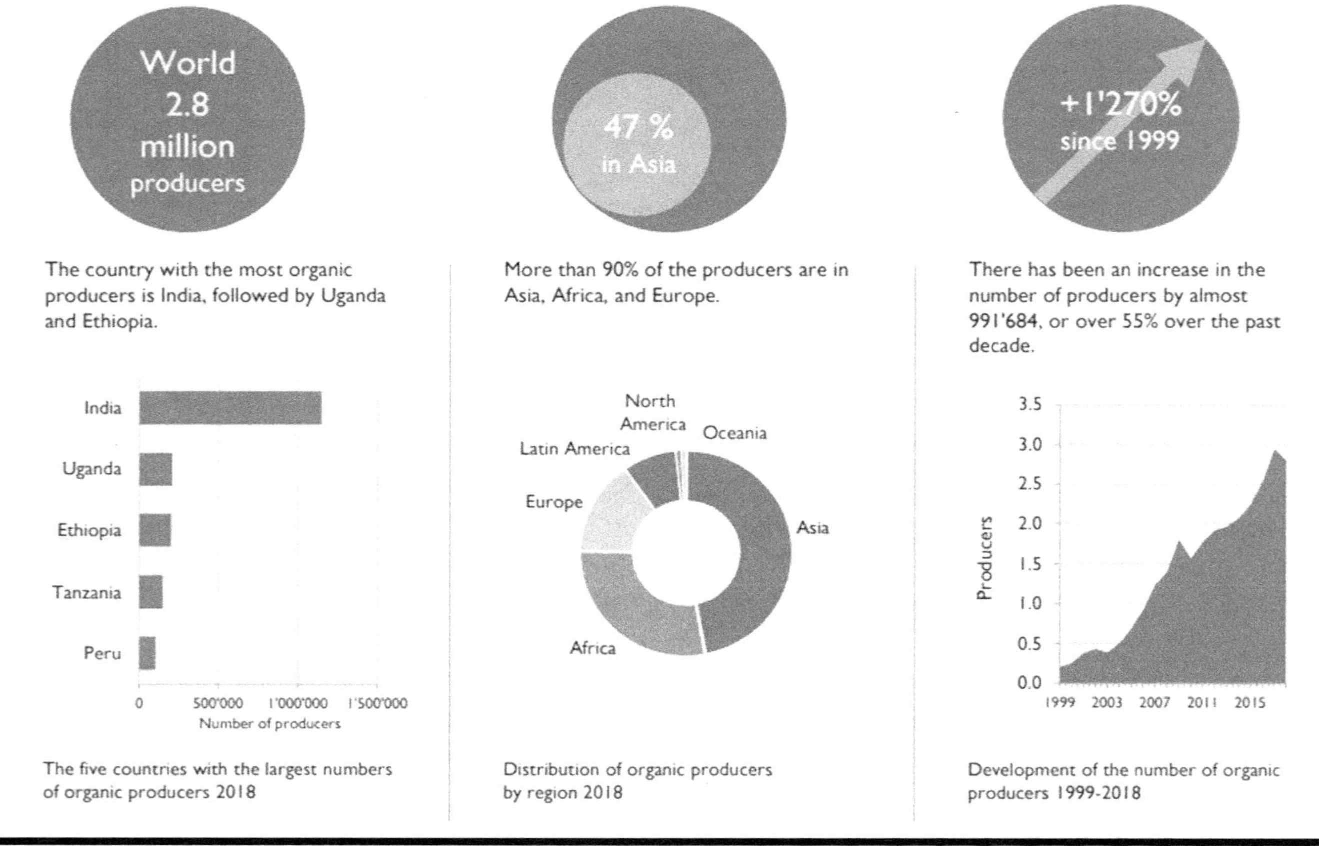

Figure 10.2 World organic agriculture.

Source: FiBL survey 2020 www.organic-world.net.

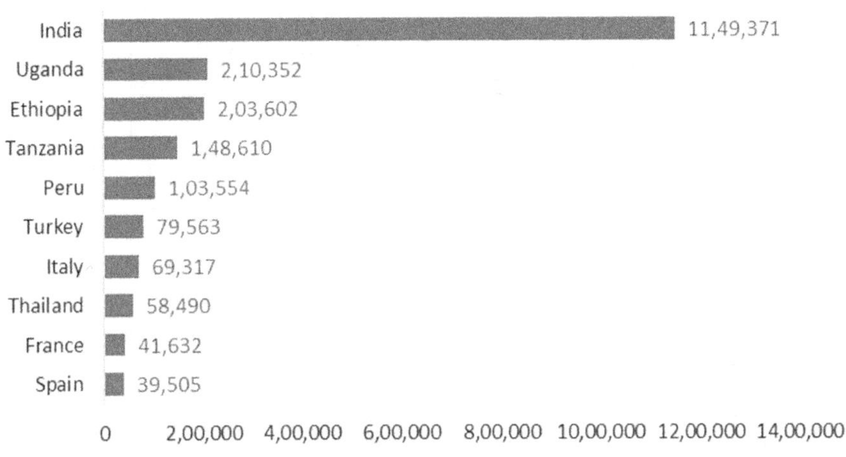

Figure 10.3 World organic producers.

Source: FiBL survey 2020 www.organic-world.net.

7.23 lakh ha in 2013–2014 to 27,77 lakh ha in 2018–2019. The state-wise Indian organic farm area registered under organic farming is given in Table 10.1 during 2018–2019.

10.1.2 Smart Organic Farming

Current farming techniques are in a state of distress, which severely affects the biodiversity and long-term sustainability of the system. The major challenges of conventional farming techniques are given in Figure 10.4.

Smart organic farming integrates conventional farming, innovation, and technologies for human welfare and satisfies the principles of ecology, health, and care of all living things. Historically, organic farming was recognized in 1905, and it reached its importance after the effects of modern agriculture in the later 1990s. The British scientist Sir Albert Howard was called the father of modern organic agriculture, and followed Indian conventional farming methods to use organic materials for farmlands. Organic farming controls energy consumption by 30.7% per unit of land by avoiding the energy required to produce synthetic fertilizers and pesticides, thereby reducing fuel consumption for transport with the usage of internal farm inputs. Although a lot of organic farming research has been performed, there is a need to address future issues, such as a healthy diet for human welfare, eco-friendly organic farms, learning economics with the market environment, and carbon organic farming models. Figure 10.5 depicts the uses of smart organic farming in the field of agriculture.

Table 10.1 State-Wise Indian Organic Farm Coverage Under Organic Certification (2018–2019)

		Organic area							
				Scheme-wise break up of total organic area of state/Union Territory					
				NPOP					
No	*State/union territory*	*Total organic area (000 ha) in 2019**	*Organic area in 2019 as % of net sown area of that state/union territory** (%)*	*NPOP %*	*In Conversion %*	*PKVY (%)*	*MoOVCDNER (%)*	*State schemes/non schemes (%)*	*Organic farming policy/ Mission/Act*
1	Madhya Pradesh	756	4.9	50.2	38.9	10.1	0	0.7	Policy,2010
2	Rajasthan	350	2	31.5	32.5	35.2	0	0.7	Policy 2017
3	Maharashtra	284	1.6	55.7	32.7	8.9	0	2.7	Policy 2013 Mission 2018
4	Andhra Pradesh	144	2.3	9.5	13	73.4	0	4.1	Draft policy 2008 CR-ZBNF 2015
5	Uttarakhand	128	18.2	15.7	13	70.2	0	1.1	Policy 2000 Act 2019
6	Odisha	118	2.6	62	19.2	17.6	0	1.2	Policy 2018
7	Karnataka	111	1.1	51.2	23.4	9.8	0	15.6	Policy 2004
8	Gujarat	103	1	58.2	32.5	1.9	0	7.3	Policy 2015
9	Uttar Pradesh	79	0.5	56.6	22.8	15.7	0	5	-
10	Sikkim	155	100	47.6	1.4	1.9	8	41.1	Policy 2004 Mission 2015
11	Chhattisgarh	71	1.5	10.3	19.4	33.6	0	36.6	Mission 2013

12	Meghalaya	56	19.5	2.9	84	1.6	11.5	–	Mission 2018
13	Kerala	54	2.7	35.5	35.4	22.9	0	6.2	Policy 2010
14	Assam	43	1.5	35.7	30.4	10.3	16.3	7.3	–
15	Jharkhand	31	2.2	9.7	69.6	16.3	0	4.4	Mission 2017
16	Tamil Nadu	30	0.6	14.4	60.5	20.8	0	4.3	Draft policy 2013
17	Telangana	28	0.6	22.9	8.8	50	0	18.2	–
18	Jammu Kashmir	26	3.4	68.3	29	2.2	0	0.5	–
19	Goa	23	18.1	45.7	11.2	43.1	0	–	Promotion scheme 2019
20	Nagaland	23	6	12	24.1	2.1	56.9	4.9	Policy 2019
21	Arunachal Pradesh	22	9.8	2.8	38.9	1.7	38.4	18.1	Policy 2014 Mission 2017
22	Manipur	19	5	1.3	27.3	3.1	65.4	2.9	Mission 2016
23	Himachal Pradesh	18	3.3	46.4	24.7	22.8	0	6.1	Organic policy 2018
24	Punjab	17	0.4	1.9	50.5	29.4	0	18.3	–
25	Mizoram	14	10	0	48.8	4.7	46.1	0.4	Act 2004 Mission 2006
26	Bihar	12	0.2	0	28.8	69.9	0	1.2	–
28	Dadar and Nagar Haveli	10	45.5	0	0	100	0	0.1	–

(Continued)

Table 10.1 (Continued)

		Organic area							
				Scheme-wise break up of total organic area of state/Union Territory					
				NPOP					
No	*State/union territory*	*Total organic area (000 ha) in 2019**	*Organic area in 2019 as % of net sown area of that state/union territory** (%)*	*NPOP %*	*In Conversion %*	*PKVY (%)*	*MoOVCDNER (%)*	*State schemes/non schemes (%)*	*Organic farming policy/ Mission/Act*
29	Andaman Nicobar	9	60	0	84.6	15.4	0	–	–
30	West Bengal	9	0.2	56.4	14.8	27.2	0	1.6	–
31	Tripura	9	3.4	2.4	27.2	11.7	58.8	–	–
32	Haryana	7	0.2	33	53.1	5.8	0	8.1	–
33	Lakshadweep	3	–	24.9	0	75.1	0	–	–
34	Chandigarh	3	–	0	0	39.4	0	60.6	–
35	Daman and Diu	1	32	0	0	98.1	0	–	–
36	Puducherry	1	–	0.5	0	99.5	0	–	–
	Total(India)	2,777	–	39.5	30.3	21.5	2.6	6.1	–

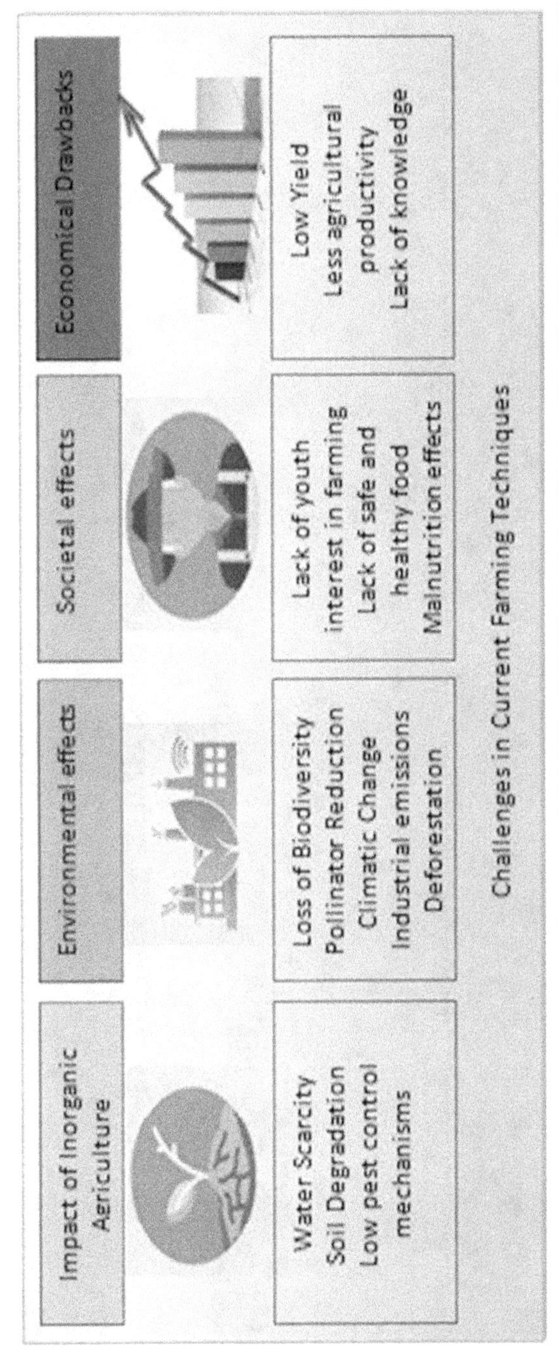

Figure 10.4 Current farming challenges.

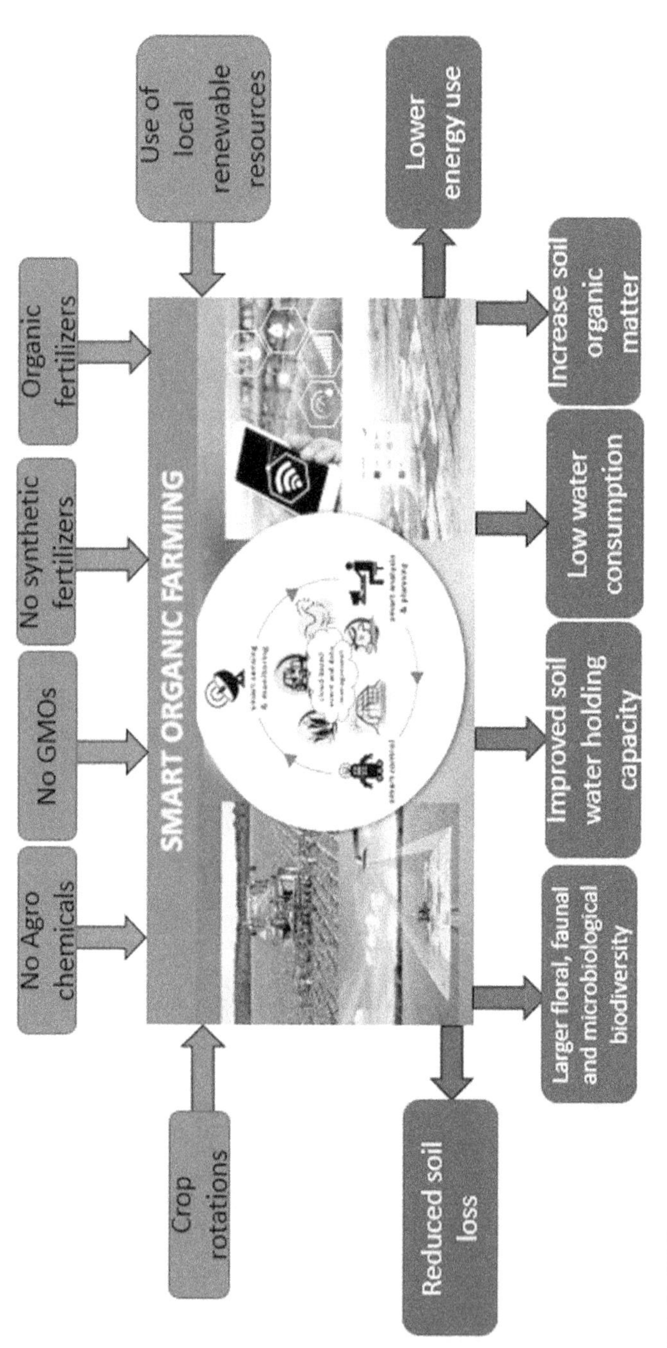

Figure 10.5 Smart organic farming.

Source: Agrotechnomarket.com.

10.1.2.1 Organic Farming Management Principles

Organic farming management concentrates heavily on low resource usage to enhance productivity. The key management principles of organic farming are [3]

1. Organic farm development
2. Organic soil conversion from inorganic farm
3. Multi-cropping
4. Crop rotation
5. Organic treatment of seed and plants
6. Enrichment of soil with manures
7. Use of organic fertilizers and microbial cultures
8. Temperature management
9. Protection of ecosystem
10. Pest control

Agricultural farms are certified as organic farms when the farming system fulfils certain minimum requirements to meet its objective as follows:

Conversion: conversion time is the time period between the initial stage of organic management to the end of organic certification. However, the conversion time is 3 years for perennial crops and 2 years for annual crops. This conversion time depends on past use of that land and its ecological conditions. The certification agency has control over the relaxation of the conversion period through verification and its objectives are to maintain biodiversity.

Mixed farming: integrated organic farming should be followed to maintain biodiversity like using poultry, fisheries, and animal husbandry with organic farming. Under the Network Project on Organic Farming, Coimbatore Agricultural University developed an integrated organic farming system model to support organic farming.

Cropping pattern: mono-cropping is avoided in organic farming to enrich soil nutrients. Crop rotation plays a major role in annual cropping and intercropping is needed for perennial crops. Green manure and fodder crops are covered by crop rotation in annual crops. Also, Kolinji (Tephrosia purpurea) crops are grown to enhance the soil for perennial crops.

Planting: cultivating species and plant varieties must adapt to that soil and climatic condition and should have high resistance to diseases and pests. Plants and seeds for organic farming must be produced from organic sources. Chemically treated seeds and genetically modified seeds/planting materials are not allowed in organic farming.

Manure Policy: Green manure crops and leguminous crops are used for enhancing soil fertility in organic farming. Biodegradable materials like animal waste and plant residue after harvest can be used as organic manure. Chemical and synthetic fertilizers are not allowed to be used in organic form as it affects soil fertility.

Pest, disease, and weed management: planting trees around the farms attracts birds, which eat insects and pests on the organic crops. To control pests and disease, pesticides are prepared from the products of animals, plants, local farms, and microorganisms (panchakavya, neem extract and cow urine spray, etc). Weeds under the plants are put as mulch around the plants to enrich soil nutrients. The organic products responsible for pest and disease control are pheromone traps, mechanical trap, neem extract, organic soap, agni asthram and EM Karaisal.

10.1.2.2 Challenges of Smart Organic Farming in India

Organic farming and organic products are more popular worldwide due to the increased need of health consciousness among people. Organic food is considered as an important health factor for upcoming generations. Some of the challenges faced by organic farming of Indian agricultural industry are as follows:

Indian farmers lack awareness of organic farming
Less marketing policies in organic farming
Less infrastructure is available in farming
High cost of the latest technologies
Inadequate agricultural practices
Less crop yield
No standard quality manure
Less government policies to promote organic farming
Unable to satisfy the market need in the global market
High demand for biomass

10.2 IoT in Smart Organic Farming

In traditional organic farming techniques, throughout the crop life farmers frequently visit the agricultural farms to get an idea about plant conditions. To overcome this, smart organic farming was introduced to reduce fieldwork by spending 70% time in monitoring and learn the crop conditions. Smart wireless devices are used to monitor crops continuously with high accuracy and detect any unwanted state occuring in the crops. Accurate reports from smart sensors are used for analysis and decision-making, which makes the environment smart, and cost-effective, and the agriculture is site-specific [4].

Recently, through prediction, the Internet of Things (IoT) is going to play an important role in the agricultural industry by connecting all devices to the internet, which will subsequently be responsible for data acquisition, cloud data analysis, decision-making, and automation of agricultural systems. These functionalities of the Internet of Things create a revolution in the agricultural sector, which is a developing industry of today's value chain.

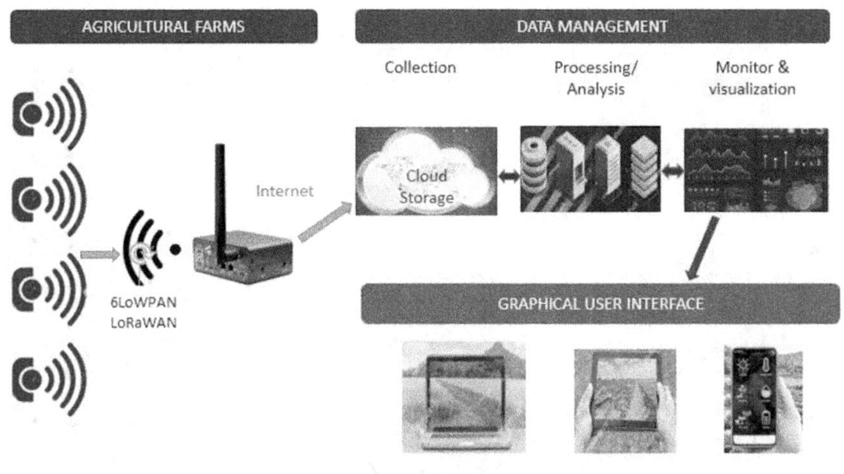

Figure 10.6 Internet of Things in smart organic farming.

Agricultural engineers and researchers are developing various methods and architectures for smart equipment to monitor and collect data about crop status at different stages. Leading Agri equipment manufacturers concentrate on developing smart sensors, communication devices, robots, and unmanned aerial vehicles (UAVs) to provide sensing data to farmers. And, also food and agriculture organizations, government bodies, and various commissions are creating policies and guidelines to regulate these agricultural technologies to produce quality food and a safe environment [5].

The general IoT-based smart organic farming architecture is shown in Figure 10.6. Here, data from agricultural farms are collected from smart sensor devices and then passed to the IoT gateway. The cloud computing environment collects the information from the gateway network, and then the end-user visualizes their results with a GUI application on a smart device with a data analytics system.

10.2.1 Applications of IoT in Organic Farming

The Internet of Things with the integration of smart sensors take the agricultural industry to the next era by providing solutions for organic farming issues like pest control, irrigation, crop yield optimization, land suitability, and drought response. However, these advanced technologies help the farming industry to improve its efficiency to an unimaginable level. The most important applications of IoT in smart farming (Figure 10.7) are as follows [6].

Monitoring Agricultural Fields: Kamilaris & Prenafeta-Boldú (2018) surveyed various machine learning algorithms used for agricultural farming, especially in monitoring

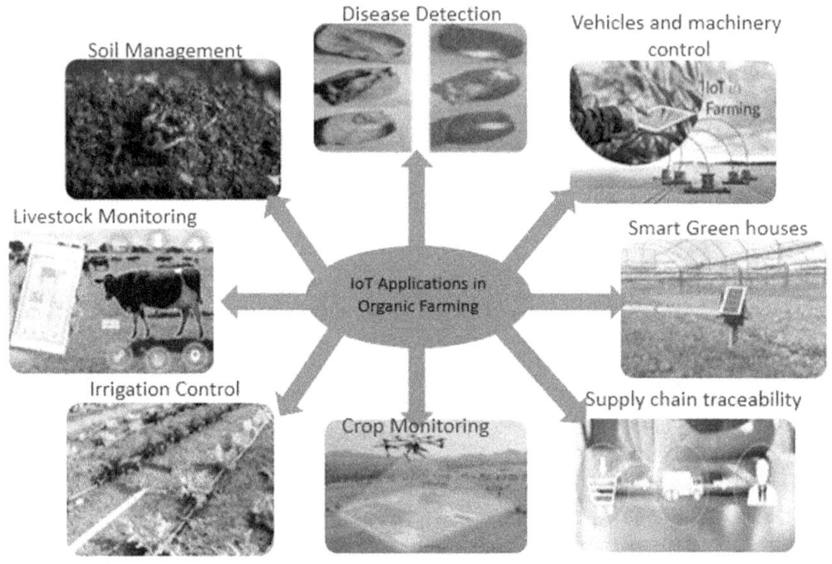

Figure 10.7 Application of IoT in smart organic farming.

the conditions of planted crops with smart sensors to enhance the crop yield and reduce the cost of crop production [7].

Lei Xiao (2010) developed a real-time farm monitoring system using smart sensors (humidity, light intensity, and temperature) devices. The monitoring system collects the agricultural condition with sensors efficiently to control the cost of agricultural production [8].

Jinhu Liao (2015) have developed agricultural monitoring system with ARM microcontroller using Zig Bee modules to collect data from farm environment and transfer it through GPRS module to cloud environment to take an accurate decision about agricultural crops, but also it enhances the profit of crop with this smart IoT technologies [9].

Smart Irrigation Control: G. Nisha(2014) developed an automated smart irrigation system for agricultural industry to optimize the use of water in the agricultural field. The automation system uses a soil moisture sensor and temperature sensor with a water quantity programming algorithm to pass water information to a microcontroller to take action about the need for water for a specific crop [10].

Srishti Rawal (2017) have developed a reliable and low power consumption irrigation monitoring system with a smart sensor placed in ad hoc mode in the farming environment. The smart sensors maintain the water level according to water reservation parameters and control the flow of excess water into a farm through moto gate valve control mechanism. IoT gateway and sensor deployment need minimal installation training for the farmers at the initial stage to use these agricultural applications [11].

Early Disease Prevention: the key challenge and requirement of farmers and agronomists is to identifydisease in the plants at an early stage to reduce heavy loss at the end of harvesting. IoT solutions with image processing help to identify crop disease at the beginning stage and specify the pest control mechanisms needed for farmers with smart sensors deployed on the crop area. Generally, the disease monitoring system uses the following aspects: a collection of sensor data, cloud data evaluation, and treatment of disease.

Advanced disease detection mechanisms use UAVs and remote sensing satellites to monitor fields, and hence, they provide a solution for a low and high efficient monitoring mechanism for the agricultural industry [12].

Soil Nutrient Enrichment: to enrich soil nutrients, an organic fertilizer is essential to enhance plant growth and fertility. The major macronutrients for the plants are NPK (nitrogen, phosphorus, potassium). Nitrogen is used to enrich the growth of plant leaves, phosphorous is used for the development of fruits, flowers, and stronger roots, potassium is used for water flow and stem development [13].

Excess use of a fertilizer or its deficiency creates a serious issue for the growth of plants and results in heavy financial loss and also affects the soil and environment. Fertilizer used in an unbalanced mode creates an imbalance in climatic conditions and soil nutrients as proved above 80 percent of deforestation arises due to poor agricultural practices [14].

Food Supply Chain and Transportation: the important issue in India is to serve quality food to chronically hungry people and to control malnutrition that arises at an early stage. Yes, we produce enough food, but the food and its transportation facility should be high quality and efficient enough to serve the people, which is an important hurdle faced by the food industry [15].

Integrating IoT with all functional areas of the food industry facilitates food safety and ensures food traceability through technology installed at the storage site and in transportation trucks. Also, configuring an online dashboard with IoT technology helps to send alerts and trigger actions at the occasion of abnormal temperature. Some of the key IoT technologies used for the food industry are listed here.

CCP Smart Tag (RC4) is a monitoring solution with IoT technology installed at the storage site and in transportation trucks to monitor the recommended environment of the food service and retail industry. CCP is responsible for automating food temperature for various food items suggested by food safety regulations. Later, the collected temperature data are communicated via the cloud platform to the end-user through GUI applications [16].

Auto Control of Agriculture Vehicles and Machines: due to the rapid development of the agricultural sector, the need for rural labor resources is high to enhance crop production, to overcome this issue automatic heavy equipment is required to fulfil the resource demand. Agricultural equipment manufacturers like Case IH, Hello tractors, and John Deere use IoT technology to produce automatically driven tractors with cloud computing functionalities. These self-driving tractors reduce the plowing overlap and avoid plowing in the same area. Also, they make an accurate turn in the field and perform weeding precisely. The target was to sell 40,000 fully automated and unmanned tractors to fulfill the requirements of the agricultural industry [17].

An important stage in agricultural production is harvesting at an accurate time, which decides profit and the success of crop production. Agriculture robots offer solutions in precise harvesting at the proper time to reduce the labor pressure. Agrobot is used to collect strawberries in the agricultural field at the correct harvesting time to reduce the loss of fruits, later packed through rural labor resources [18].

10.3 Smart Organic Farming Education and Curriculum Development

Smart organic farming education and curriculum development should be developed to enrich the organic farming concepts to the next generation. Also, the climatic condition and food production varies in different region results in different crop yield and return on investment for the same food crop [19].

10.3.1 Developing an Organic Farming Syllabus

To provide training on recent technologies to farmers, farm managers, and policymakers in smart organic farming, the syllabus should be framed and training should be given to all agricultural communities. Funds from NGOs and Government can be used to train farmers and certify the organic agriculture workforce. The next-generation students should adopt smart organic farming as their careers to cultivate organic foods. The syllabus should be multidisciplinary to integrate branches, such as agronomy, sensor technology, sensing approach, and biological sciences with a goal of enhancing crop yields with low-cost farm inputs, good farm management and informed decisions, and sustainable organic cultivation techniques. The objective of this curriculum is to solve the upcoming challenges in future smart organic agriculture [20].

Farmers adopt the new technology when it is easy to learn and use, profitable, trialled or testable, compatible with their farm area, and visible to all organic farmers.

Opportunities should be given to organic farmers to learn new technology through training, conferences, and workshops, and encourage farmers to trial the new digital agricultural technology.

10.4 AI in Smart Farming

Artificial intelligence in smart farming created a revolution in agricultural industry. Machine learning (ML) methods have the capability to process an enormous amount of data from sensing environment of various agricultural operational applications. The two major categories of machine learning techniques are supervised and unsupervised learning. The supervised learning methods maps input to output data in order to predict the missing outputs from test data samples. But, unsupervised learning identifies the hidden patterns with unlabeled data sets [21].

Machine learning supports various applications of smart farming like disease detection, yield prediction, weed detection, species recognition and livestock management. Finally, artificial intelligence helps farmers to achieve higher crop yield through precise cultivation with limited resource utilization. Table 10.2 provides the use of various artificila intelligence algorithms in smart farming applications.

10.5 Smart Organic Farming Privacy and Security

This section discusses the possible cyber-attacks in smart organic farming and presents the possible issues when more smart devices generate sensitive data with the Internet of Things. The common threats in smart devices are data leakage, unsecured Wi-Fi connection, spyware, weakly encrypted apps, and phishing attacks.

According to the Kaspersky Security Network (KSN) report, India takes the 11th position in the world with 2,299,682 cyber-attacks in the first quarter of 2020 when compared to 854,782 attack detected in the fourth quarter of 2019.

Major security and privacy challenges in smart organic farming arise with an adaptation of cloud, big data, and sensor technologies, which has a greater possibility for cyber-attacks. Smart organic farming faces technological issues as wireless smart devices are heterogeneous and ubiquitous in nature. The human challenges arise in the need for security by confidentiality, authentication, integrity, and end-to-end security in a secure organic smart farming environment [34].

Any smart farming system with the Internet of Things should satisfy the integrity, availability, and confidentiality goals. However, smart farming with IoT faces computational problems with its smart devices. There is a need for lightweight security solutions, key management, and security policies for smart organic farming systems and we need antivirus to recover infected devices, anti-rootkit, firewall, and

Table 10.2 AI in Smart Farming

AI Application in Smart Farming	*Crop*	*Discussion*	*Feature Extraction*	*AI Algorithm/Models*	*Accuracy*	*References*
Crop yield prediction	Coffee	To count the total number of coffee fruits on the single side of its branch using a non-destructive method	Color features	Support vector machine (SVM)	Semi-ripe: 68.25–85.36% Ripe/ overripe: 82.54–87.83%	[22]
	Tomatoes	Detection of tomatoes in the image from unmanned aerial vehicles (UAVs)	RGB (high spatial resolution)	Expectation maximization (EM) self-organizing map (SOM)	Precision: 0.9191	[23]
	Wheat	Yield potential prediction	Soil parameters and remote sensing vegetation indices	Counter-propagation artificial neural networks Supervised Kohonen Networks	CP-ANN 78.3% SKN was 81.65%	[24]
	Rice	Yield prediction and rice development stage prediction	Surface weather, and soil physico-chemical data	Support vector machine	Results based on tillering, heading, and milk stage	[25]
Disease detection	Wheat	Water stressed wheat canopies detection	Spectral reflectance and fluorescence features	Least-squares support vectors machine (LSSVM) with sensor fusion	98.7%	[26]
	Rice	Bakanae disease	color and morphological traits	Support vector machine	87.9%	[27]

	Strawberry	Thrips detection	Hue, saturation, and intensity	Support vector machine	Mean percent error (MSE)<2.25%.	[28]
	Silybum marianum	Smut fungus Microbotyum silybum	Leaf spectra images	XY-fusion network	Y-F (95.16%).	[29]
Weed detection	Grassland	Rumex and Urtica detection	Images of weeds	Support vector machine	Rumex 97.9% Urtica 94.65%	[30]
	Silybum marianum	Weed detection	High-resolution images from UAV	self-organizing maps Supervised Kohonen Network (SKN)	98.87%	[31]
Species recognition	White beans, red beans, and soybean	Recognizing the morphological similarity OF white beans, red beans, and soybean,	Leaf vein patterns	Deep convolutional neural network (CNN)	Soybean: 98.8% White bean: 90.2% Red bean: 98.3%	[32]
Livestock management	Cattle	Accurate behavioral monitoring through sensors	Grazing, ruminating,resting, walking	Ensemble classification with tree learner	96%	[33]
	Pig	Tracks pig movement with depth video cameras	3D motion data	Gaussian mixture models (GMMs)	89%	[33]

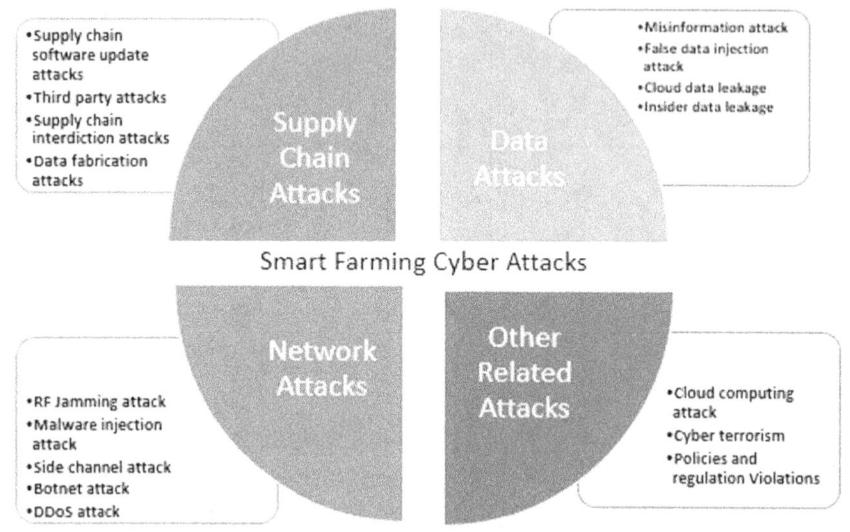

Figure 10.8 Smart organic farming privacy and security.

Source: [3].

total control of all smart devices. Figure 10.8 represents the various attacks on data from different agricultural application.

The information from the smart sensor of the Internet of Things and other smart agricultural machines are traveled over the internet faces data privacy and ownership issues, in the situation of data attacks. The various categories of data attacks are data theft from smart devices, which do not follow confidentiality standards, data attacks from stakeholders, unethical sale of farmer data to reduce their profit in agribusiness, and use of drones and cameras for unattended access of sensitive data to use against farmers [35].

The data integrity and data availability on demand are of utmost important in smart farming as it is needed for analysis and decision-making with agricultural data. Intentional or unintentional data falsification causes severe effects in smart farming automation systems. False data causes a dangerous threat to human life and agri-products, results in over dosage of pesticides to farm plants [36].

Supply chain attacks

A food supply chain is an integrated approach with IoT and smart agricultural devices where agricultural products face consequences due to data delays, environmental factors, climatic conditions from agricultural farms, and data center to farmers. Nevertheless, each stage of the supply chain faces security vulnerabilities, and it is a challenge to make the supply chain network highly secure [37].

Phishing attacks of the supplier data faces major security vulnerabilities at supplier chain applications. To provide security to supply chain mechanisms it needs security awareness among farmers and proper authentication techniques.

Network attack

The smart agricultural ecosystem contains various smart IoT smart devices prone to network attacks by malicious software called the Botnet of Things. Infected IoT devices in smart farms can infect other devices in the network withcybercriminals. However, smart IoT devices are not built with a highly secure mechanism and there is a need for highly efficient cybersecurity defence methods.

Smart devices in agricultural farms use the 802.11 protocol for network formation in IoT networks. Attackers utilize BSSID and SSID to make an auto connection with an access point and persuade a wireless client to attach with the reliable access point. Wireless access protocols are vulnerable to an evil twin access point attack in creating a rogue access point in IoT networks.

Malware injection attacks in a cloud platform redirects the farmer's request to the hacker's site to make it a malicious code resulting in stealing or eavesdropping of farmer's data. Here, attackers install infected service implementation module in SaaS, PaaS, or in IaaS Module. The Amazon Web Services cloud computing platform was attacked by a German researcher in 2011 to crash its servers.

Other related cyber-attacks

Agricultural technology providers introduce complex software licence agreements to take control of farmers' data and share it with smart agricultural farming providers. The farming equipment like drones, sensors, tractors, and agri-machines contain software licences, which are not mostly discussed with farmers at the time of sale. The General Data Protection Regulation (GDPR, 2018) in Europe in 2018 helps to regulate farmers to access and use data created at their agricultural farms.

The sensitive data from the IoT agricultural farms are moved to cloud environments, which are susceptible to a data breach or damaged by attackers. It is important for the cloud service provider to protect the farmers' data with secure encryption methods to maintain data integrity.

With the anonymity of the internet, hackers can exploit the access point of farmers to breach heterogeneous data from networks of smart devices in an agricultural environment. The crime number is growing rapidly in agricultural application with less enforcement law for data security [37].

10.6 Future Aspects of Smart Organic Farming

Agriculture is predominant in India with 67% of its population depending on farming and other related activities. Organic farming is considered as an indigenous practice in India that is carried out by rural and farming communities. The awareness of health and environment makes the Indian organic food industry grow at a rapid rate and provides opportunities to various companies associated with the organic food industry. It is noteworthy that globally the need for Indian organic products has started to increase significantly due to the use of toxic fertilizer for agricultural products in conventional farming. The safety and quality of foods are the key objectives of organic farming for the long-term sustainability of the system and also it ensures a profitable livelihood option.

Introduction of the latest technologies for smart organic farming, such as sensor technologies, the Internet of Things, UAVs, and AI optimize the use of labor and increases the quality and quantity of organic food production. The technologies used for future smart organic farming are given in Figure 10.9.

Smart organic farming is expected to play a key role in building relationships between small- and large-scale farmers globally. Smart organic farming predicts a future where all our agricultural food products are grown using organic fertilizers and biopesticides with the adoption of the latest technologies. More importantly, the support of the Internet of Things in organic farming enhances food safety and encourages the environmental benefits of smart organic farming, leading to sustainable development.

10.7 Conclusion

India is a land of agriculture, which fulfils the demands of rapid growing population with the introduction of genetically modified high-yielding seeds and fertilizers. In recent years, the growth of productivity in the agricultural industry remains stagnant, resulting in a decline of net income to Indian farmers. Needless to say, the Indian agricultural industry is under a state of distress due to negative environmental effects like the emission of greenhouse gases and contamination of surface and ground water. Organic farming is an innate farming method of India practiced by major rural and farming communities with the aim of being free from chemicals and pesticides, such as fungicides and insecticides that are harmful for human health. The global market for organic products is under rapid development in both developed and developing countries. It is evident that smart organic farming is benign to obtain enough sustainable healthy food and eco-friendly environment. Therefore, it is important for the organic farming industry to introduce innovation with new technologies like the Internet of Things, cloud computing, and artificial intelligence to attain a higher level of healthier farm production and a high net income to organic farmers.

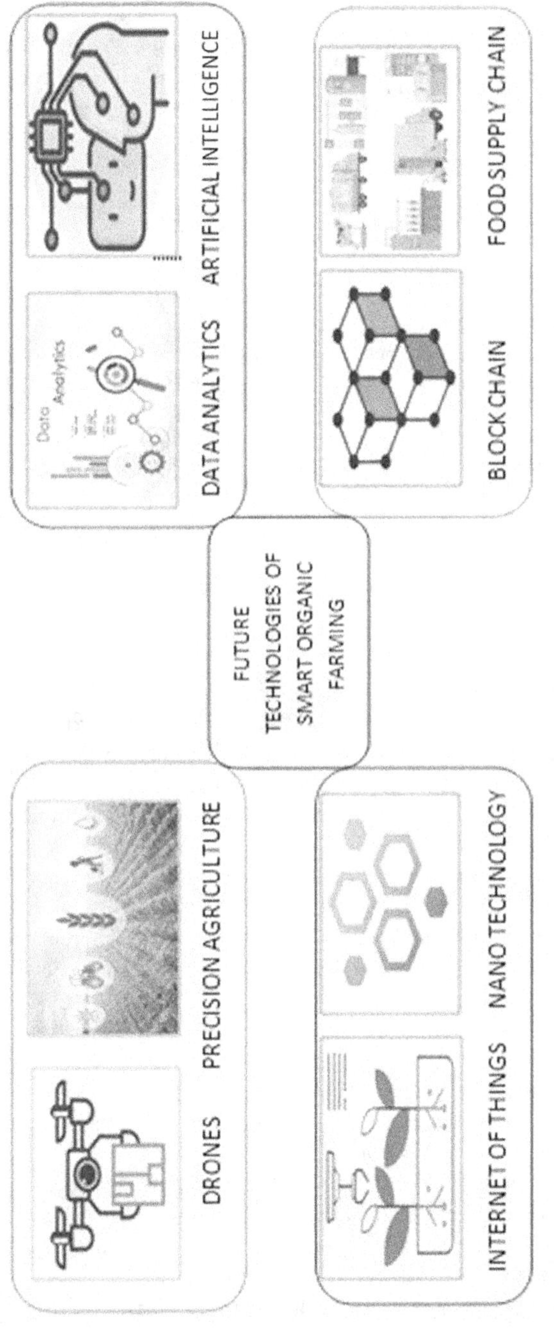

Figure 10.9 Future technologies of smart organic farming.

References

1. N. Akseer, S. Al-Gashm, S. Mehta, A. Mokdad, and Z. A. Bhutta, "Global and regional trends in the nutritional status of young people: A critical and neglected age group: Global and regional trends in the nutritional status of young people," *Ann. N. Y. Acad. Sci.*, vol. 1393, no. 1, pp. 3–20, Apr. 2017. doi: 10.1111/nyas.13336
2. E. Said Mohamed, Aa. Belal, S. Kotb Abd-Elmabod, M. A. El-Shirbeny, A. Gad, and M. B. Zahran, "Smart farming for improving agricultural management," *Egypt. J. Remote Sens. Space Sci.*, vol. 24, no. 3, pp. 971–981, Dec. 2021. doi: 10.1016/j.ejrs.2021.08.007
3. M. Gupta, M. Abdelsalam, S. Khorsandroo, and S. Mittal, "Security and privacy in smart farming: Challenges and opportunities," *IEEE Access*, pp. 1–1, 2020. doi: 10.1109/ACCESS.2020.2975142
4. R. Rajesh, C. Annadurai, and K Nirmaladevi "Low power device coordination in internet of things environment using analytic hierarchy process model," *Concurr. Computa. Pract. Exp.*, vol. 33, no. 7, pp. 1–1, 2021. doi.org/10.1002/cpe.5022
5. V. Seufert, N. Ramankutty, and T. Mayerhofer, "What is this thing called organic? – How organic farming is codified in regulations," *Food Policy*, vol. 68, pp. 10–20, Apr. 2017. doi: 10.1016/j.foodpol.2016.12.009
6. V. B. Hans, and R. Rao, "Organic farming for sustainable development in India," 2018.
7. A. Kamilaris, and F. X. Prenafeta-Boldú, "Deep learning in agriculture: A survey," *Comput. Electron. Agric.*, vol. 147, pp. 70–90, Apr. 2018. doi: 10.1016/j.compag.2018.02.016
8. L. Xiao, and L. Guo, "The realization of precision agriculture monitoring system based on wireless sensor network," in *2010 International Conference on Computer and Communication Technologies in Agriculture Engineering*, Chengdu, China, Jun. 2010, pp. 89–92. doi: 10.1109/CCTAE.2010.5544354
9. J. Liao, Q. Zhang, Y. Fang, and X. Xu, "The remote monitoring system design of farmland based on ZigBee and GPRS," in *Proceedings of the 4th International Conference on Mechatronics, Materials, Chemistry and Computer Engineering 2015*, Xi'an, China, 2015. doi: 10.2991/icmmcce-15.2015.460
10. G. Nisha, and J. Megala. "Wireless sensor network based automated irrigation and crop field monitoring system," in 2014 Sixth International Conference on Advanced Computing (ICoAC), 2014, pp. 189–194.
11. S. Rawal, "IOT based smart irrigation system," *Int. J. Comput. Appl.*, vol. 159, no. 8, pp. 10–16.
12. I. Bhakta, S. Phadikar, and K. Majumder, "State-of-the-art technologies in precision agriculture: A systematic review," *J. Sci. Food Agric.*, vol. 99, no. 11, pp. 4878–4888, Aug. 2019. doi: 10.1002/jsfa.9693
13. T. Bernauer, and E. Meins, "Technological revolution meets policy and the market: Explaining cross-national differences in agricultural biotechnology regulation," *Eur. J. Polit. Res.*, vol. 42, no. 5, pp. 643–683, Aug. 2003. doi: 10.1111/1475-6765.00099
14. D. Glaroudis, A. Iossifides, and P. Chatzimisios, "Survey, comparison and research challenges of IoT application protocols for smart farming," *Comput. Netw.*, vol. 168, pp. 107037, Feb. 2020. doi: 10.1016/j.comnet.2019.107037

15. T. W. Crowther *et al.*, "Sensitivity of global soil carbon stocks to combined nutrient enrichment," *Ecol. Lett.*, vol. 22, no. 6, pp. 936–945, Jun. 2019. doi: 10.1111/ele.13258
16. K. Sakadevan, and M.-L. Nguyen, "Livestock production and its impact on nutrient pollution and greenhouse gas emissions," *Adv. Agron.*, vol. 141, Elsevier, 2017, pp. 147–184. doi: 10.1016/bs.agron.2016.10.002
17. A. Gharehgozli, E. Iakovou, Y. Chang, and R. Swaney, "Trends in global E-food supply chain and implications for transport: Literature review and research directions," *Res. Transp. Bus. Manag.*, vol. 25, pp. 2–14, Dec. 2017. doi: 10.1016/j.rtbm.2017.10.002
18. M. Ayaz, M. Ammad-Uddin, Z. Sharif, A. Mansour, and E.-H. M. Aggoune, "Internet-of-Things (IoT)-based smart agriculture: Toward making the fields talk," *IEEE Access*, vol. 7, pp. 129551–129583, 2019. doi: 10.1109/ACCESS.2019.2932609
19. M. Li, K. Imou, K. Wakabayashi, and S. Yokoyama, "Review of research on agricultural vehicle autonomous guidance," *Biol Eng.*, vol. 2, pp. 1–6, 2009.
20. R. Rathinam, P. Kasinathan, U. Govindarajan, V. K Ramachandaramurthy, U. Subramaniam, and S. Garido, "Cybernetics approaches in intelligent systems for crops disease detection with the aid of IoT," *Int. J. Intell. Syst.*, vol. 36, no. 11, pp. 6550–6580, 2021.
21. S. Sigamani, and R. Venkatesan, "Air quality index prediction with influence of meteorological parameters using machine learning model for IoT application," *Arab. J. Geosci.*, vol. 15, no. 4, pp. 340, Feb. 2022. doi: 10.1007/s12517-022-09578-2
22. J. Senthilnath, A. Dokania, M. Kandukuri, K.N. Ramesh, G. Anand, and S. N. Omkar, "Detection of tomatoes using spectral-spatial methods in remotely sensed RGB images captured by UAV," *Biosyst. Eng.*, vol. 146, pp. 16–32, 2016. ISSN 1537-5110, https://doi.org/10.1016/j.biosystemseng.2015.12.003
23. N. Fjodorova, M. Vracko, A. Jezierska, and M. Novič, "Counter propagation artificial neural network categorical models for prediction of carcinogenicity for non-congeneric chemicals," *SAR and QSAR in Environ. Res.*, vol. 21, pp. 57–75, 2010. doi: 10.1080/10629360903563250
24. K. K. Paidipati, C. Chesneau, B. M. Nayana, K. Rohith Kumar, K. Polisetty, and C. Kurangi. "Prediction of rice cultivation in India—support vector regression approach with various Kernels for non-linear patterns," *Agri Eng.*, vol. 3, no. 2, pp. 182–198, 2021. https://doi.org/10.3390/agriengineering3020012
25. D. Moshou, X. E. Pantazi, D. Kateris, and I. Gravalos, "Water stress detection based on optical multisensor fusion with a least squares support vector machine classifier," *Biosyst. Eng.*, vol. 117, pp. 15–22, 2014. ISSN 1537-5110, https://doi.org/10.1016/j.biosystemseng.2013.07.008
26. C. Chung, K. Huang, S. Chen, M. Lai, Y. Chen, and Y. Kuo, "Detecting Bakanae disease in rice seedlings by machine vision," *Comput. Electron. Agric.*, vol. 121, pp. 404–411, 2016. ISSN 0168-1699, https://doi.org/10.1016/j.compag.2016.01.008
27. M.A. Ebrahimi, M.H. Khoshtaghaza, S. Minaei, and B. Jamshidi, "Vision-based pest detection based on SVM classification method," *Comput. Electron. Agric.*, vol. 137, pp. 52–58, 2017. ISSN 0168-1699, https://doi.org/10.1016/j.compag.2017.03.016

28. X.E. Pantazi, A.A. Tamouridou, T.K. Alexandridis, A.L. Lagopodi, G. Kontouris, and D. Moshou, "Detection of Silybum marianum infection with Microbotryum silybum using NIR field spectroscopy," *Comput. Electron. Agric.*, vol. 137, pp. 130–137, 2017. ISSN 0168-1699, https://doi.org/10.1016/j.compag.2017.03.017
29. A. Binch *et al.* "Controlled comparison of machine vision algorithms for Rumex and Urtica detection in grassland" *Comput. Electron. Agric.*, 2017.
30. X.E. Pantazi, A.A. Tamouridou, T.K. Alexandridis, A.L. Lagopodi, J. Kashefi, and D. Moshou "Evaluation of hierarchical self-organising maps for weed mapping using UAS multispectral imagery," *Comput. Electron. Agric.*, vol. 139, pp. 224–230, 2017. doi: 10.1016/j.compag.2017.05.026
31. S. Zhu, L. Zhou, C. Zhang, Y. Bao, B. Wu, H. Chu, Y. Yu, Y. He, and L. Feng. "Identification of Soybean varieties using hyperspectral imaging coupled with convolutional neural network," *Sensors*, vol. 19, no. 19, pp. 4065, 2019. https://doi.org/10.3390/s19194065
32. D. S. Dutta, R. Rawnsley, G. Bishop-Hurley, J. Hills, G. Timms, and D. Henry. "Dynamic cattle behavioural classification using supervised ensemble classifiers," *Comput. Electron. Agric.*, vol. 111, pp. 18–28, 2015.
33. L. Zhang, H. Gray, X. Ye, L. Collins, and N. Allinson. "Automatic individual pig detection and tracking in pig farms," *Sensors*, vol. 19, no. 5, pp. 1188, 2019. https://doi.org/10.3390/s19051188
34. R. Mahmoud, T. Yousuf, F. Aloul, and I. Zualkernan, "Internet of Things (IoT) Security: Current status, challenges and prospective measures," in 2015 10th International Conference for Internet Technology and Secured Transactions (ICITST), 2015, pp. 336–341.
35. S. Hu, S. Huang, J. Huang, and J. Su, "Blockchain and edge computing technology enabling organic agricultural supply chain: A framework solution to trust crisis," *Comput. Ind. Eng.*, vol. 153, pp. 107079, Mar. 2021. doi: 10.1016/j.cie.2020.107079
36. K. Demestichas, N. Peppes, and T. Alexakis, "Survey on security threats in agricultural IoT and smart farming," *Sensors*, vol. 20, no. 22, pp. 6458, Nov. 2020. doi: 10.3390/s20226458
37. S. Sontowski *et al.*, "Cyber attacks on smart farming infrastructure," in *2020 IEEE 6th International Conference on Collaboration and Internet Computing (CIC)*, Atlanta, GA, USA, Dec. 2020, pp. 135–143. doi: 10.1109/CIC50333.2020.00025

Chapter 11

Predictive Analysis of Toxic Ions and Water Quality Based on Sensor Data Using LSTM and ARIMA Models

Vallidevi Krishnamurthy, P.R. Joe Dhanith, R. Sujithra Kanmani, Surendiran Balasubramanian, and N. Rekha

11.1 Introduction

Water resources and bodies are polluted by various man-made pollutants from factories and industries. One of the major water-pollution-causing industries apart from the mining and chemical industries is the textile industries, whose effluents prevent photosynthesis of deep water plants as well as infect consumers of the water source. Among the various contaminants, dyes are more harmful as they are visible even at low concentration and can be classified into acute and chronic [1]. Such pollutants like anionic and cationic dye effluents are monitored and removed using hollow tubes, nano-tubes, and ultra filtration hollow fiber membranes [2–5].

A sensor is an electronic device that detects, measures, or occasionally reacts to a physical quantity and transmits the obtained data to other electronic devices into an electrically measurable entity. They are utilized in numerous applications, including

DOI: 10.1201/9781003469612-11

broadcasting on radio and television and voice recognition. The usage of a sensor to find hazardous ions in wastewater will be discussed in this chapter. The methods that are currently utilized to treat wastewater only assist in identifying and regulating the level of toxicity in the wastewater; as a result, they do not offer a long-term or reliable solution for the purification of water. It is crucial to manage the quantity of toxicity and prevent it from rising in the water sources since toxic water from home and industrial outlets is extremely detrimental to living things as well as the environment.

Modern technologies could foresee the quantity of hazardous ions, which are likely to increase in the future alongside the installation of sensors that detects them. This helps the wastewater treatment to be done in a way that will reduce toxic ion levels in the future and prevent the presence of toxic ions from increasing.

The technologies applied in recent years include sensors such as nano adsorbents, biosensors, spectroscopy, electrodes, and chemicals, CNTs, to identify traces of harmful ions like lead, cadmium, mercury, and other heavy metals, which helps to apply the appropriate treatment method. Due to significant advancements in CNT and graphene nanomaterials as well as the morphology and hydrophilicity of membranes, novel water purification methods using membrane technology for desalination, nanofiltration, ultrafiltration, reverse osmosis, front osmosis, and separation have also been developed recently. Despite the fact that this approach is constrained by membrane fouling, research has improved the antifouling properties of membranes [2–4, 6–8] because deposited foulants reduce production flux and raise costs.

After a thorough study of the sensor data, recent technologies such as Bayesian networks, artificial neural networks, support vector machines, machine learning, and artificial intelligence could be utilized to predict the concentration of these ions in the future.

This chapter also discusses the usage of various ion-detection sensors and how data from these sensors could be collected over a few years in a region to anticipate water toxicity involving more precise treatment options.

11.2 Sensors Which Detect the Toxic Ions Present in the Wastewater

The advancement of technology has led to the development of sensors that can identify various harmful ions in wastewater. Smartphones can also be used as detectors, with the output signal being identified by a high-quality camera [9]. Heavy metal ions as well as synthetic cerium-based oxides and ions are also treated as contaminants. Lead, cadmium, chromium, mercury, iron, arsenic, organic dyes, heavy metals, cobalt, nickel, nitrogen, zinc, and copper are just a few of the harmful ions that are commonly found in wastewater. These ions have significant effect when

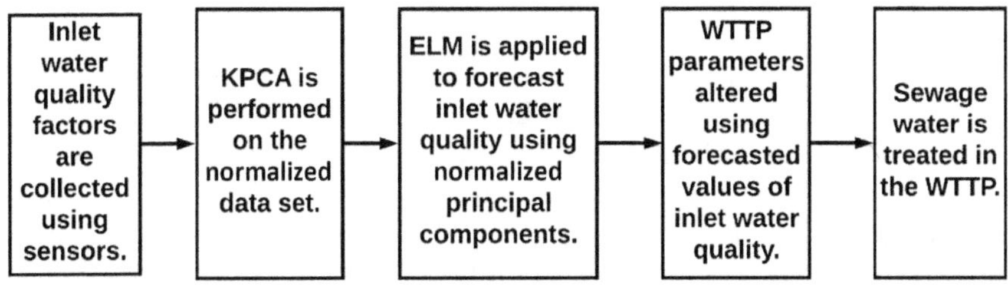

Figure 11.1 Flowchart to depict the process.

they are disposed into the water bodies. The negative impacts of these ions present in wastewater could be prevented by using various techniques of ultrafiltration membranes [5, 10]. As we have seen, dye removal is equally crucial, the use of innovative PPSU/SnO_2 (tin oxide) mixed matrix HFM (hollow fiber membranes) is the most efficient removal technique [5].

Toxic ions have negative consequences on living things, as demonstrated in Figure 11.1. The sensors used to predict the presence of the most prevalent hazardous ions in waste water from various streams are covered in this section. We restrict our discussion to the detection of lead, cadmium, chromium, mercury, and heavy metals in general, as well as the minute quantities of nitrogen, zinc, and copper. While the sensors can occasionally be employed to detect the presence of several ions.

11.2.1 Detection of Lead (Pb)

Lead is one of the normally found heavy metals because it is almost present in everything we are exposed to on a daily basis, including hair dye, contaminated air, paint, and fertilizers. This substance causes irreparable harm to the mental system and blood-related ailments when it comes into contact with the human body [11].

Lead is found as a major pollutant in contaminated water especially by factories and is absorbed by the body and accumulates in the bones. Hence, it is crucial to identify the presence of lead and predict the potential consequences of its proportions.

According to French standards, there can be no more than 10 $\mu g\ l^{-1}$ of lead in drinking water [12]. Fluorescent sensors, nanoparticles, field effect transistors, and other electrochemical detection techniques are a few of the sensors used to identify levels of lead in water. One such highly effective technique is the creation of a Pb^{2+} ion selective membrane electrode based on a composite cation exchanger made of poly(3,4-ethylenedioxythiophene) and poly(styrenesulfonate) (PEDOT:PSS) Zr(IV) monothiophosphate [13]. Let us examine these in more detail.

11.2.1.1 Ion Selective Field Effect Transistor

These sensors are built using high electron mobility transistors coated with lead ion selective membrane, which has improved lead ion sensitivity. The detection threshold is close to 10-10 M. As it is used directly in drinking water and tap sources, the sensor dynamically monitors changes in lead concentrations [14]. The sensor combines a high sensitivity electrical double layer gated field effect transistor with an ion selective membrane. High electron concentrations at the AlGaN and GaN junction form the transistor's active region, which has a high conductivity. Between the two metal plates of the gate electrodes and the transistor channel, the test solution serves as the liquid dielectric. At the transmission channel's gate electrode and interface, the mobile electron gets concentrated. When it comes to electrical double layers, the dielectric current of the transistor changes through modulation. This sensor operates by measuring the difference in current when the gate bias is applied versus when it is not. As lead ions touch the ion selective membrane coated on the transmission channel, the capacitance changes, leading to a change in the drain current, which assists in ion detection.

11.2.1.2 Fluorescent Aptamers

An aptamer which is a form of synthetic oligonucleotide, has the capacity to bind to specific target molecules [15]. The guanine (G)-quadruplexes having high affinity towards leads ions are formed due to the presence of lead ions in wastewater. Many fluorescence detection methods are now available in relation to the synthesis of these specific molecules [15].

11.2.1.3 Potentiometric Sensors

Using 1,10-dibenzyl-1,10-diaza-18crown-6 (DBzDA18C) as the membrane carrier, Mousavi *et al.* created a PVC membrane lead (II) ion selective electrode that demonstrated Nernstian response for lead ion over a broad concentration range [11].

11.2.1.4 Modified Electrodes

Lead can be detected using anodic stripping voltammetry electrodes. It is an extremely sensitive detection method based on electrochemical measurements, analogous to polarography. Two steps include applying voltage to scan the electrode and electrolytic chemical deposition of chemical species on an inert electrode whose surface is at constant potential [12].

11.2.1.5 Sensors Using Gold Nanoparticles

Gold nanoparticles which are highly inert exhibit sable oxidation states with large surface-to-volume ratio and unique optical and electrochemical properties.

So functionalized gold nanoparticles sense the cadmium ions throughout various mechanisms as cadmium induces aggregation, Cd assisted chelation [16].

11.2.2 Detection of Cadmium

Cadmium is an inevitable byproduct of industries, which is an extract of zinc, lead, and copper, and is used in electroplating and nuclear reactors and batteries. The contamination happens through the alloys of this in pesticides or the underground water. Cadmium leads to bone defects (osteoporosis) and renal malfunction. There are many sensors developed for the detection of the cadmium ions of which some are discussed below [11]. The human cardiovascular system and the gastrointestinal and reproductive systems drastically being affected may lead gradually to cancer [11]. Network-type carbon spheres of potassium hydroxide (KOH) with oxygen-rich groups can be adopted for the removal of cadmium (II) heavy metal [10].

11.2.2.1 Modified Electrode

A PVC membrane electrode coated with [1, 1-bicyclohexyl]-1, 1, 2, 2-tetrol as a membrane carrier detects the presence of the cadmium ion if a high range of concentrations is present [11].

11.2.2.2 Novel Modified Carbon Paste Electrode

Carbon paste electrode modified with lanthanum tungstate ion exchanger is used to detect the Cd (II) ions [17]. The Cd (II)-CPE has lanthanum tungstate ion exchanger as an ionophore. The distribution coefficient of the lanthanum tungstate cation exchanger determines the level of the ions present in the water, by taking into account that the highest amount of adsorbed ion was Cd based on the observations made [17].

11.2.2.3 Laser-Induced Breakdown Spectroscopy

Initially aluminum is used as a cathode with a platinum wire wound around is placed inside a glass beaker holding the aqueous solution, which deposits the ions C_0 in the solution after passing a voltage of 5 V. The water is then analyzed by LIBS method with assistance of electrodeposition. Using the already observed curves and the noted value of the C_0 the presence of cadmium is determined [18].

11.2.2.4 Nanoabsorbent Sensors

It is one of the most effective, efficient, and economic method. The main property of an adsorbent is the high porosity and large surface for absorption. Particles like clay minerals, zeolites, activated carbon, are used to remove the ions from water. Graphene has adsorption capacity for Cd^{2+} ions and hence can be used to remove cadmium [19].

11.2.3 Detection Chromium

Chromium is an essential nutrient for human beings and the shortage might cause diabetes along with heart disruptions and difference in metabolism. But increased presence of the ion, has adverse effects on health causing skin diseases. The main causes for the generation of Cr are steel, leather, and textile-manufacturing industries [11].

11.2.3.1 Ion Selective Electrodes

An electrode covered with a PVC-based membrane of 4-dimethylaminoazobenzene is found to have high Nernstian potentiometrics in the detection of the chromium ions in water. Another electrode is PVC-based membrane of 3,10-c-meso-3,5,7,7,10,12,14,14-octamethyl-1,4,8,11 tetra aza cyclo tetradecane diperchlorate with sodium tetraphenyl borate (STB) as an anion excluder and dibutyl phthalate (DBP), dibutyl butylphosphonate (DBBP), tris(2ethylhexyl) phosphate (TEP), and tributyl phosphate (TBP) also was found to be a Cr(III) selective electrode [11].

11.2.3.2 Gold (Au) Nanoparticle

Cr can be detected using the sensing mechanism of chelation of Cr ions, which leads to the aggregation of gold nanoparticles, which is done using functional agents such as citrate [16].

11.2.3.3 Biosensor

The biosensor is based on the measurement of the activity of urease, which the metal inherits. Chromium ion inherits *Dolichos uniflorus*. The inherited and the non-inherited activities are studied in phosphate buffer. And further chemical reactions conducted reveal the presence of the ion. The sensor used is the modified sol gel process. The sensor is calibrated to detect chromium after studying the reaction of the sol gel technique with urease [20].

11.2.3.4 Microbial Fuel Cell

A three-stage single-chambered microbial fuel cell biosensor inoculated with *Exiguobacterium aestuarii* was developed for real time and also continuous monitoring of the presence of chromium was performed. The voltage of the sensor after a period of time of running indicates the amount of chromium in the wastewater [21].

11.2.3.5 Cu-S Nanospheres

This method employs adsorption of the ions as they are found in very trace contents. The nanospheres are loaded with *Phyllostachys pubescent*, which is cheap and has high adsorption capabilities. After the adsorption, the concentration is determined by flame atomic adsorption spectrometer (FAAS) [22].

11.2.4 Detection of Mercury (Hg)

This element has been used for many purposes like thermometers and light bulbs. Other sources include natural sources like volcanic eruption.

Mercury leads to many health complications. Mercury remains accumulated in the living organisms for a very long time. Organs like kidneys, pancreas, and sweat glands are adversely affected.

11.2.4.1 Field Effect Transistor

An extended field transistor coated with mercury ion selective membrane is employed. This was found to have ultrahigh sensitivity. The extended gate is constructed by integrating a PVC-based ion-sensor with it, in which epoxy resins are poured and treated for an hour at 125°. The electrical characteristics of extended gate Hg-ISMFET are recorded. The current gain recorded is used to detect the mercury ion based on the voltage and the capacitance recorded [23].

11.2.4.2 Gold Nanoparticles

The aggregation of gold nanoparticles when exposed to the ion is used to detect the ion. The amalgamation of Hg-Au and the color change happens due to the aggregation of the mercury ions by the functionalizing agents like papain, polyvinylpyrrolidone [16].

11.2.4.3 Fluorescent Sensors

The fluorescent sensors are based on aptamers, which are artificial oligonucleotide sequences with the ability to bind to a specific target molecule. A particular aptamer is developed to detect the presence of the mercury ion. Hg detected by competitive binding of the T-rich aptamer that is initially been labeled to fluorophore gets linked to its complementary DNA. In the presence of mercury, the aptamer gets separated from the DNA, and the mercury ions bind to the aptamer, which leads to the formation of T-Hg^{2+} having spiral structure and being easily detectable [15].

11.2.4.4 EDTA-Carbon Paste Modified Electrode

EDTA plays an important role in the complexion of metal ions. Volta metric determination method is used by the carbon paste modified electrode. The electrode is modified using the ethylene diamine tetra acetic acid and square wave voltammetry technique is used for detection. The electrolysis is conducted by dipping it in the test solution. After the Hg ions bind with the EDTA on the electrode, it is removed and placed in a voltammetry cell having NaCl as thed electrolyte and the voltammeter measurements are compared to the already made measurements to find the presence of the mercury ion.

11.2.5 Heavy Metals

Although there is particular definition to the elements that fall under this category, they refer to the metals and metalloids of density higher than 5 g/cm^3 [24]. That increases the toxicity and cause harm to the lives of animals and plants directly or indirectly. Some of them include lead, cadmium, mercury, zinc, iron, manganese, cobalt, arsenic, iron, copper, which gradually find their way into the ecosystem and cause harm to the living species. Arsenic causes serious health hazards. Pressure-driven membrane filtration processes like ultrafiltration, reverse osmosis, nanofiltration are used for their removal [3]. Heavy metals are generally present in their cationic form.

Sensors that can be used to detect the toxic ions in wastewater include chemical sensors, electrochemical sensors, optical sensors, biological sensors, sensors involving the nanoadsorbents, sensors using the organic matter [26]. Table. 11.1 gives the list of some of the toxic ions and the sensors used to detect them, and the type of technology or concept that is used to remove them.

As detection of these toxic ions from the earth is necessary, it is also important to predict their concentration in the future so as to stop the worst from occurring. The prediction of this concentration level will help mankind to gain awareness and take the necessary precautionary steps to stop the increase in the concentrations of the toxic ions. The predictions are done by employing machine learning and data analytics, to the data collected from the sensors. These algorithms will help in getting a rough idea of where the purity of the water is headed. The following paragraphs contain the detailed explanation of the major machine learning algorithms that are used in the present day. With the help of these predictions it is possible to prevent the toxic ions in harming the environment.

11.3 Machine Learning to Predict/Detect Pollutants

Acute shortage of water in daily life is a prime concern all over the world today. Wisely conserving our water resources with judicious usage of the water supply

Table 11.1 Toxic Ions and Their Sensors

Toxic ion	*Sensor used*	*Method used*
Lead	Electrochemical sensors	Anodic stripping voltammetry [26]
	Gold nanoparticles	Adsorption of particular molecules [27]
	Lead ion coated high electron mobility transistor	Sensitivity of the ion in the high electron mobility channel [28]
	Ultra-sensitive sensors	Laser-induced breakdown spectroscopy [29]
	Graphene and carbon nanotubes	Removal using different types of absorbents [19]
Cadmium	Biosensor-single chamber microbial fuel cell	Based on the reaction towards the bacteria [30]
	Aptamers	Ability of binding of aptamers to target molecule [15]
	Carbon paste electrode	Study of the difference in the electrode potential [17]
Chromium	Phyllostachys pubescens powder	Absorption of toxic ions [22]
	Ion selective electrodes	Analytic information collected from the electrolysis [11]
	Biosensor	Measurement of the activity of urease, which is inhibited by heavy metal ions [20]
	Microbial fuel biosensor	Atomic absorption spectrometry [21]
Mercury	Electrochemical sensors	Anodic stripping voltammetry [12]
	EDTA-CPE modified electrode	Voltammetric determination of mercury ions (square wave voltammetry) [24]
	Ion-selective sensors	Changes in the transistor drain current [23]
Iron	Ion-selective electrodes	Analytic information collected from the electrolysis [11]
	Nanoparticles	Morphology of Fe_3O_4 and Fe3O4@ZnO was observed by TEM [31]
Arsenic	Electrochemical sensors	Anodic stripping voltammetry [32]
	Fluorescent optic sensor	Highly sensitive enzymatic catalysis system [33]

(Continued)

Table 11.1 (Continued)

Toxic ion	*Sensor used*	*Method used*
Organic dye	Malaysian plant, Scirpus grossus	Phytoremediation [34]
	Auto calibration biosensors	Calibrated to detect the particular dye [35]
	Laccase biosensor	Reagent-less continuous online analysis [36]
Nitrogen or ammonia	Biosensor-single chamber microbial fuel cell	Based on the reaction towards the bacteria [30]
	Ion chromatography	Gas chromatography (GC), thin layer chromatography (TLC) and liquid chromatography (LC) [37]
	Biosensors	Biosensing the activated sludge and getting one or more information from many activated sludge process which is in the form of substrate addition
Zinc	Chemically modified carbon paste sensors	Reaction with chemicals like Thiosemicarbazide and Acetaldehyde Thiosemicarbazone Complexes [38]
	Colorimetric sensors	immobilizing potassium ferri cyanide and diphenyl amine on Whatman filter paper [39]
Nickel	Ion-selective electrodes	Analytic information collected from the electrolysis[11]
Other heavy metals	Biosensors based on luminescent bacteria	Conversion of concentration of target solution to electrical solution
	Eukaryotic cell lines	Sensitive layer in cell based sensor technology
	Carbon nanotubes	Stripping voltammetry [40]
	A double-microbial fuel cell heavy metals Toxicity Sensor	Study of the aerobic/anaerobic activated sludge [41]
	Chemical-sensor and biosensor	Different methods for different sensors [25]
	Optic sensors	Different methods for different ions [24]
	Biosensor	Atomic absorption and emission spectroscopy, plasma mass spectroscopy

is one solution to address the problem, as well as treatment of the wastewater so that it can be recycled or reused is another. However, for this treatment, a meticulous analysis into the type of water pollutant and prediction of the quality of water samples using previously fed data is very much required. Machine learning algorithms and models are implemented to do the same. Water pollutants are of different types. Various ML algorithms are chosen and opted suitably as follows [42–44].

11.3.1 Sewage Treatment

11.3.1.1 Inlet Wastewater

High levels of pollutants in water (sewage) adversely affects human health and wildlife survival. Purifying this polluted water periodically is necessary. To treat sewage water, it is crucial to forecast the quality of the inlet water supply. Water quality cannot be tested efficiently in a short time. Misinterpreted data will affect the purification of the next set of polluted water. Thus, past observed data of inlet water quality can be used to adjust the parameters of the WTTP. For this, a water sample is collected from the inlet supply and taken through two processes:

(1) KPCA: principal components are extracted from the inlet water sample with kernel by reducing dimensionality and eliminating linear correlation between data thus increasing accuracy.

    Here covariance matrix C is obtained by nonlinearly mapping the elements of the data matrix in feature space. The score vector of *k*th element (observation) is obtained by projecting the element onto eigen vectors. Thus the general rule for selecting the main element is that the ratio between the sum of eigen values of principal components and the sum of eigen values of all components must exceed the threshold of principal components E.

(2) ELM: algorithms such as ANN forecast purity of future samples. The wastewater treatment plant's performance parameters (WTTP) in order to operate in an efficient and economically viable manner are tweaked according to the forecast of the inlet water quality components, which is obtained as stated in Figure 11.1, which depicts the flow through which the treatment of water proceeds before the treatment plant.

Comparison between the models

The KPCA-ELM model performed better than the PCA-ELM model, ELM model and BPNN algorithm in terms of accuracy, i.e., with lower error values, to predict the BOD and COD levels in the inlet water supply.

Compared to PCA, the KPCA predicts the nonlinear characteristics of data (as opposed to the spatial distribution of the original data) [44].

11.3.1.2 Integrated Food Waste and Wastewater Treatment

Biodegradable waste faced a major problem for disposal. With various acts that prohibits them from being dumped into wastelands and oceans, wastewater treatment plants treat the waste with methods that are likely to produce the optimal solution to resolve the problem.

A large amount of food waste leachate with high moisture content is available for all recycling processes to be used in the wastewater treatment systems. Thus to treat using conventional wastewater treatment plants (WTTP) is not advisable. This is because the high concentration of total-nitrogen (T-N) might impact the effluent water quality.

Two machine learning models are applied to predict T-N concentration of effluent from a wastewater treatment plant.

(1) artificial neural networks (ANNs) and
(2) support vector machines (SVMs).

The ANN structure, called a multilayer perception network, consists of three distinctive layers: input, hidden, and output with linked-nodes and functions. SVM generalizes the classification or regression problems. As a whole, MATLAB is used for building ANN and SVM models to predict effluent T-N in the WWTP (wastewater treatment plant).

Comparison between the models

The models' efficiencies were measured using three criterions: coefficient of determination (R2), Nash–Sutcliff efficiency (NSE), and relative efficiency criteria (REC) for the training and validation. Latin-hypercube one-factor-at-a-time (LH-OAT) is applied to analyze the sensitivity of the models to varying datasets. Further a pattern search algorithm was instrumented to select and use the most optimized parameters for the model.

Even though both models could be effectively applied to the 1-day interval prediction of T-N concentration of effluent, the SVM model showed a higher accuracy while predicting the training stage as well as the validation stage. Thus, the sensitivity analysis produced a result that the ANN model is a superior model for 1-day interval T-N concentration prediction in terms of the cause-and-effect relationship between T-N concentration and modeling input values to integrate food waste and wastewater treatment [44].

11.3.2 Activated Sludge

Activated sludge is nothing but aerated sludge containing anaerobic bacteria to help break it down. This is formed primarily due to increased human exploitation of resources, i.e., from excess transformation of soluble constituents and solids from secondary wastewater treatment plants. Activated sludge plays a beneficial role in adding nutrient content to the soil. However, activated sludge impacts the purity level of water bodies adversely and increases management costs. Thus methods to improve procedures to stabilize sludge are urgently needed.

Pretreatment methods based on ultrasonic, thermal and ozonization are followed by supplementary treatment processes for overall mineralization and stabilization of the sludge.

Electrochemical process opposite to the biological, aerobic, and anaerobic methods in order to stabilize sludge is a quicker process that does not depend on environmental factors.

The electrochemical technique is also proved to be more versatile, energy efficient, safe, and component selective, amenable to automation and cost effective. Both electro-oxidation and electrodegradation methods are used. Wastewater treatment plants also showed to increase efficiency in negating the influence of high T-N concentration and low chemical oxygen demand. The electro-oxidation method is used to decrease the organic compound content and remove the available microorganisms in the activated sludge of the sewage. The sensitivity analysis showed that the parameters of the electrolytic component such as electrode type and pH of the solution was used to increase the sensitivity desirably.

The processes to conduct the electrochemical method is depicted in Figure 11.2.

The first is indirect oxidation of activated sludge through free radicals, particularly OH• with a high oxidation power as non-selective oxidants. The continuous generation of OH• radicals takes place on the anodic surface followed by indirect oxidation. The next two mechanisms are sludge destruction on the electrode surface through direct oxidation and destruction by oxidants, which are electrically generated.

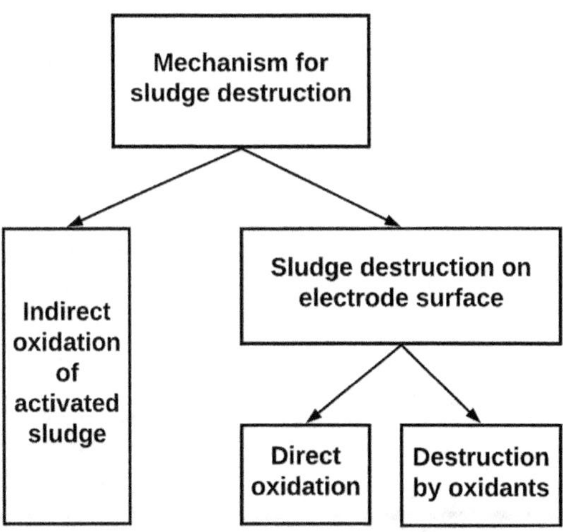

Figure 11.2 Depicting various mechanisms for sludge destruction.

For the multivariate system where not much information about the parameters is known, artificial intelligence tools and simulation processes are used as stated below.

(1) Artificial neural networks (ANN): useful tool for modeling of complex non-linear processes with no knowledge of the mechanism that obtained the outputs.
(2) Support vector machine (SVM): accurate method of classification and regression. It requires the user to choose very few parameters.

Comparison between the models as shown in Figure 11.3
A comparison between the two modeling methods (artificial neural networks and support vector regression) is drawn. Both models are optimized with genetic algorithm and are applied for benzene isopropylation on Hbeta catalytic process. The result that ANN obtains is slightly better due to the reduced time it takes and because it is suitable for online optimal control procedures [45].

11.3.3 Chemical Impurities

11.3.3.1 Phenol and its Compounds

Phenol and its derivatives are included in the major water pollutants' list. They are toxic even at low concentrations due to its toxicity, pungent odor, and carcinogenic properties. The adsorption process compared to other procedures, such as ion exchange, reverse osmosis, chemical oxidation, precipitation and distillation, is the

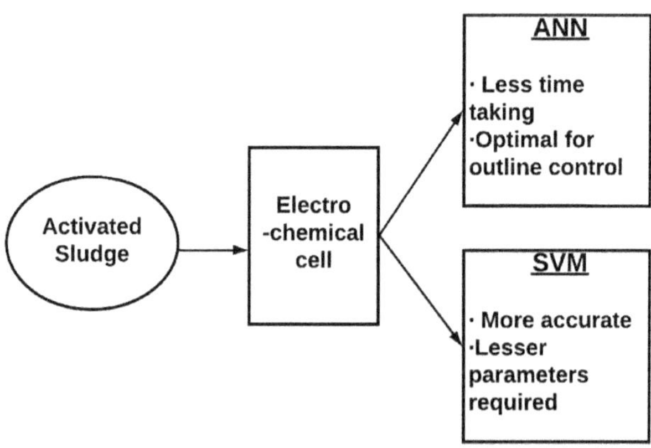

Figure 11.3 Depicting the flow of activated sludge for treatment.

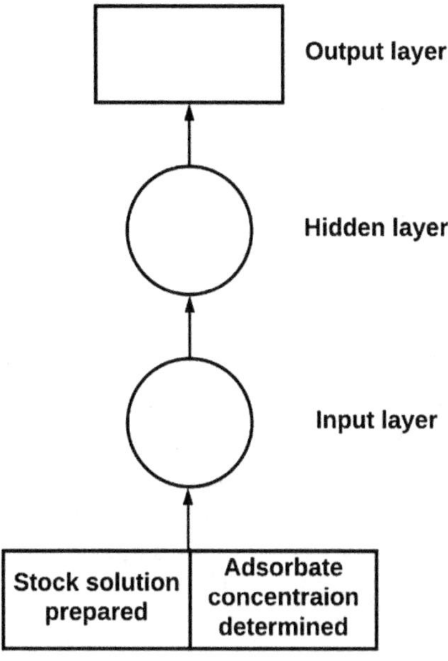

Figure 11.4 Depicting the flow through ANN.

most cost-effective method of treatment of phenol containing water and wastewater even at low concentration levels.

Artificial neural networks (ANN) is a tool used to model the adsorption process. Its three-layer model uses an optimized version of the back-propagation (BP) algorithm for the four-step and six-step pressure swing adsorption (PSA) cycle. The performance of the ANN model was checked using mean square error (MSE) and verified that it was a better model than the regular multiple-regression model. The input parameters used for the ANN model was contact time of test water, initial dye(aura mine O) concentration, agitation speed, temperature, initial solution pH and activated carbon mass. ANN's most vital feature is its ability to determine randomly, the complex relationship between data set elements without detailed knowledge of the mechanism itself. It could predict accurately the pollutant removal efficiency (in competitive adsorption) of the sample, which is the most important parameter when it comes to wastewater treatment using adsorption. That parameter depends on factors like adsorbent concentration, contact time and agitation speed. The ANN model has three layers as shown in Figure 11.4: input, hidden, and output layers.

The data is sent through the input layer, a number of (depending on complexity) hidden layers and finally the output layer. The training and validation of the model was done using MATLAB [45].

11.3.3.2 Soluble Fluoride

Ground water is an important source of portable water in many areas across the world. High fluoride concentrations pollute the water bodies and thus prediction of the extent of its contamination of groundwater for planning and management is necessary. Numerical models that used to be employed to practically view the problem led to a large difference between real and predicted values. Artificial intelligence bridged this gap. The solubility of fluoride as a pollutant was analyzed using ML algorithms and thus three are compared:

(1) SVM: estimates the relation between data, i.e., classification and regression.
(2) MLP: the input layer distributes information to the hidden layer from where it is passed on to the output layer. The last two layers process data by multiplying weights (factors) with the data.
(3) ELM: fast learning technique with high performance and uses single-hidden layer feed-forward networks (SLFNs) as opposed to ANN. It determines all parameters required analytically. This is a mostly error-free technique.

Comparison between the models

Low root mean square error values for all techniques proved the successful prediction yields of all three methods. However, ELM was the fastest learning method and SVM had the highest computation time [46].

11.3.3.3 Acid Mine Drainage (AMD)

Acid mine drainage/acid rock drainage is a global problem that poses a great hazard to human health as well as having environmental implications. Once AMD is formed, water transports the toxic substances into the environment and contaminates it. This study presents machine learning techniques to develop models to predict AMD quality using historical monitoring data of a mine site. The ML techniques include artificial neural networks (ANN), support vector machine with polynomial (SVM-poly) and radial base function (SVM-RBF) kernels, model tree (M5P) and K-nearest neighbors (K-NN) [47]. The ML algorithms work on the data as shown in Figure 11.5.

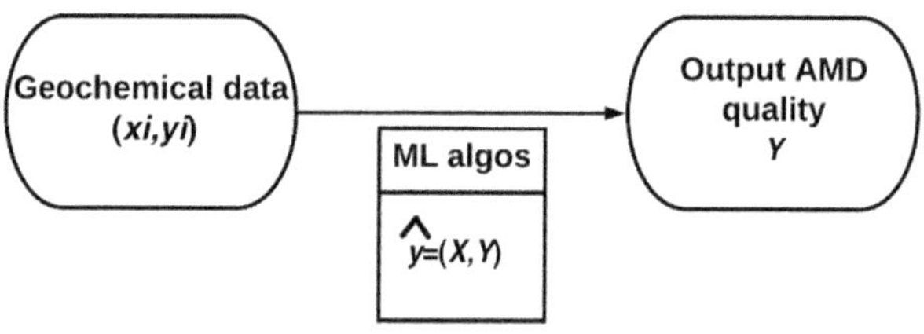

Figure 11.5 Depicting input and output of the ML algorithms implemented.

The ML techniques have been used to build models that predict future drainage quality levels based on existing data.

Comparison between the models

The accuracy with which each model predicts future quality of drainage is used to compare between the techniques. It compares between observed and predicted values for each machine learning technique. The prediction accuracy of each machine learning technique was evaluated using the root mean squared error (RMSE), mean absolute error (MAE), root relative squared error (RRSE), and relative absolute error (RAE). The lesser the error, the better considered the technique.

Copper concentrations are predicted by the methods and it is found that SVM-poly is the best technique (both in terms of predictive accuracy and uncertainty evaluation methods), followed by SVM-RBF, ANN, M5P, and K-NN techniques. K-NN was the poorest performing model. Values over and above the prediction line for the two methods, respectively, suggested that the K-NN technique overestimates and M5P underestimates the overall predictions. Thus the ML techniques proved useful for AMD chemistry and the results can be used to identify AMD management alternatives. Also, sustainable mine wastes management methods can be integrated.

11.3.4 Eutrophication

The enrichment of water in terms of nutrients because of various sources such as manual activities such as industrial activities, agriculture production, energy generation, fishing and tourism, is called eutrophication. It increases the total nitrogen (T-N) and phosphorous content leading to an exponential growth of algae which in turn leads to reduction in the quality of water. Depletion of dissolved oxygen and a resulting loss of flora and fauna in the water body is a major effect. It is essential to keep track of eutrophication levels in water bodies to make use of the water resource. Field sampling and analysis using features like BOD, COD, and DOC using multivariate regression analysis was time consuming.

The support vector machine is very useful in depicting the non-linearity between indicator and input with minimal errors. The grid-search (GS) algorithm fed to the machine and eutrophic conditions of a sample is predicted with maximum possible accuracy. Sample sets are collected at standard depths and excitation-emission spectroscopy provides a 3D contour plot with the data depicting the eutrophication levels. The parallel factor analysis (PARAFAC) is used to further simulate the model and this data set is split as a training set and validation set for the SVM. Finally the SVM generates the real-time marine eutophication status of waters using easily measured parameters [48].

11.3.5 Machine-Learning-Based Predictive Analysis of Toxic Ions

Water is an important resource required for various purposes in day-to-day life as well as for industries so has to be used carefully. Polluted water sources that cannot be used can be treated as per desire based on its pollution content and cause. The pollutants are detected categorically using sensors.

Further basic machine learning algorithms and techniques, such as ANN, SVM, and ELM primarily are used to predict the pollution content of a water sample based on the past results detected by sensors. This prediction helps to alter the parameters of the treatment plants used to recycle the water as per requirement. Based on the pollutant properties and the technique most optimal to remove them, various ML algorithms are chosen and applied. Various methods are also applied for their removal. One such method is the usage of polyetherimide membrane with bentonite clay for the removal of metal ions [54]. Pollutants have been categorized into non-chemical and chemical divisions to analyze the ML algorithms in the respective fields of pollutants more accurately.

11.4 Water Quality Prediction Using LSTM and ARIMA

Predicting water quality is a key and essential activity for the management and prevention of water environment pollution. The fluidity of water causes water quality trends across several stretches of the same river to be consistent. The current water quality prediction algorithms do not take into account how similar the water quality is between sections, thus they cannot make use of the correlation between the water quality of each part to deeply capture information. This work develops a machine-learning-based water quality prediction model in order to handle this problem, taking into account the PH and dissolved oxygen of the research object's aqueous environment to do the study this LSTM, SVM, neural network, and the autoregressive integrated moving average (ARIMA) model is implemented.

Water quality has a direct effect on the ecology and public health. Water is used for a variety of things, such as drinking, farming, and industry. The rise of water sports and entertainment has recently attracted a lot of tourists. Rivers have been used for the expansion of human cultures more often than other water sources due to their accessibility. Occasionally, other water sources—like groundwater and the ocean—can help with problems.

The earth sciences are currently conducting extensive research on river water quality. Two aspects are considered while determining the river's quality: defining the transmission of contaminants and assessing the components of water quality.

In the earth sciences, research on water quality is a hot topic. While analyzing the quality of rivers, two factors are taken into account: defining the transmission of pollutants and evaluating the elements of water quality.

It has been recommended to monitor dissolved oxygen (DO), pH, temperature, turbidity, among other water quality indicators. Governments have built hydrometric stations along rivers that connect urban and rural regions, agro-industrial projects, industrial estates, and rivers that cross dam reservoirs in order to do this.

The purpose of this research is to use machine learning techniques to forecast several aspects of water quality.

11.5 Literature Survey

Haghiabi *et al.* discuss the effectiveness of artificial intelligence techniques for forecasting water quality elements of the Tireh River in southwest Iran [49]. These methods include support vector machines, group approach of data management, and artificial neural networks (ANN) (SVM). Analyzing the ANN and SVM findings revealed that both models perform well in predicting the various elements of water quality.

Srivastava and Kumar (2013) discuss the water quality index, giving a single value that, depending on a number of quality parameters, indicates the total water quality at a specific location and time [50]. An index's goal is to simplify complicated data on water quality into information that the general public can use. In this, a formula will be presented for calculating the water quality. Nine water quality factors are used in the standard formula to determine the water quality index.

Zhu *et al.* (2019) discuss the work based on 353 datasets of adsorption tests from the literature, the adsorption of six heavy metals (lead, cadmium, nickel, arsenic, copper, and zinc) on 44 biochars was predicted using artificial neural networks (ANN) and random forests (RF) [51]. Biochars' pHH2O contributed 66% of the biochar properties. However, the biochars' surface area only offered 2% of the adsorption efficiency. The RF models, on the other hand, were more generalizable than ANN models.

Asadollah *et al.* (2021) explores extra tree regression (ETR), a new ensemble machine learning model for forecasting monthly WQI values at the Lam Tsuen River in Hong Kong, in this study [52]. The performance of the traditional standalone models, support vector regression (SVR) and decision tree regression, is contrasted with that of the ETR model (DTR).

Chen *et al.* (2020) suggest the effectiveness of machine learning models for predicting water quality may depend on both the models themselves and the parameters in the data set that were selected for training the learning models [53]. In order to further lower prediction costs and increase prediction efficiency, the learning models should additionally identify the essential elements.

In Tan *et al.* (2012) [54], river water is investigated. Quality measurement data, after LS-SVM model training for water quality parameters BP and RBF networks are used in the same sample as a monitoring system for prediction. The small sample situation with noise and a least-squares support vector machine, according to an experimental approach more effectively addresses the needs of water than multi-layer BP.

Sakaa *et al.* (2022) use a hybrid artificial intelligence model called sequential minimal optimization-support vector machine and random forest as a benchmark model for predicting water quality values [55]. Utilizing a variety of statistical indicators and graphical representations, the prediction performance of various input data combinations is assessed. Results demonstrate that during the dry season in the downstream northeastern half of the basin, less than 40% of samples were found to have low-quality water.

In Muharemi *et al.* (2019) [56], various methods for spotting changes or anomalies in time series data on water quality are described. Additionally, several difficulties encountered while working with time series data are discussed, and a solution is suggested. The F-score measure is used to evaluate performance.

In Rosero-Montalvo *et al.* (2020) [57], a wireless sensor network (WSN) system for assessing river water quality is presented. We use the Tahuando River in Ibarra, Ecuador, specifically, as a case study. By creating data reports into an interactive user interface, this study's primary objective is to ascertain the river's condition over its whole course. To achieve this, we employ a variety of sensors to gather data on turbidity, temperature, water quality, pH, and other variables. Data gathered is used in an IoT sensor.

In Solanki *et al.* (2015) [58], the major goal is to make reasonably accurate forecasts for variable data. The Chaskaman River, which is close to Nasik, Maharashtra, India, was the subject of study using secondary data obtained from a third party and entered into the WEKA tool. To assess the prediction error rate, use metrics like mean square error and mean absolute error.

In Shkurin (2016) [59], the goal of the research that informed this paper was to show engineers and scientists working in the environmental field examples of how these models could be applied to environmental tasks. The STREAMES (STreamREAch Management, an Expert System) project, which initially aimed to produce tools for improving the quality of European rivers, produced the water quality data used in the research [59].

Shah *et al.* (2021) discuss whether artificial intelligence's (AI) unique characteristics can provide a thorough understanding of the growing water quality challenges [60]. The current study examines the accuracy of monthly total dissolved solids (TDS) and specific conductivity (EC) modeling using gene expression programming (GEP), artificial neural networks (ANN), and linear regression models (LRM) in the upper Indus River at two outlet stations.

Wang *et al.* (2017) [61] assessed the surface water quality of the Ebinur Lake Watershed. The WQI method was employed to evaluate the river's water quality. For each sampling point in the Ebinur Lake Watershed in October 2016, values for pH, HCO3, TP, TN, BOD, NH3 +-N, iron, copper, zinc, volatile phenol, DO, TDS, Cl, SO4 2, Na, Ca, Mg, COD, PO4 3, and Cr were taken into consideration.

Uddin *et al.* (2022) [62] estimated coastal WQIs in Cork Harbour, the current study used a recently created enhanced WQI model. Instead of repeatedly using SI and weight values to reduce model uncertainty, the research aims to identify the most reliable and robust machine learning (ML) algorithm(s) to forecast WQIs at each monitoring point. Models excelled over competing models.

11.6 Experiment

Dataset Utilized: The dataset, which includes the above parameters, is taken for different years and will help us to predict the quality of water. For this study, the dataset includes 1300 observations comprising pH, dissolved oxygen, temperature, turbidity, conductivity, TDS.

pH: pH is a crucial factor in determining the acid-base balance of water. Moreover, it shows if the water is acidic or alkaline. The maximum pH allowed range, according to the WHO, is between 6.5 and 8.5. The range of the most recent investigation was 6.52 to 6.83, which is within the range of WHO criteria.

Turbidity: the amount of solid stuff present in the suspended state determines how turbid the water is. The test is used to determine the quality of waste discharge with regard to colloidal matter and measures the light-emitting capabilities of water. The Wondo Genet Campus's mean turbidity value (0.98 NTU) is less than the WHO-recommended threshold of 5.00 NTU.

Conductivity: clean water is a good insulator and poor conductor of electrical electricity. The electrical conductivity of water is improved by an increase in ion concentration. The electrical conductivity of water is typically determined by the amount of dissolved particles present. The ability of a solution to convey current through its ionic process is measured by electrical conductivity (EC). According to WHO guidelines, the EC value should not be more than 400 S/cm.

Solids (total dissolved solids, or TDS): water may dissolve a variety of inorganic and certain organic minerals or salts, including bicarbonates, chlorides, magnesium, sulphates, potassium, calcium, and sodium. These minerals gave the water an undesirable flavor and muted color. This is a crucial variable while using water. Water with a high TDS rating is one that has a high mineral content. The recommended TDS level for drinking purposes is 500 mg/l, with a maximum limit of 1000 mg/l.

11.7 Methodology

The method in this research is based on LSTM and Arima model that predicts the quality of water by analyzing the predicted and expected value of dissolved oxygen and ph. Hence, giving us indepth knowledge of water quality by the means of time series. By analyzing previous data, the algorithm may determine the pH and dissolved oxygen for the coming year. The system consists of two modules: the Arima model and LSTM neural network.

11.7.1 LSTM Model Working

A unique type of recurrent neural network (RNN) that can recognize long-term trends is the LSTM neural network. Hochreiter and Schmidhuber made the initial suggestion. It has been successfully used to solve a variety of issues and is now extensively used.

Time-series features with extremely lengthy intervals and delays in the time series can be processed and predicted using the LSTM network. The gradient vanishing and explosion issues that frequently arise in conventional recurrent neural networks can also be successfully addressed by the LSTM network.

An input gate, an output gate, and an ignore gate are components of the LSTM model that are used to alter memory. The forget gate is primarily used to determine which memories in the memory unit should be kept and which memories can be forgotten, which can be described by the equations. The input gate and output gate are primarily used to control the input features and output contents. It can be given as

$$\text{hid}_1 = H\left(Wt_{\text{hid}x} + Wt_t + Wt_{\text{hid}x}\text{hid}_{t-1} + b_{\text{hid}}\right) \tag{11.1}$$

$$P_t = Wt_{\text{hid}y}Wt_{\text{hid}x}y_{t-1} + b_y \tag{11.2}$$

where

$X = (x1, x2, \ldots\ldots, xn)$ denotes the input time series

hid = $(h1, h2, \ldots\ldots, hn)$ denotes the hidden state of memory cells

$Y = (y1, y2, \ldots\ldots, yn)$ denotes the output time series

Wt denotes weighted matrices

b denotes the bias vectors

11.7.2 ARIMA Model Working

A variety of sectors use ARIMA models. It is frequently used to anticipate demand, such as to predict future demand. This is so that managers have solid guidelines to follow when making choices, thanks to the model. Additionally, based on the past pH, ARIMA models can be used to forecast the future pH of water. Be aware that even though they could aid you in predicting future shifts in water quality and it is given by

$$\text{ARIMA } (p, d, q)\ (P, D, Q)\ m \tag{11.3}$$

where,
p is the order of the AR term, no. of lags used as a predictor
d is the no. of differencing required to make the time series data stationary, if $d = 0$ then it is stationary
q is the order of the MA term, no. of lagged predicted errors which go into the ARIMA model
m is the number of periods in each season
(P, D, Q) represent the periodic part of the time series that is (p, d, q)

11.7.3 Performance Criteria

The water quality prediction is essentially a regression problem. In the present study, the mean absolute error (MAE), root-mean-square error (RMSE), and coefficient of determination (R2) were used to quantitatively evaluate the model prediction effect.

The mean absolute error (MAE) is a statistic that examines the mean amount of mistakes in a set of predictions without taking into account their direction. It evaluates the accuracy of continuous values. In other words, the MAE is the mean of the absolute data points representing the differences between the predicted and relevant observations. It is a linear score, which means that all individual variances are equally weighted in the mean.

The root mean square error (RMSE) is a quadratic scoring mechanism that determines the error's mean magnitude. In other words, the difference between the prediction and actual values is squared and then averaged over the sample value. Finally, the average's square root value is computed. The MAE and RMSE can be combined to diagnose forecast error variation. The RMSE will always be greater than or equal to the MAE; the bigger the difference, the greater the variation in the single errors in the sample. If the RMSE equals the MAE, then all errors are of identical magnitude. Both the MAE and the RMSE have a range of 0 to infinity. The lower the value, the better the performance.

When forecasting the outcome of a given event, the coefficient of determination is a statistical measurement that assesses how variations in one variable may be explained by differences in another one. In other words, this coefficient, often known as r-squared (or r2), evaluates the strength of the linear relationship between two variables and is frequently used by investors when conducting trend analysis.

Table 11.2 Performance Evaluation of Various Models

ARIMA MODEL	Test MSE: 0.031 Test RMSE:0.189 R2score:0.96
LSTM MODEL	TEST MSE: 0.61 TEST RMSE:0.823 R2 score:1.20
Neural Network	TEST MSE:0.76 TEST RMSE:0.73 R2 score:1.03
SVM Model	TEST MSE:0.83 TEST RMSE:0.67 The R2 score: 0.7

Source: [63].

11.7.4 Results

The study uses a dataset for the purpose of water quality prediction. The dataset consists of 1300 files for five attributes. After preprocessing the data to obtain features, we move on to decide models for implementing. The model was trained over several epochs. The experiments are carried out on Google Collaboratory notebooks with support of GPU.

The water quality prediction is essentially a regression problem. In the present study, the mean absolute error (MAE), root-mean-square error (RMSE), and coefficient of determination (R2) were used to quantitatively evaluate the model prediction. The first model implemented was the ARIMA model, which gave good performance as shown in the Table 11.2 and Figures 11.6–11.8.

11.8 Conclusion

In this chapter, the LSTM ARIMA neural network model is used. In order to classify the water quality, the LSTM model, a linear approach model, is used to predict the water quality. This research illustrates that all four implemented models showed good accuracy but the best accuracy was seen by the ARIMA model as the predicted and expected value are very near to each other.

This model may convert passive water environment risk emergency treatment to automatic prediction, greatly enhancing the ability of relevant departments to predict water environment danger. Many applicants can use this research to apply it to real-world situations.

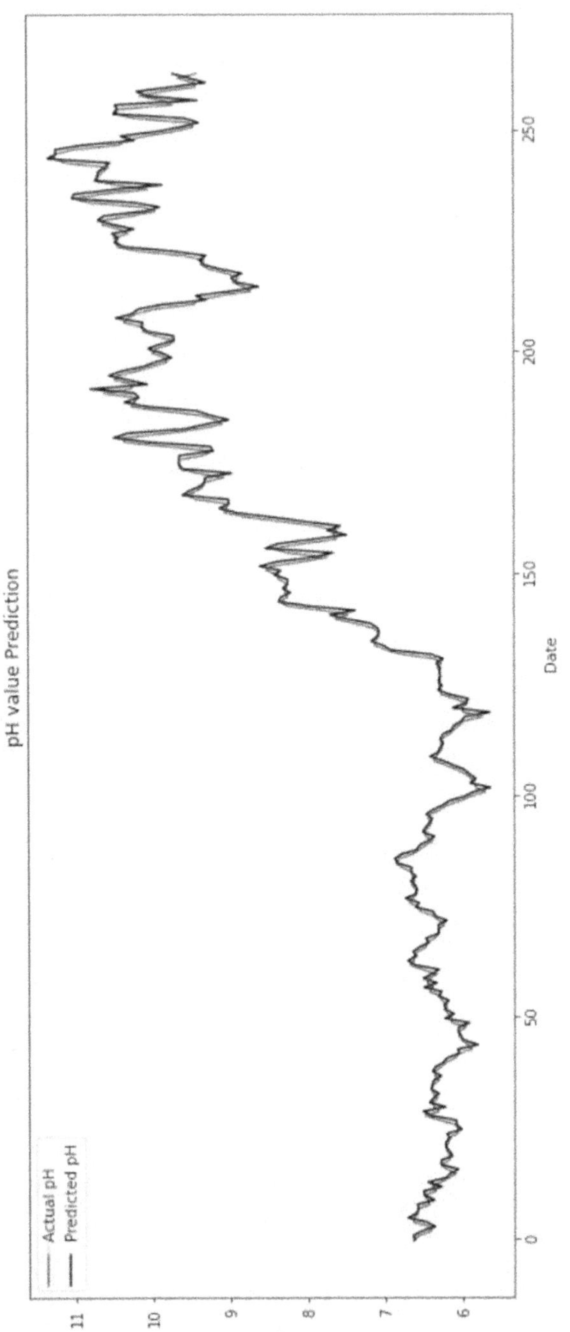

Figure 11.6 pH value prediction.

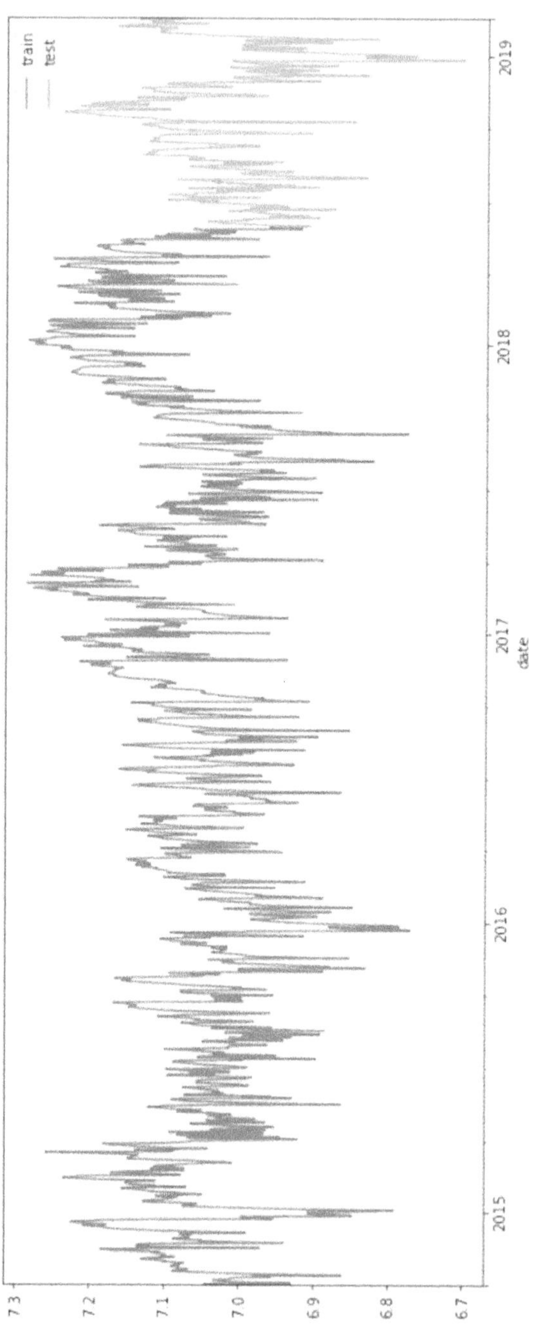

Figure 11.7 Testing and training accuracy of ARIMA.

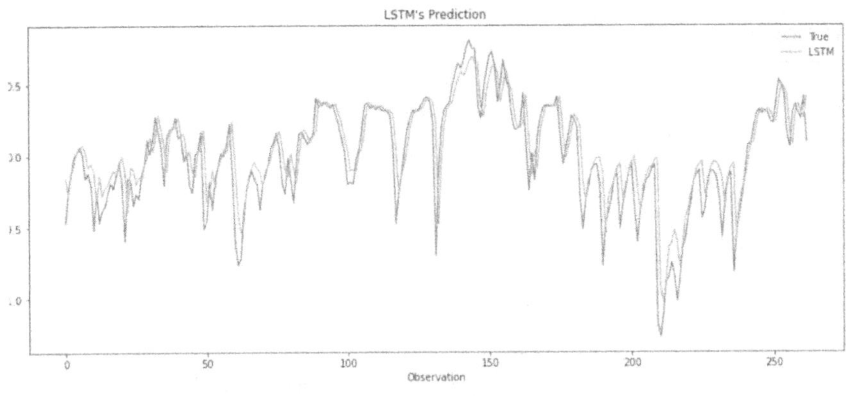

Figure 11.8 Testing and training accuracy of LSTM.

Data from one site was used for this prediction. But in the future, more site data might be added, which would make this research more beneficial because it would cover all aspects of predicting water quality.

References

[1] Mashkoor, F., et al., Exploring the reusability of synthetically contaminated wastewater containing crystal violet dye using Tectona grandis Sawdust as a very low-cost adsorbent. *Scientific Reports*, 2018. 8(1): p. 8314.

[2] Ranjbar-Mohammadi, M., M. Rahimdokht, and E. Pajootan, Low cost hydrogels based on gum Tragacanth and TiO2 nanoparticles: Characterization and RBFNN modelling of methylene blue dye removal. *Int J Biol Macromol*, 2019. 134: p. 967–975.

[3] Hebbar, R. S., et al., Fabrication of polyetherimide nanocomposite membrane with amine functionalised halloysite nanotubes for effective removal of cationic dye effluents. *Journal of the Taiwan Institute of Chemical Engineers*, 2018. 93: p. 42–53.

[4] Olivera, S., et al., Open ended tube like hollow bio-carbon derived from banana fibre for removal of anionic and cationic dyes. *Desalination and Water Treatment*, 2018. 132: p. 298–306.

[5] Isloor, A. M., et al., Novel polyphenylsulfone (PPSU)/nano tin oxide (SnO2) mixed matrix ultrafiltration hollow fiber membranes: Fabrication, characterization and toxic dyes removal from aqueous solutions. *Reactive and Functional Polymers*, 2019. 139: p. 170–180.

[6] Hebbar, R. S., et al., Removal of metal ions and humic acids through polyetherimide membrane with grafted bentonite clay. *Scientific Reports*, 2018. 8(1): p. 4665.

[7] Madhura, L., et al., Membrane technology for water purification. *Environmental Chemistry Letters*, 2017. 16(2): p. 343–365.

[8] Rezakazemi, M., et al., Fouling-resistant membranes for water reuse. *Environmental Chemistry Letters*, 2018. 16(3): p. 715–763.
[9] Kanchi, S., et al., Smartphone based bioanalytical and diagnosis applications: A review. *Biosensors and Bioelectronics*, 2018. 102: p. 136–149.
[10] Olivera, S., et al., Oxygen enriched network-type carbon spheres for multipurpose water purification applications. *Environmental Technology & Innovation*, 2018. 12: p. 160–171.
[11] Gupta, V. K., et al., Electrochemical analysis of some toxic metals by ion-selective electrodes. *Critical Reviews in Analytical Chemistry*, 2011. 41: p. 282–313.
[12] March, G., T. D. Nguyen, and B. Piro, Modified electrodes used for electrochemical detection of metal ions in environmental analysis. *Biosensors*, 2015. 5: p. 241–275.
[13] Inamuddin, et al., Synthesis and characterisation of poly(3,4-ethylenedioxythiophene)-poly(styrenesulfonate) (PEDOT:PSS) Zr(IV) monothiophosphate composite cation exchanger: Analytical application as lead ion selective membrane electrode. *International Journal of Environmental Analytical Chemistry*, 2015. 95(4): p. 312–323.
[14] Chen, Y.-T., et al., High-field modulated ion-selective field-effect-transistor (FET) sensors with sensitivity higher than the ideal Nernst sensitivity, *Scientific Reports*, 2018. 8(1): p. 1–11.
[15] De Acha, N., C. Elosúa, J. M. Corres, and F. J. Arregui. Fluorescent sensors for the detection of heavy metal ions in aqueous media. *Sensors*. 2019; 19(3):599. https://doi.org/10.3390/s19030599
[16] Priyadarshini, E. and N. Pradhan, Gold nanoparticles as efficient sensors in colorimetric detection of toxic metal ions: A review. *Sensors and Actuators, B: Chemical*, 2017. 238: p. 888–902.
[17] Aglan, R. F., M. M. Hamed, and H. M. Saleh, Selective and sensitive determination of Cd (II) ions in various samples using a novel modified carbon paste electrode. *Journal of Analytical Science & Technology*, 2019. p. 4.
[18] Zhao, F., et al., Ultra-sensitive detection of heavy metal ions in tap water by laser-induced breakdown spectroscopy with the assistance of electrical-deposition. *Analytical Methods*, 2010. 2: p. 408–414.
[19] Grozdanov, A., et al., Removal of Heavy Metal Ions from Wastewater Using Bio- and Nanosorbents, In Proceedings of the International Conference on Microplastic Pollution in the Mediterranean Sea, 2018: p. 239–244.
[20] Nepomuscene, N. J., D. Daniel, and A. Krastanov, Biosensor to detect chromium in wastewater. *Biotechnology and Biotechnological Equipment*, 2007. 21: p. 377–381.
[21] Wu, L-C, G-H. Wang, T-H Tsai, S-Y Lo, C-Y Cheng, and Y-C Chung. Three-stage single-chambered microbial fuel cell biosensor inoculated with Exiguobacterium aestuarii YC211 for continuous Chromium (VI) measurement. Sensors. 2019. 19(6):1418. https://doi.org/10.3390/s19061418
[22] Ai, T., Jiang, X. and Liu, Q. Chromium removal from industrial wastewater using Phyllostachys pubescens biomass loaded Cu-S nanospheres. Open Chem, 2018. 16: p. 842–852.
[23] Sukesan, R., et al., Instant mercury Ion detection in industrial waste water with a microchip using extended gate Field-E ffect transistors and a portable device. *Sensors*, 2019. 20.

[24] Mayr, T. and O. S. Wolfbeis, Optical sensors for the determination of heavy metal ions. 2002. Ph.D. Thes: p. 140.

[25] Abdullah, N. A., et al., Chemical and biosensor technologies for wastewater quality management. *International Journal of Advanced Research and Publications*, 2017. 1: p. 1–10.

[26] Bahadori, A., M. Al-Haddabi, and H. B. Vuthaluru, Simple predictive tool estimates sodium adsorption ratio for evaluation of potential infiltration problems using reclaimed wastewater. *Communications in Soil Science and Plant Analysis*, 2012. 43(19): p. 2492–2503.

[27] Li, D., S. Liu. Remote monitoring of water quality for intensive fish culture. In: Mukhopadhyay, S., Mason, A. (eds) *Smart Sensors for Real-Time Water Quality Monitoring. Smart Sensors, Measurement and Instrumentation*, vol 4. Springer, Berlin, Heidelberg, 2013. https://doi.org/10.1007/978-3-642-37006-9_10

[28] Gebicki, J., Application of electrochemical sensors and sensor matrixes for measurement of odorous chemical compounds. *TrAC – Trends in Analytical Chemistry*, 2016. 77: p. 1–13.

[29] Manufacturing, F., (12) United States Patent. 2002. p. 1.

[30] Logroño, W., et al., A terrestrial single chamber microbial fuel cell-based biosensor for biochemical oxygen demand of synthetic rice washed wastewater. *Sensors*, 2016. 16(1): p. 101.

[31] Li, J., et al., Highly selective fluorescent chemosensor for detection of Fe 3 + based on Fe 3 O 4 @ ZnO. Nature Publishing Group, 2016: p. 1–8.

[32] Bahadori, A., M. Al-Haddabi, and H. B. Vuthaluru, Simple predictive tool estimates sodium adsorption ratio for evaluation of potential infiltration problems using reclaimed wastewater. *Communications in Soil Science and Plant Analysis*, 2012. 43: p. 2492–2503.

[33] Li, J., et al., Improvements in the decision making for Cleaner Production by data mining: Case study of vanadium extraction industry using weak acid leaching process. *Journal of Cleaner Production*, 2017. 143: p. 582–597.

[34] Metilena, R., A. Sisa, and M. Scirpus, Treatment of Methylene blue in wastewater using Scirpus grossus. *Malaysian Journal of Analytical Science*, 2017. 21: p. 182–187.

[35] Vanrolleghem, P. A. and D. S. Lee, On-line monitoring equipment for wastewater treatment processes: State of the art. *Water Science and Technology*, 2003. 47: p. 1–34.

[36] Yashas, S. R., et al., Laccase biosensor: Green technique for quantification of phenols in wastewater (a review). *Oriental Journal of Chemistry*, 2018. 34: p. 631–637.

[37] Michalski, R., Ion chromatography applications in wastewater analysis. *Separations*, 2018. 5: p. 16.

[38] Yosry M Issa, K. A. and R. R Ami, A modified carbon paste sensor for determination of Zn in vitamin and waste water using Thiosemicarbazide and Acetaldehyde Thiosemicarbazone complexes. *Journal of Biosensors & Bioelectronics*, 2015: p. 06, 176.

[39] Sharma, R. D. and S. Amlathe, Quantitative determination and removal of zinc using disposable colorimetric sensors: An appropriate alternative to optodes. *Journal of Chemical and Pharmaceutical Research*, 2012. 4: p. 1097–1105.

[40] Wang, T. and W. Yue, Carbon nanotubes heavy metal detection with stripping voltammetry: A review paper. *Electroanalysis*, 2017. 29: p. 2178–2189.

[41] XIE, T., et al., A double-microbial fuel cell heavy metals toxicity sensor. *DEStech Transactions on Environment, Energy and Earth Sciences*, 2018: p. 5–10.

[42] Aghav, R. M., S. Kumar, and S. N. Mukherjee, Artificial neural network modeling in competitive adsorption of phenol and resorcinol from water environment using some carbonaceous adsorbents. *Journal of Hazardous Materials*, 2011. 188: p. 67–77.

[43] Ingildsen, P., Realising full-scale control in wastewater treatment systems using in situ nutrient sensors. Department of Industrial Electrical Engineering and Automation, 2002. PhD: p. 365.

[44] Özdemir, Ş. K., et al., Highly sensitive detection of nanoparticles with a self-referenced and self-heterodyned whispering-gallery Raman microlaser. *Proceedings of the National Academy of Sciences*, 2014. 111: p. E3836–E3844.

[45] Curteanu, S., et al., Electro-Oxidation method applied for activated sludge treatment: Experiment and simulation based on supervised machine learning methods. *Industrial & Engineering Chemistry Research*, 2014. 53(12): p. 4902–4912.

[46] Barzegar, R., et al., Comparison of machine learning models for predicting fluoride contamination in groundwater. *Stochastic Environmental Research and Risk Assessment*, 2017. 31: p. 2705–2718.

[47] Betrie, G. D., et al., Predicting copper concentrations in acid mine drainage: A comparative analysis of five machine learning techniques. *Environmental Monitoring and Assessment*, 2013. 185(5): p. 4171–82.

[48] Kong, X., et al., Real-time eutrophication status evaluation of coastal waters using support vector machine with grid search algorithm. *Marine Pollution Bulletin*, 2017. 119: p. 307–319.

[49] Haghiabi, A. H., Nasrolahi, A. H., and Parsaie, A. Water quality prediction using machine learning methods. *Water Quality Research Journal*, 2018. 53(1): p. 3–13.

[50] Srivastava, G., and Kumar, P. Water quality index with missing parameters. *International Journal of research in Engineering and Technology*, 2013. 2(4): p. 609–614.

[51] Zhu, X., Wang, X., and Ok, Y. S. The application of machine learning methodsfor prediction of metals orption onto biochars. *Journal of Hazardous Materials*, 2019. 378, 120727.

[52] Asadollah, S. B. H. S., Sharafati, A., Motta, D., and Yaseen, Z. M. (2021). River water quality index prediction and uncertainty analysis: A comparative study of machine learning models. *Journal of Environmental Chemical Engineering*, 2021. 9(1): p. 104599.

[53] Chen, K., Chen, H., Zhou, C., Huang, Y., Qi, X., Shen, R., ... and Ren, H. Comparative analysis of surface water quality prediction performance and identification of key water parameters using different machine learning models based on bigdata. *Water Research*, 2020. 171: p. 115454.

[54] Tan, G., Yan, J., Gao, C., and Yang, S. Prediction of water quality time series data based on least square ssupport vector machine. *Procedia Engineering*, 2012. 31: p. 1194–1199.

[55] Sakaa, B., Elbeltagi, A., Boudibi, S., Chaffaï, H., Islam, A. R. M., Kulimushi, L. C., ... and Wong, Y. J. Water quality index modeling using random forest and improved SMO algorithm forsupport vector machine in SafSaf river basin. *Environmental Science and Pollution Research*, 2022: p. 1–18.

[56] Muharemi, F., Logofătu, D., and Leon, F. Machine learning approaches for anomaly detection of water quality on a real-world data set. *Journal of Information and Telecommunication*, 2019. 3(3): p. 294–307.

[57] Rosero-Montalvo, P. D., López-Batista, V. F., Riascos, J. A., & Peluffo-Ordóñez, D. H. Intelligent WSN system for water quality analysis using machine learning algorithms: A case study (Tahuando river from Ecuador). *RemoteSensing*, 2020. 12(12): p. 1988.

[58] Solanki, A., Agrawal, H., and Khare, K. Predictive analysis of water quality parameters usingdeep learning. *International Journal of Computer Applications*, 2015. 125(9): p. 0975–8887.

[59] Shkurin, A. Water quality analysis using machine learning algorithms. Bachelor's Thesis in Environmental Engineering. Mamk University Applied Science. Finland. 2015.

[60] Shah, M. I., Alaloul, W. S., Alqahtani, A., Aldrees, A., Musarat, M. A., and Javed, M. F. Predictive modeling approach for surface water quality: development and comparison of machine learning models. *Sustainability*, 2021. 13(14): p. 7515.

[61] Wang, X., Zhang, F., and Ding, J. Evaluation ofwater quality based on a machine learning algorithm and water quality index forthe Ebinur Lake Watershed, China. *Scientific Reports*, 2017. 7(1): p. 1–18.

[62] Uddin, M. G., Nash, S., Diganta, M. T. M., Rahman, A., and Olbert, A. I. Robust machine learning algorithmsfor predicting coastal water quality index. *Journal of Environmental Management*, 2022. 321: p. 115923.

[63] Hafeez, S., Wong, M. S., Ho, H. C., Nazeer, M., Nichol, J., Abbas, S., … and Pun, L. Comparison of machine learning algorithms for retrieval of water quality indicators in case-II waters: A case study of Hong Kong. *Remote Sensing*, 2019. 11(6): p. 617.

Chapter 12

Digital Representation of Agriculture Forms

K. Rajasathiya

12.1 Introduction

Agriculture is incorporating virtual technology more and more. Global technology companies, local startup businesses, and the government of the United Kingdom are offering and supporting a wide range of solutions aimed at creating the "clever farmer," from farm management apps to milking robots and from self-operating tractors to soil disease detection drones. The first is agrodrone1, a startup company that operates globally and has developed a sprayer drone for crop safety in orchards and vineyards. Recent research has focused on three study areas: (a) power and governance; (b) knowledge, competence, and learning; and (c) the use of digital technology, knowledge transfer, and public–private partnerships in the field of digital agriculture. First, inside the present day literature on clever farming traits, a whole lot has been written approximately huge-scale farming tasks in anglophone international locations (e.g., United States of America). Second, transparent conductive films (TCFs) are essential parts of a variety of technologies, including flat panel displays, light-emitting diodes, and solar cells, despite the fact that a lot of scholarly attention has been paid to a variety of technologies and despite the fact that studies have looked at drone technologies in agriculture. Farms and agricultural operations operate very differently today than they did a few decades ago. With the use of features like sensors, GPS, intelligence-powered gadgets, and other modern

 DOI: 10.1201/9781003469612-12

agricultural technology, the field is able to redefine the significance of agricultural technology.

- More productive crops.
- Less pesticide, water, and fertilizer consumption will result in cheaper food costs.
- Reduced negative effects on natural ecosystems.
- Fewer chemical spills into rivers and groundwater.
- Enhanced worker safety.
- Reliable management of natural resources is made possible by robotic technology.

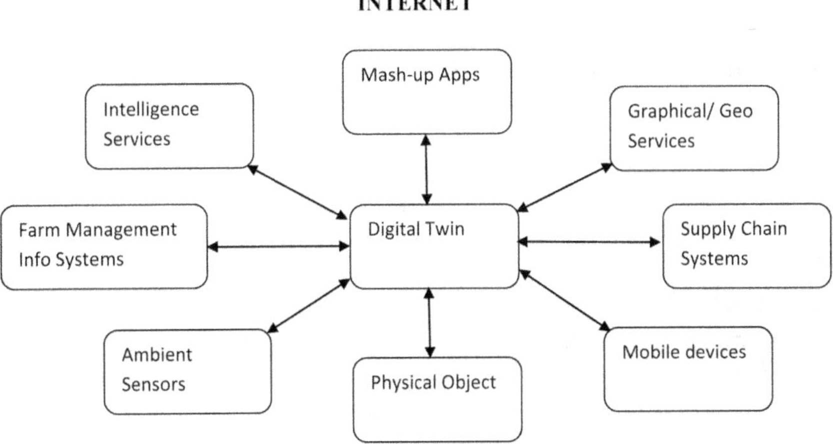

The impact of technology on agriculture

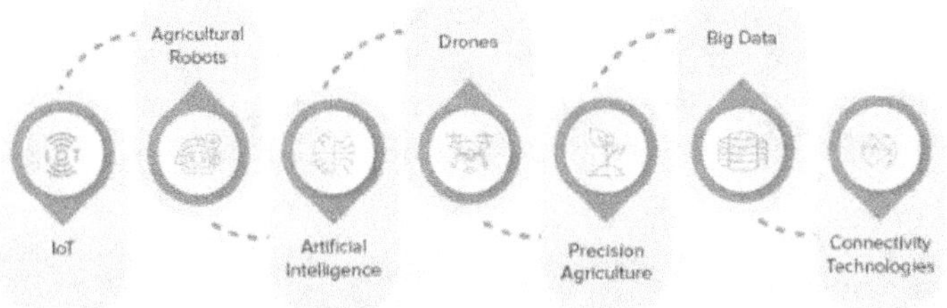

Internet of Thongs (IoT)

An Internet of Things (IoT) gadget is made up of sensors that gather farm data and provide accurate, real-time information via applications or other channels. These Internet of Things sensors facilitate agricultural tasks, such as the following: sensing of soil temperature and humidity, tracking of plants and animals, and remote farm monitoring systems.

Along with this, a number of firms are developing cutting-edge linked sensors that link IoT gadgets with robotics, drones, and computer imaging to improve the accuracy and agility of farming operations.

Robots for Farming

Farmers frequently struggle with a labor shortage, particularly in large enterprises. The state-of-the-art technology that is overcoming this issue in agriculture is agricultural robots.

To assist farmers with physically taxing tasks like harvesting, plucking produce, planting, spraying, and weeding, several businesses have started producing robots.

Smart agricultural tools, such as autonomous and semi-autonomous tractors for harvesting, as well as robots for livestock management tasks like incubators, automated weighing scales, auto feeders, and milking machines are some of the robots they are purchasing to observe how technology is affecting agriculture.

Artificial Intelligence

Intelligent chatbots used in agricultural AI provide advice and suggestions to farmers. While chatbots operate at a more granular level, AI and ML algorithms can be used to more broadly diagnose and track abnormalities and diseases in plants and livestock.

Drones

It is challenging to strike a balance between raising farm productivity and cutting costs. Drones, however, are one piece of contemporary agricultural technology that is helping farmers get over this hurdle. The equipment gathers raw data, which is later transformed into useable data for agricultural monitoring. Drones that are equipped with cameras aid in surveying and aerial imaging information about maximizing the use of pesticides, seeds, water, livestock tracking, grazing, and geofencing monitoring.

Precision Agriculture

A farm field is divided into various sections, each with its own unique soil characteristics, sunlight exposure, and slopes. Thus, treating every aspect of the farm the same might result in inefficiency and increased time and resource waste.

Precision-based agricultural technology utilization has been developed by the industry to address this problem. They are compiling data using AI-powered analysis and drones.

Big Data

The limited technological progress in agriculture is being transformed by big data approaches, which translate all farm data into useful insights. The following are some examples of how agricultural technology is used.

Laying the groundwork for the farming season to come by researching data on crop production, land use, irrigation, agricultural prices, weather predictions, and crop diseases

To gather data about farm operations, data about weather events, water cycles, farm machinery, and crop quantity and quality are used.

Technology for Connectivity

Smart connectivity technologies like 5G, LPWAN, or satellite-based communication are necessary for smart farming to function. The use of devices like robots, IoT devices, sensors, and other items that require quick, real-time communication is made possible by these connectivity technological breakthroughs in agriculture.

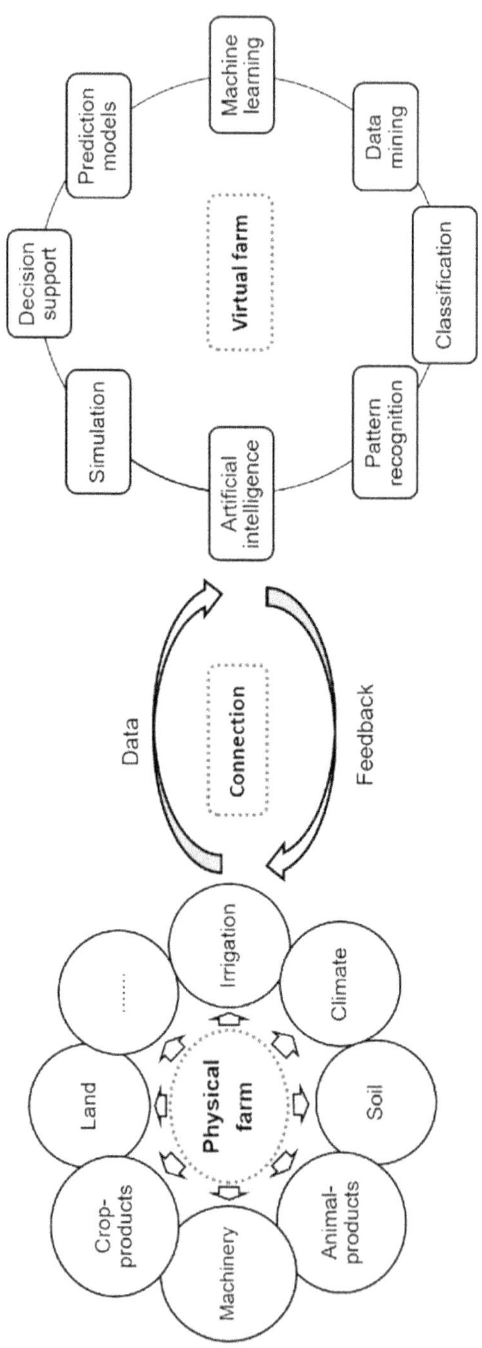
Physical farm
Land
Crop-products
Machinery
Animal-products
Soil
Climate
Irrigation
......
Data
Connection
Feedback
Virtual farm
Artificial intelligence
Simulation
Decision support
Prediction models
Machine learning
Data mining
Classification
Pattern recognition

12.1.1 Uses of Digital Representation in Agriculture

Digital farming techniques can provide farmers with useful information on how to

(I) utilize seeds, fertilizers, pesticides, and irrigation management techniques;
(II) safeguard the environment;
(III) oversee crop, pest, and weather monitoring;
(IV) satisfy market expectations and address business circumstances.

12.2 Agriculture

Definition
"The art and science of cultivating the soil, raising animals, and growing plants are collectively referred to as agriculture". It consists of the instruction of plant and animal things that people can use and the markets where they are sold.

12.2.1 Grain Farming

Definition
"Cultivation of cereals for human consumption and animal in addition to other uses, such as the creation of starch and bio fuel industrially". Purity of the seed is important. The cost of sowing ranges from 22.5 kilograms per hectare (20 kilograms per acre) and more. The depth of sowing, which is typically 1 to 3 inches or 2.5 to 7.5 cm, might be significantly less in advantageous places. To make the soil more conducive to planting, a variety of plowing tools and instruments are hired.

Plant Protection
The earth should be rolled in the spring to solidify the soil across the roots because winter plants are frequently disturbed by frost. The crop is typically harrowed in the spring to aerate the soil and kill young weeds if the soil has crusted due to heavy rainfall followed by surface drying.

Harvesting
Threshing machines had been powered first with the aid of men or animals, frequently the use of treadmills were followed by internal combustion and steam engines.The integrate cleans the grain, chops the standing grain, separates the grain from the straw and chaff, and empties the grain into bags or grain reservoirs.

Fungus Diseases

The leader damage in the rust fungus group is caused by black rust. Cereal blooms turn yellow instead of green due to black rust. The grain produced is tiny, shriveled, and has a low weight that is consistent with a bushel.

12.2.2 Livestock Ranching

Definition

"The practice of "ranching" involves leaving animals in a confined pasture to graze on grass". Ranching is the act of jogging a massive farm, specifically breeding livestock on a ranch, to elevate them. Normally, ranchers are involved in raising grazing animals such as sheep and cattle. Farming is the exercise of elevating dwelling organisms for the cultivation of natural sources. This consists of meals within the shape of meat, produce, grains, eggs, or dairy, in addition to other assets like natural fibers, plant oils, and rubber.

12.2.3 Mediterranean Agriculture

Definition

"The techniques and practices used to cultivate animals and plants in areas with Mediterranean temperatures. It is a type of agriculture developed in an environment similar to that of the Mediterranean".

Elements of Mediterranean Agriculture

The four important elements are growing of cereals and vegetables, vineyards, and orchards. Olives, pomegranates, oranges, figs, pears, grapes, and other produce are examples of what is grown. Four additives form a strategic compound in Mediterranean agriculture:

(i) Rainfed annual plants based totally at the winter rain.
(ii) Everlasting crops, treecrops surviving the dry summer.
(iii) Transhumance keeping off the dry summer time.
(iv) Irrigation compensating the lacking (summer season) precipitation.

Mediterranean Agriculture Climate

On occasion known as the dry summer time temperate weather, mediterranean climates are classified under the köppen (rounded) weather system as cs. The subtypes are

- *Hot-summer season mediterranean weather (csa),*
- *Warm-summer season mediterranean weather (csb),*
- *Cold-summer time mediterranean weather (csb).*

First	*Second*	*Third*
A (Tropical)	f (Rainforest) m (Monsoon) w (Savanna, dry winter) s (Savanna, dry summer)	
B (Dry)	W (Arid Desert) S (Semi-Arid or steppe)	h (Hot) k (Cold)
C (Temperate)	w (Dry winter) f (No dry season) s (Dry summer)	a (Hot summer) b (Warm summer) c (Cold summer)
D (Continental)	w (Dry winter) f (No dry season) s (Dry summer)	a (Hot summer) b (Warm summer) c (Cold summer) d (Very cold winter)
E (Polar)		T (Tundra) F (Ice cap)

12.2.4 Commercial Gardening and Fruit Farming

Definition

"Commercial horticulture" refers to the production and management of ornamental flowers and turfgrass, as well as the sale of fruits and vegetables. There are two types of farming:

12.2.4.1 Intensive Farming

"High labor and financial inputs in comparison to the size of the farms". It is defined by performance: higher meat and dairy production from fewer animals in smaller spaces, and higher agricultural yields from smaller farms.

Advantages:

- Restricts cultivation to specific places, freeing up space for other uses.
- The most productive type of farming in terms of output.
- Capable of feeding and sustaining large human populations.

Disadvantages:

- Intensive farming is most viable in temperate regions; intensive farming is not conceivable in a desert, for example, without irrigation.
- Subsistence farmers may benefit from intensive agriculture, but commercial farmers may not.
- Intensive cattle agriculture can unfold pollution and can be perceived as inhumane.
- Extensive farming calls for financial and bodily investments a few farmers cannot afford.
- If incorrectly managed, extensive crop cultivation can produce pollutants and greatly deteriorate soil.
- Traditional farming methods are preferred by cultural practices over more sophisticated farming techniques.

Plantation Agriculture
Plantations take up a lot of space, yet they aim for maximum profits solely through economies of scale.

12.2.4.2 Extensive Farming

Smaller labor inputs in comparison to the amount of land being farmed are the reverse of intensive farming.

12.3 Types of Agriculture

"In India, agriculture significantly contributes to the increase of American revenue. New farming strategies assist farmers in growing their earnings". Agriculture in India is classified into several types. The sort of crop planted in fields serves as the foundation for categorization. The varieties of agriculture mentioned above are mentioned in element below.

- *Subsistence farming:* it is a form of farming practiced by small farmers' families. They use much less generation for farming and greater manpower.
- *Shifting cultivation*: it is a type of farming practiced in forests on the countries of India, America, Africa, and others.
- *Intensive Farming:* rotational grazing is one of the fine examples of intensive farming. Right here, a massive range of manufacturing is performed on a small a part of the land.
- *Pastoralism:* it is associated with animal husbandry. In this technique, domesticated animals get left in a massive a part of the land for grazing.

Types of Agriculture	
Shifting Cultivation	Concerned with the rotation of vegetation
Subsistence Farming	This kind refers back to the exercise of elevating farm animals or growing plants best for self and no longer for exchange functions.
Pastoralism	Related to animals herding
Intensive Farming	Makes a speciality of increasing the input and the ensuing output in line with unit of agricultural land

12.4 Usage of Digital Technologies

Virtual farming includes using numerous technologies, including sensors, drones, precision irrigation structures, GPS-guided machinery, and data analytics and machine learning tools. Twin technology is used for agriculture forming.

Definition

"A digital twin is a representation of a real-world object in the digital realm. It employs real-time data supplied from sensors on the object to simulate behavior and track actions across the course of the object's lifecycle".

A virtual replica of an object or device from the actual world is called a digital twin. A software model or enclosed software item that replicates a special real object, system, company, person, or other abstraction is known as a digital dual implementation. For a composite picture across certain real-world entities, such as a power plant or a community, and their connected approaches, data from a few virtual twins may be combined.

12.5 Types of Twins and Digital Twins

Identical (monozygotic)

"*Twins that originated from a single embryo are known as monozygotic twins*". Etymologically, the time period is derived from mono, meaning "one", and zygotic, from zygote (from zugoun, that means "to sign up for"). The zygote splits into equal halves, every independently developing into a character with the identical intercourse and genetic constitution as the other.

Fraternal (dizygotic)

"Fraternal twins or non identical twins" are other names for dizygotic twins. Its root words are zygote, which means "egg," and di, which means "two." Internationally, the cost of variable dizygotic twinning greatly. Dyzygotic twins are genetically unique, and as a result, they are no longer identical.

12.5.1 Digital Twin Types

Component Twins

Component twins are virtual representations of an individual a part of a machine or product, such as a tools or screw. In preference to virtually modeling all of the person components of a product, although, factor twins are commonly used to model imperative parts, along with the ones under precise strain or warmth.

Asset Twins

Machine twins, also referred to as unit twins, are digital representations of systems of merchandize working collectively. While asset twins version real-international products created from many parts, machine twins version these individual products as additives of a bigger machine.

System Twins

Machine twins, also referred to as unit twins, are digital representations of systems of merchandize working collectively. While asset twins version real-international products created from many parts, machine twins version these individual products as additives of a bigger machine.

Process Twins

Manner twins are digital representations of systems working together. For instance, while a machine dual might version a production line, a method dual may want to model the whole manufacturing unit all the manner all the way down to the employees working the machines on the factory floor.

Why and How to Design Digital Twins?

A virtual dual layout is created by collecting data and constructing computational models to evaluate it. This will contain a real-time interface between the virtual version and a physical object to communicate and receive comments and facts.

Data

To construct a virtual model that can represent the behaviors or states of the real world object or system, a virtual dual needs information about the thing or system in question. These statistics may also pertain to a product's lifecycle and include design specs, manufacturing methods, or engineering records. It is able to also include manufacturing records including system, substances, elements, techniques, and exceptional control. Records also can be associated with operation, consisting of real-time comments, historical analysis, and renovation information. Different information utilized in digital twin layout can encompass commercial information or end-of-life strategies.

Modeling

Once the data has been collected, it can be utilized to develop computational analytical patterns to expose running outcomes, anticipate states such as weariness, and decide behaviors. Simulation in engineering, physics, chemistry, statistics, device learning, synthetic intelligence, commercial business good judgment, or goals can all be used to prescribe actions. These models may be displayed through three-D representations and augmented truth modeling so one can resource human expertise of the conclusions.

Linking

Digital twin findings may be combined to generate an outline, for example, by including device twin findings into a production line dual, which may then inform a virtual twin at manufacturing scale. It is possible to enable smart business applications for real globally functional qualities and improvements by using connected digital twins in this way.

Benefit

Additionally globally functional, digital twins can be utilized to prototype before production, reducing product flaws and speeding up the time to market. Other uses for the digital twin could be process improvements, including monitoring staffing levels in relation to production or coordinating a supply chain with needs for manufacturing or preservation.

The aim is to achieve increased availability and reliability through modeling and monitoring to improve performance as a whole. Additionally, they can reduce the likelihood of accidents and unexpected downtime due to failure, save maintenance costs by anticipating failure, and ensure that production needs are not compromised by planning maintenance, repairs, and the ordering of replacement parts. Digital twin can also offer persevered enhancements with the aid of analyzing customization fashions and making sure products have the best overall performance through checking out in real time.

12.6 Working

A digital twin is an object that replicates the behavior and states of a real-world object across the course of its lifetime in a virtual environment. The decoupling of physical flows from its planning and manipulation is made possible by the use of virtual twins, which is an effective method for farm management.

Smart farming can be improved using digital twins. The framework includes a model for implementing digital twin systems in line with the Internet of Things structure (IoT-A), a reference architecture for IoT structures, as well as a control version built primarily on a standard structure technique.

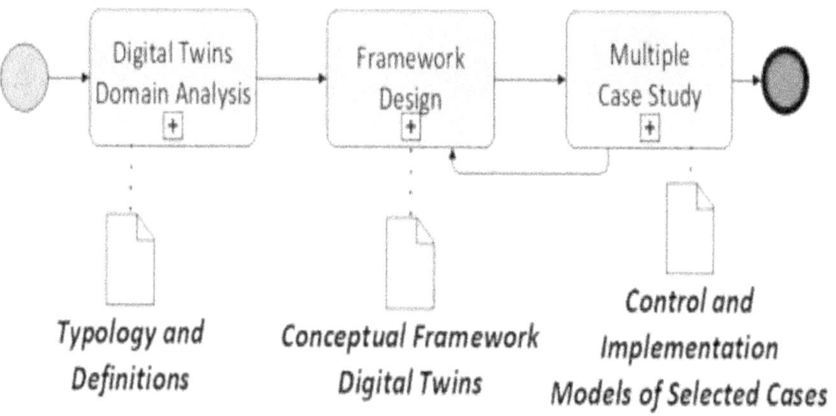

First, the case studies that only briefly discuss digital dual without providing a more detailed definition were disregarded from the study. Digital twins are a relatively new idea, and as a result, they are used in many different contexts. As a result, key definitions and viewpoints on virtual twins in the literature have been idnetified, particularly in the context of product lifecycle management and the net of factors.

Second, a conceptual framework for creating and implementing digital twins in farm management altered. This work focuses on providing a solid conceptual foundation for the use of virtual twins in smart farming, including the management aspects of their use to plan, observe, manage, and improve farm strategies.

Thirdly, the conceptual framework becomes implemented to instances. The steps of the protocol are as follows:

(1) Case study approach
(2) Data collection planning
(3) Data collecting for the analyzed instance
(4) Data analysis of the collected data
(5) Reporting.

Smart products and clever flowers generate massive quantities of statistics concerning their usage and effectiveness. The performance virtual twin captures this statistic from merchandize and flowers in operation and analyzes it to provide actionable perception for informed selection making. With the aid of leveraging overall performance virtual twins, organizations can

- Create new commercial enterprise opportunities
- Advantage insight to enhance digital models
- Capture, aggregate, and analyze operational statistics
- Enhance product and manufacturing gadget performance

Where it is Used?

Manufacturing
Manufacturing might become more efficient and productive thanks to virtual twins, which would also speed up throughput.

Automotive
To acquire and analyze operational data from a car in order to study its reputation in real time and inform product improvements is one way that digital twins are utilized in the automotive industry.

Retail
Virtual twin is used in the retail industry outside of manufacturing and business to mimic and improve the customer experience, no matter what degree level of a shopping center or for specialty shops.

Healthcare
Digital dual has benefited science in areas including de-risking techniques, surgical procedure training, and organ donation. Structures have also been developed to mimic how patients move around hospitals, detect potential infection hotspots, and identify potential contact dangers.

Disaster Management

In recent years, there has been a global impact from international weather trade, but virtual twin can help to prevent this by leveraging the thoughtful deployment of intelligent infrastructures, emergency action strategies, and tracking changes in the weather.

Smart Cities

Additionally, digital dual can assist cities to grow and be more economical, as well as environmentally and socially sustainable. The interaction of digital technology with couture is known as digital fashion. The nexus of technology and human representation is human AI, where the importance of human value is highlighted and strengthened by technology and the potential for discovering design. For example, timely solutions to issues may be informed with the aid of real-time records from digital twins to permit assets, including to respond to a crisis in hospitals.

Advantages of Technology in Agriculture

- Farmers who lack practical knowledge are unable to safely handle machinery.
- The time is cut down.
- Water is used to irrigate crops.
- In contrast, machines are useful for planting seeds.
- Technique for irrigation.
- Using artificial fertilizers in the soil.
- A pest control method involving chemicals.
- They raise both the price and demand for the goods.
- Better pricing and marketing exposure.
- E-commerce and internet trade amenities.
- Additionally, raise the soil's fertility.
- Reduce your use of fertilizers and water to help keep costs down.
- Low dumping of waste materials and pollutants into the ocean.
- Lessen your ecological footprint.
- Suitable for a campus setting.

Disadvantages of Technology in Agriculture

- Farmers who lack practical understanding cannot operate machines safely.
- Although maintenance is highly expensive.
- Machine overuse could harm the ecosystem.
- Driverless agricultural equipment is also a barrier to accessing the technology.
- Boost the scouting initiatives.

- Robotic machines are unable to influence a people's culture; therefore, we must manually create their programmes.

12.7 Applications in Agriculture

Integrating Digital Platform

The first step toward a digital representation of any industry is the creation of a digital platform that connects individuals with the tools and information they use. This pertains to agriculture and involves storing and connecting information on stakeholders, a lot of data, and economic analyses.

Workflow Engine

Using distinct operations, farming is strictly regulated. We can classify movements and their location along the value chain from farm to consumer by precisely describing the stairs in this complex system.

Analytics Automation

By automating analytics via the Internet of Things (IoT) and other technology-driven techniques, we may gather orders of magnitude more information from sensors and other sources without the need for human participation.

Soil: The Productive Asset

Soil is a useful resource in agriculture. It is a constant in all land-based agriculture (hydroponics, of course, is a little variation). We must measure and comprehend as much as we can about the composition and capability of the soil where crops grow in response to an agricultural digital dual call.

Digital Product Descriptions

If we are to simulate consequences during a growing season, we have to know the whole thing so that we are able to approximately predict the ability of a crop and can tweak those inputs to discover our best bet at an excellent harvest.

Weather Prediction

With the use of artificial intelligence (AI) and other technologies, many organizations are investing heavily in the modeling and prediction of climate change. We can use those predictive models as a component of our agricultural digital twin.

Constant Stress Recognition

These pressures tend to occur frequently in the same places and are consistent drags on agricultural productivity. Picking them out and grading them is necessary for dealing with them. We will plan, study, improve, and model how we raise our crops. At some point in the future, we will help farming become a sustainable activity by increasing yields, reducing strain on water resources and soil quality, and other related goals.

12.8 Digital Agriculture

Using digital tools to manage crops, animals, and other agricultural operations is referred to as "digital agriculture." Farmers have the opportunity to increase output, minimize risk, and amass long-term financial savings thanks to digital agriculture. It is viewed by agricultural researchers as a data-gathering technology with the potential to accelerate data collecting and processing, improving predictive capacities for crop management and animal behavior and productivity. Digital agriculture, smart farming, or e-agricultural are terms used to describe tools used in agriculture that collect, store, analyze, and distribute electronic data and/or information. Digital farming is already transforming the landscape of agricultural research by enabling the integration of many more data sources for analysis.

The capacity to make judgements in real-time and use new sensors that provide instantaneous information have completely transformed the landscape since the introduction of smart phones and iPads, according to Plaut. "We now have access to data from a much wider variety of sources than ever before, and we can use that data to form conclusions". The power of digital agriculture resides in that.

Digital twins, which are virtual versions of real-world objects like fields, animals, or machinery, can be enhanced with information from sensors and cameras on the ground and use Cloud, Edge, AI, and IoT technologies to spread seeds and fertilizer precisely, use less harmful pesticides, and optimize water use.

The digital agriculture system accurately gathers data from external sources, such as meteorological data. In order to assist the farmer in making wiser decisions, the combined data is analyzed and interpreted. Farmers may get real-time feedback on the results of their decisions thanks to robotics and advanced equipment, which swiftly and accurately put these judgments into action.

Precision weather forecasts, yield mapping, variable-rate fertilizer, herbicide, and water applications, as well as GPS guiding systems, are a few examples of digital agriculture technologies.

Systems for digital agriculture provide a distinctive viewpoint that enhances the capability for making and using decisions. Robots are used by the current generation of farmers to plant and harvest crops.

1. **Technologies for Digital Agriculture Reduce Your Vulnerabilities To Climate Change**

Accurate weather forecasts made possible by digital agriculture can help farmers better adapt to weather conditions caused by climate change. By foreseeing weather patterns, digital agriculture might decrease climate change risks. Because of the information that is disseminated to communities via radio broadcasts and mobile communications, farmers can navigate the distribution of rainfall and make better judgments on various livelihood methods. Agriculture producers who are not familiar with digital technology can benefit from the bulletins and notifications provided by extension services, which are digitally enabled services.

2. Technology Risks in Digital Agriculture

A significant concern is the potential use of resources in digital agriculture. Despite the excitement and enthusiasm surrounding the adoption of digital technology in agriculture, caution is absolutely necessary, especially in light of unanticipated consequences related to the probable extinction of natural systems.

12.8.1 Agriculture Technology and Digital

Digital technologies including the Internet, mobile devices, data analytics, artificial intelligence, and digitally delivered services and apps are changing agriculture and the food system.

At various points along the agri-food value chain: remote satellite data and in-situ sensors improve accuracy and reduce the cost of monitoring crop growth and the quality of land or water; digital logistics services and traceability technologies have the potential to streamline agri-food supply chains and provide consumers with trustworthy information.

Governments can use digital technologies to boost the effectiveness and efficiency of current policies and programmes as well as create new ones. For instance, the expense of monitoring various agricultural activities is drastically reduced by the availability of free, high-quality satellite photography. In addition to ensuring environmental regulations are followed, digital technologies allow for the automation of administrative processes for agriculture and the provision of additional government services, such as in conjunction with advising or extension services.

Digital technologies can facilitate trade in agricultural and food products by expanding markets for private sector suppliers, enabling new ways for governments to monitor and ensure compliance with standards, and enabling quicker and more efficient border procedures—all of which are essential for perishable goods.

These advantages result from both direct adoption of technologies by sector actors, such as service providers, and indirect adoption of technologies by governments to produce better policies.

12.8.1.1 Ways of Technology to Improve Agriculture

- Internet of Things (IoT), which enables remote and real-time agriculture field monitoring.
- Robots for farming, which are capable of weeding, harvesting, and other agricultural activities.
- Artificial intelligence (AI), which can examine data and offer perceptions for making decisions.
- Drones, which can precisely survey and spray crops.
- To optimize inputs and outputs, precision agriculture uses GPS and sensors.
- The use of big data to manage opportunities and dangers for farmers.
- Communications and information sharing between farmers and stakeholders are made possible by connectivity technologies.

12.8.2 Adoption of Digital Agriculture's Effects

Efficiency

By reducing the cost of data replication, transit, tracking, verification, and search, digital technology modifies economic activity. Digital technology will increase productivity across the entire agricultural value chain as a result of these declining costs.

On-Farm Efficiency

Precision agricultural technologies on-farm can reduce the amount of input needed for a given output. The physical capital allocation efficiency inside and between farms can be enhanced through digital agriculture. Armies with limited financial resources can more easily get equipment to increase productivity.

The increased farmer knowledge brought about by digital agriculture increases labor productivity. E-extension (the electronic transmission of traditional agricultural extension services) makes it possible to spread farming knowledge at a reasonable cost.

Through a reduction in labor requirements, digital agriculture increases labor productivity. Precision agriculture's inherent automation—from "milking robots on dairy farms to greenhouses with automated climate control"—can improve crop and livestock management by needing less labor.

Off-Farm/Market Efficiency

Digital agriculture technologies can improve the effectiveness of agricultural markets in addition to increasing farm output. Mobile devices, online ICTs, e-commerce platforms, digital payment systems, and other agricultural technology can help to alleviate market failures and lower transaction costs throughout the value chain.

- Matching customers and sellers can help reduce information asymmetry.
- Reducing the cost of transactions in business markets
- Reducing the cost of transactions for government services

Equity

Digital agriculture has the potential to create a fairer agri-food value chain. Because it reduces transaction costs and information asymmetries, digital technology can help smallholder farmers get market access in a variety of ways:

- financial participation
- market penetration
- possible injustices brought on by digital agriculture

12.8.3 Impact of Digital Agriculture on Climate and Food Security

Change Mitigation

A thorough analysis of previously published papers revealed that the effects of climate change on food security include decreased agricultural yields, slowed animal growth rates, and decreased livestock output. This is due to the fact that it has warmed the earth, changed precipitation patterns, and made extreme events more frequent.

12.8.3.1 Impact of Climate Change on Agriculture

Establishing sustainable food and farming systems, eliminating environmental pollution, boosting yields, distributing food fairly and equitably, and lowering malnutrition are concerns that are important for achieving food security.

In addition to increased profitability brought about by improved management techniques and the advancement of agricultural information systems and digital agriculture offers a number of other advantages, such as improved product quality, increased sustainability, reduced food safety and risk in the management system thanks to product traceability, environmental protection, and rural development. The major goal of this study was to identify the constraints and challenges in various sections on food production and management, take into consideration these advantages, and present an overview of digital agriculture's role in increasing food security as a climate-smart technology.

It is anticipated that various aspects of agriculture, such as crop productivity, soil characteristics, biodiversity, aquaculture, livestock, and fisheries, will be significantly impacted by climate change.

Table 12.1 Weather-Related Factors' Effects on Crop Growth and Development

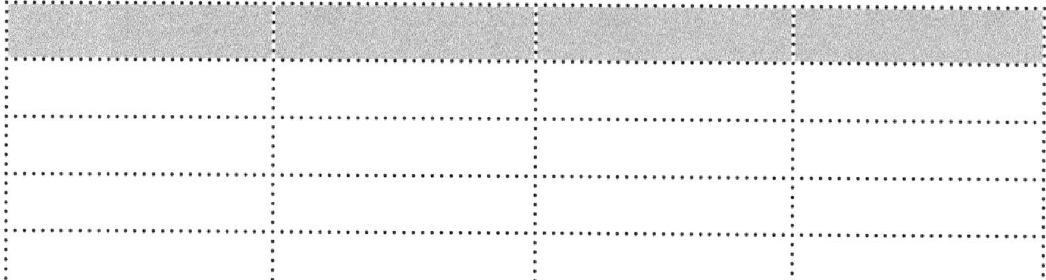

12.8.3.2 Impact the Production of Crops

Crop production may be impacted by changes in soil temperature and moisture, precipitation patterns, pests and diseases, extreme weather events like droughts, floods, and storms occurring more frequently and severely, and changes in water availability. These modifications may result in lessened food security and farmer financial losses. The environment also affects the development, yield, and quality of agricultural crops, which in turn impacts the interactions between plants and microbes. Table 12.1 shows the effects of weather-related factors on crop growth and development.

12.8.3.3 Impact on Soil

Climate, relief, parent material, and creatures are only a few of the numerous elements that interact intricately to generate soils over time. The climate component, however, has the greatest impact on soil development and management, having a significant impact on soil structure, stability, water retention, fertility status, and erosion .The primary projected effects of climate change include higher temperatures and climatic extremes, such as excessive rainfall, drought conditions, frosty conditions, storms, and sea level rise, which represent serious challenges to the soil in terms of erosion, compactness, soil healthiness, and productivity.

An increasingly important step in combating climate change is improving and protecting soil carbon, which has numerous environmental and socioeconomic advantages.

According to the International Union for Conservation of Nature (IUCN), soil stores more carbon than biomass (vegetation) and the atmosphere combined, making it an important carbon sink along with oceans and forests.

By managing and maintaining carbon-rich soil ecosystems and enhancing soil health by halting land degradation, it is possible to accomplish a cost-effective triple win for people, the environment, and the climate.

12.8.3.4 Climate Change's Impacts on Fish and Livestock

Increased human competition for limited natural resources, changing feed crop species and quality, disease outbreaks, and heat stress will all have a significant influence on animal productivity and efficiency. Winter warming can encourage the occurrence and spread of cattle diseases, which can last all year and get worse as temperatures rise. The crucial component that predominantly impacts animals is heat stress, which has an impact on their ability to reproduce, produce milk and meat, and overall health.

Methane and nitrous oxide are the two most significant greenhouse gases produced by animal agriculture. Enteric fermentation and the storage of manure are the main sources of methane, a gas that contributes to global warming 28 times more than carbon dioxide. The worldwide warming potential of nitrous

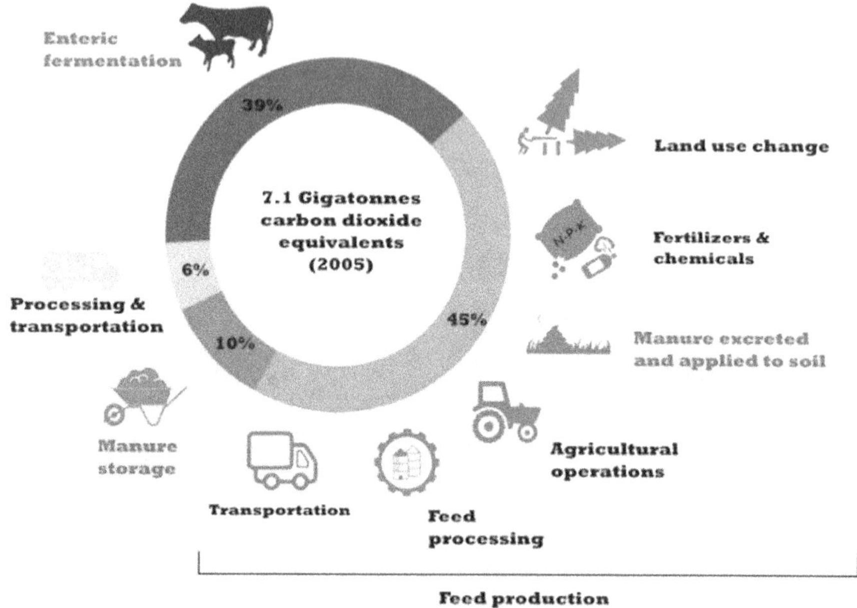

Figure 12.1 Livestock emissions by source. Red represents direct emissions from cattle.

oxide, which results from the storage of manure and the application of organic and inorganic fertilizers, is 265 times more than that of carbon dioxide. Figure 12.1 shows livestock emissions by source. Red represents direct emissions from cattle.

Goglio *et al.* (2018) claim that soil dynamics (such as decaying plant remains, soil organic matter mineralizing, changes in land use, etc.), the production of synthetic fertilizers and pesticides, and the use of fossil fuels in on-farm agricultural operations are the main causes of soil carbon dioxide emissions. Nitrous oxide emissions are created when organic and inorganic fertilizers are given to the soil.

12.8.3.5 Live Stock Mitigation Strategies

The vast heterogeneity of the agricultural sector must be taken into account when assessing the overall sustainability of a mitigation plan, which can vary across various livestock systems, species, and climates.

Methane emissions during enteric fermentation may decrease, but there may be an increase in greenhouse gas emissions from applied manure. Decreased direct nitrous oxide emissions during storage may result in increased nitrate leaching and ammonia volatilization upon field application.

Mitigation potential of various strategies

Strategies	*Category*	*Potential mitigating effect**	
		Methane	*Nitrous oxide*
Fermentation of enteric bacteria	Grass quality	Low to medium	†
	Feed processing	Low	Low
	Concentrate inclusion	Low to medium	†
	Dietary lipids	Medium	†
	Electrons receptors	High	†
	Ionophores	Low	†
	Methanogenic inhibitors	Low	†
Manure storage	Solid-liquid separation	High	Low
	Anaerobic digestion	High	High
	Decreased storage time	High	High
	Frequent manure removal	High	High
	Phase feeding	‡	Low
	Reduced dietary protein	‡	Medium
	Nitrification inhibitors	‡	Medium to high
	No grazing on wet soil	Low	Medium
	Increased productivity	High	High
Animal management	Genetic selection	High	‡
	Animal health	Low to medium	Low to medium
	Increase reproductive eff.	Low to medium	Low to medium
	Reduced animal mortality	Low to medium	Low to medium
	Housing structures	Medium to high	Medium to high

12.8.4 *Digitalization of Agriculture in India*

The term "digitalization of agriculture" refers to the integration of slice-edge digital technologies, similar as artificial intelligence (AI), robots, unmanned aeronautics systems, detectors, and communication networks, into the ranch product system. India is always attempting to create and put into effect rules that will increase the sustainability of its agriculture sector. Partnerships between corporations and the government can aid in the development of a smart agriculture industry given India's dynamic corporate structure.

To accomplish goals like raising farmer incomes and promoting sustainable growth, the Indian government is putting numerous policies and initiatives into practice. Therefore, for the widespread adoption of digital agriculture in India, a multi-stakeholder strategy would be needed, with the government playing a large facilitation role in the ecosystem. Increased agricultural production, lower greenhouse gas emissions, better farmer and family health, and increased food security are the aims of digital agriculture.

Not just agriculture, but all economic sectors, are going digital. Farmers are the last to use digitization; therefore, the government should invest time and resources into spreading the word about its advantages, the private sector should band together, and the top software businesses in India should assist. Utilization simplicity, accessibility, familiarity, acceptability, and popularization of advantages for farmers who are fast to pick up on and value the advantages.

12.8.5 New Frontiers in Indian Agriculture

By using data to produce actionable intelligence and substantive added value, digital farming goes beyond the simple presence and availability of data. Precision farming and smart farming are being combined in digital farming.

A definition of digital farming is "consistent application of precision farming and smart farming methods, internal and external farm networking, and use of web-based data platforms in conjunction with Big Data analyses."

12.8.6 Digital Farming Strategies (ICRISAT)

(i) *Collaboration with T-Hub*

The relationship includes numerous programmes, with agriculture as the primary domain and technology innovation having a big impact. Among the areas to be investigated are the establishment of an agricultural accelerator hub, partnerships and synergies across the innovation spectrum, sponsorship of agricultural-related programmes, and ways in which the agriculture-focused T-Hub Accelerator initiative may accept ICRISAT as a partner.

(ii) *ICRISAT: Revolutionizing Agriculture with ICT*

Farmers and other participants in the agriculture value chain require a lot of information. To make agriculture a successful business and appealing to young people, information and communication technologies (ICTs) will be essential for knowledge exchange, targeted advice, market integration, and access to financing.

12.8.7 Mobile Application for Digital Farmers

It offers farmers with daily news and also displays the current market rate. It also displays government programmes for farmers.

It displays notifications for chatting, product bidding, schemes, and updates. It offers free information about organic farming and animal farming in order to promote understanding.

The most difficulties in digital farming

1. Accessibility in remote places.
2. Ignorance of the many farm production functions.
3. Individual management zone size.
4. Entry barriers for new businesses.
5. Inadequate scalability and configuration issues.
6. The benefits are not immediately obvious.

12.8.7.1 How to Overcome the Difficulties in Digital Farming

In order to meet these demands, the agriculture sector must overcome a number of obstacles, including the limited amount of arable land, the effects of climate change, the growing water shortage, the cost and availability of energy, particularly energy derived from fossil fuels, and the effects of urbanisation on the availability of rural labour.

When it comes to overcoming these obstacles and ensuring that expenses are kept to a minimum, technology will be crucial. This is already evident with the introduction of precision agriculture to the milking business. Figure 12.2 shows the digital agriculture revolution.

12.8.7.2 Digitalization of Rural and Agricultural Sectors

The primary goal of this work is to identify, categorize, and characterize the potential implications of digital technologies in rural regions and agriculture described in the scientific literature rather than to provide a theoretical analysis of technological innovation per se building on current ideas.

12.8.8 The Future of Indian Agriculture

With higher agricultural yields and improved sustainability due to less water and pesticide use, the current developments in digital farming technologies will speed expansion. Digital innovations that are modernizing agricultural value chains include artificial intelligence (AI) and machine learning (ML), remote sensing, big data, block chain, and the Internet of Things (IoT).

Figure 12.2 The digital agriculture revolution.

India is expected to foster the use of digital agriculture through public–private partnerships (PPP). It is likely that the Department of Agriculture's data pool, which will be created as part of the National Agri Stack, will be crucial.

The Ministry of Agriculture & Farmers Welfare has developed significant digital tools to encourage farmers to use technology, such as the following:

National Agriculture Market (eNAM): To provide a single national market for agricultural goods, the National Agriculture Market (eNAM) brings together all of India's Agricultural Produce Market Committee (APMC) mandis. By giving farmers adequate returns on their investments, eNAM enables farmers to sell commodities directly to customers without the aid of brokers or mediators.

Transfer of Direct Benefits (DBT) Central Agricultural Portal: All agricultural programs in the country are centralized in the DBT Agri Portal. The website provides government funds to assist farmers in purchasing cutting-edge farm equipment. Figure 12.3 depicts smart farming: robotics, the IoT and agriculture and Figure 12.4 shows the future of farming.

12.8.8.1 Application of Digital Agriculture

Farmers may gather, visualize, and assess crop and soil health concerns at different stages of production in an efficient and practical way thanks to remote sensing, soil sensors, unmanned aerial surveys, market insights, and other technological interventions.

Artificial intelligence/machine learning (AI/ML) algorithms can generate current actionable insights to assist with soil screening, increase crop output, lessen pests, and lessen farmer workload.

Figure 12.3 Smart farming: Robotics, the IoT, and agriculture.

You can track your food and track accurate, tamper-proof information about farms, inventories, and transactions with the aid of block chain technology. As a result, farmers are no longer needed to record and store important information using files or paperwork.

12.8.8.2 Benefits of Digital Agriculture

Since they have immediate access to a comprehensive digital analysis of their farms, farmers may take the proper action and refrain from using excessive amounts of pesticides, fertilizers, and water. Additional benefits are as follows:

- Reduces the use of chemicals in agricultural production, which reduces production costs and increases productivity in agriculture, and encourages water resources to be used effectively and efficiently.
- Improves the socioeconomic standing of farmers and has a smaller negative environmental and ecological impact.
- Improves workplace safety.

12.8.9 India's Digital Agriculture Implementation

The following actions could be taken in India to make digital agriculture successful:

- Affordable technology
- Portable equipment
- Platforms for renting and sharing farm equipment and machinery
- Academic assistance

Figure 12.4 The future of farming.

12.8.10 Case Study: Smart Farming

Digital twins provide great promise for advancing sustainable and productive farming practices. Without accurate and current information regarding farm activities, modern agricultural manufacturing is not possible. More and more farms must rely on digital technologies like smart devices, complex analytics, and sensing and monitoring equipment. Agricultural manufacturing is changing fast in the direction of clever farming systems, driven with the aid of the speedy tempo of technical advancements such using the cloud, the net of factors, massive records, gadget learning, augmented reality and robotics. This chapter aims to contribute to clearing up this gap by way of analyzing how virtual twins can strengthen clever farming. More specifically, the goals are threefold:

1. To define the concept of virtual twins and present a typology;
2. Recommending a conceptual framework, or a set of rules, for creating and enforcing digital twins;
3. To use a few case examples to apply and validate the conceptual framework to intelligent farming.

Methodology

A conceptual framework for the creation and application of virtual twins in farm management is the design artifact that resulted from this paper's development. Virtual twins are a very novel and complicated idea. An excellent way to gain a deeper knowledge of such complicated events, which cannot be researched outside

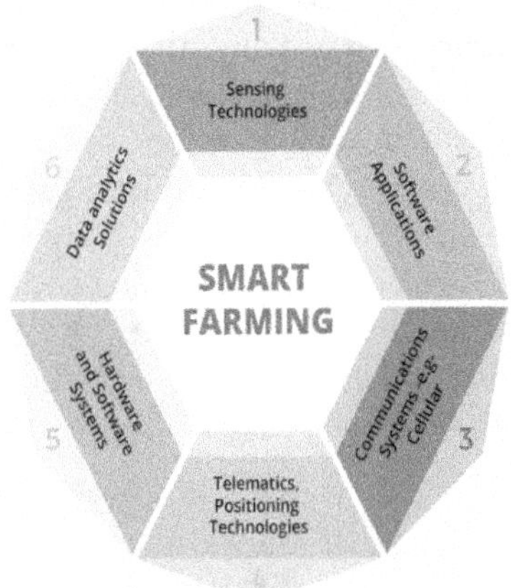

Figure 12.5 The future of agriculture.

of their rich, real-world context, is through case studies. Figure 12.5 shows the future of agriculture.

12.9 Conclusion

With cutting-edge work covering a number of use-cases, including plants, robots, and aquaponics, enabled by technology such as the internet of matter, machine learning, and cyber-bodily systems, the digital twin is an effective concept with a promising future in agriculture. To overcome present-day challenges and difficult circumstances, the development of fresh approaches, agriculturally specific definitions, and the application of supporting technology may be essential. The virtual dual offers an intriguing chance to automate dynamic structural modeling, simulation, and modeling, as well as an intriguing chance to achieve genuine digitalization in challenging areas like agriculture. It is in the national interest to adopt a holistic ecosystem approach to solving the issues the Indian agriculture sector faces in order to accomplish objectives like doubling farmer incomes and achieving sustainable growth. Therefore, a multi-stakeholder approach will be necessary for digital agriculture to be widely adopted in India, with the government playing a key enabling role in the ecosystem.

Bibliography

1. Bakken, B. R., & Hernes, T. (2004). The implementation of e-farming: A Norwegian case study. *Journal of Agricultural Technology*, 10(2), 143–154.
2. Brown, K. W., & Townsend, R. (2014). The use of digital technologies in agriculture: A systematic review. *Journal of Agricultural Systems Research*, 1(1), 1–14.
3. Burchell, K., & Zapata, P. (2015). Understanding the role of digital technologies in agricultural development. *Journal of Agricultural Education and Extension*, 21(3), 231–244.
4. Carlsson, M., & Åberg, M. (2017). Digitalization in agriculture: A review of the literature. *Journal of Agricultural Technology*, 13(2), 63–76.
5. Colledge, L., & Fennell, R. (2019). The impact of digital technologies on agricultural decision-making: A systematic review. *Journal of Agricultural Economics*, 70(2), 247–265.
6. Crutzen, J., & Swartjes, F. (2017). Digital agriculture and the future of farming: A review of the literature. *Journal of Agricultural Systems Research*, 3(2), 1–14.
7. De Jesus, M., & Fuchs, S. (2019). Smart farming and the role of digital technologies in agriculture: A review of the literature. *Agriculture and Food Security*, 8(1), 1–15.
8. Dijksterhuis, G., & van der Werf, W. (2018). The role of digital technologies in sustainable agriculture: A review of the literature. *Journal of Cleaner Production*, 172, 3341–3352.
9. Fletes-Santos, E., & Lopez-Montano, J. (2019). The impact of digital technologies on agricultural productivity: A systematic review. *Journal of Agricultural Education and Extension*, 25(2), 141–156.
10. Gadallah, F., & Rodriguez-Diaz, J. A. (2018). The use of digital technologies in agriculture: A systematic review and meta-analysis. *Journal of Agricultural Technology*, 14(3), 157–172.
11. Giannakopoulos, G., & Triantafyllou, E. (2019). Digital agriculture and its potential for sustainable agriculture: A systematic review. *Sustainability: Science, Practice and Policy*, 15(1), 1–15.
12. Hernandez-Ceron, M., & Rodriguez-Diaz, J. A. (2020). The impact of digital technologies on agricultural sustainability: A systematic review and meta-analysis. *Journal of Agricultural Systems Research*, 6(1), 1–14.

13. Jansen-Hoogland, G., & Bergers, P. J. (2017). Digitalization in agriculture: Opportunities and challenges for sustainable agriculture in the Netherlands. *Journal of Agricultural Systems Research*, 2(2), 1–14.
14. Kahiluoto, H., & Juga, J. (2019). The role of digital technologies in sustainable agriculture: A systematic review and meta-analysis. *Journal of Cleaner Production*, 235, 1218–1228.
15. Leege, P., & Sturm-Meyer, K. (2019). Digital agriculture and its potential for sustainable agriculture: A systematic review and meta-analysis. *Agriculture and Food Security*, 8(2), 1–15.
16. Lipscomb-McGee, M., & O'Connor, T.-A. (2020). The impact of digital technologies on agricultural sustainability: A systematic review and meta-analysis. *Journal of Agricultural Systems Research*, 7(1), 1–14.

Chapter 13

Accelerators for Clustering Applications in Machine Learning

Mrinalika Durairaju, Mrinalini Durairaju, Vallidevi Krishnamurthy, V. Sakthivel, and Surendiran Balasubramanian

13.1 Introduction

In recent years, machine learning (ML) has become an essential tool for organizations across various industries to process, analyze, and extract insights from large amounts of data.

Machine learning is a subfield of artificial intelligence cantered on algorithms that can learn to complete tasks using data. ML is currently experiencing a renaissance thanks to recent computational, theoretical, and practical advances in the field. Recently, various ML approaches have been successfully demonstrated in real-world tasks for computer vision [1], anomaly detection [2], and language translation [3]. Those wishing to use ML are able to access an increasingly wide variety of high-performance computing resources. For example, computing clusters at universities and other research institutions are now widely accessible, along with a variety of web-based services. The greater availability of large datasets and the variety of new open-source high-level software libraries has also helped to improve the reach of existing ML techniques and to facilitate rapid prototyping of new algorithms. These technical advances go hand-in-hand with new real-world applications that motivate ML research and funding in the public and private sectors.

 DOI: 10.1201/9781003469612-13

In the big data era, clustering algorithms have been successfully applied in many different fields [4], including machine learning, pattern recognition, image analysis, information retrieval, bioinformatics, and data compression [5]. However, in the big data era, it poses significant challenges to improve the processing performance due to huge data volume and a wide variety of applications. This problem has seriously dragged down the efficiency and performance of the algorithm, consequently, acceleration techniques have become a hot research topic during the past few years.

As datasets have grown larger and more complex, the demand for computing power has also increased. To address this challenge, organizations have turned to accelerators, specialized hardware devices designed to perform specific types of computations more efficiently than general-purpose CPUs.

Accelerators have become increasingly important for cluster applications in machine learning. These hardware devices are designed to perform specific types of computations more efficiently than general-purpose CPUs. By using accelerators, organizations can significantly improve the performance and energy efficiency of their machine learning applications, as well as handle larger and more complex workloads.

13.2 Clustering Technique

Clustering is a versatile machine learning method that finds application in a wide range of domains, such as natural language processing, image and speech recognition, and recommendation systems. Its ability to group similar items, like images or products, is particularly useful in tasks such as organizing image databases and providing product recommendations in e-commerce platforms.

Clustering is a key analysis method in machine learning that leverages unsupervised techniques to group data points based on similarity metrics and assign unique cluster-IDs to each cluster. This simplifies the processing of large and complex datasets, allowing for the identification of patterns and relationships among data points. Clustering is utilized for identifying patterns and structures in both labeled and unlabeled datasets, making it an important technique in machine learning. The technique of clustering is commonly utilized for analyzing statistical data.

The clustering technique can be better understood with an example from real-life situations, such as a visit to a shopping mall. In such cases, similar items are often grouped together, such as t-shirts in one section and trousers in another, to facilitate easy access and retrieval of items. We often observe that in grocery stores, items are organized into separate sections for ease of access. For instance, in the fruit section, apples, bananas, mangoes, and other fruits are placed in separate groups. This process of grouping is similar to the clustering technique, which groups data points based on similarity metrics. Another example of clustering is the grouping of documents according to the topic.

Some of the most common uses of this technique are the following:

1. Statistical data analysis
2. Market segmentation
3. Image segmentation
4. Social network analysis
5. Anomaly detection, etc.

In addition to the aforementioned applications, clustering is utilized in various fields such as e-commerce and entertainment. For instance, Amazon leverages this technique to suggest products based on the users' previous search and purchase history, while Netflix utilizes clustering to recommend movies and TV shows based on users' watch history and preferences.

Clustering algorithms are developed for implementing different clustering techniques. Clustering algorithms work by identifying similarities between data points and grouping them together based on those similarities. The goal of clustering is to group data points together so that points within the same cluster are as similar as possible to each other, while points in different clusters are as dissimilar as possible.

There are many different clustering algorithms available in machine learning, each with its own strengths and weaknesses. Some common clustering algorithms include *k*-means, PAM, and DSCAN.

Figure 13.1 is an illustration that demonstrates how the clustering algorithm operates. It displays various patterns that are grouped into different clusters based on their similar properties.

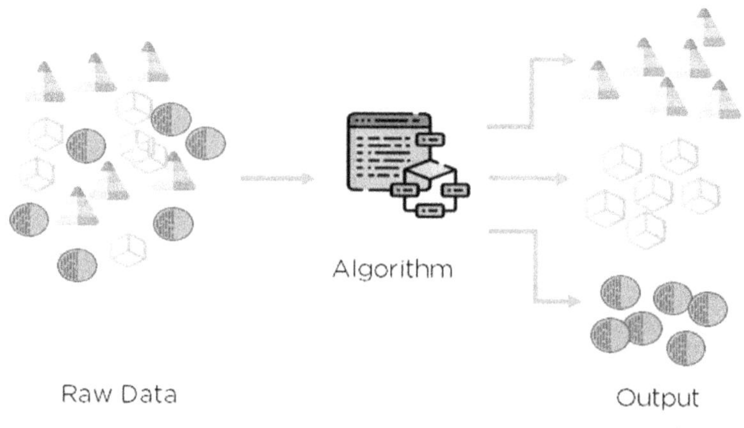

Figure 13.1 Illustration of clustering algorithm.

13.3 Types of Clustering Methods

In machine learning, clustering methods can be categorized into two main types: hard clustering, where data points belong to only one group, and soft clustering, where data points can belong to multiple groups. While these are the two primary categories, there are also other clustering approaches utilized in machine learning. Here are some of the main clustering methods employed in the field:

1. Partitioning clustering
2. Density-based clustering
3. Distribution model-based clustering
4. Hierarchical clustering
5. Grid-based clustering
6. Fuzzy clustering

13.3.1 Partitioning-Based Clustering

Partitioning clustering is a type of clustering that involves dividing data into non-hierarchical groups. It is also known as centroid-based clustering, and the *k*-means clustering algorithm is a popular example. This method partitions the dataset into a set of *k* groups, where *k* defines the number of predefined groups. Each cluster should consist of at least one data object, and each data object should be classified into exactly one cluster. The cluster center is created in such a way that the distance between the data points of one cluster is minimum as compared to the data points of other clusters. These methods optimize a targeted benchmark similarity function, with distance being a significant parameter to consider.

Two popular techniques are the following:

1. *k*-means clustering (understand *k*-means clustering from here in detail)
2. CLARANS (clustering large datasets based upon randomized search)

In addition, partitioning clustering algorithms are non-hierarchical methods that are typically applied to static datasets. They aim to identify groups within the data by optimizing an objective function, which often involves minimizing the distance between data points within each cluster. The quality of the partition is improved iteratively through optimization techniques.

Partitioning-based clustering (Figure 13.2) is highly efficient in terms of simplicity, proficiency, and it is easy to deploy, and computes all attainable clusters synchronously.

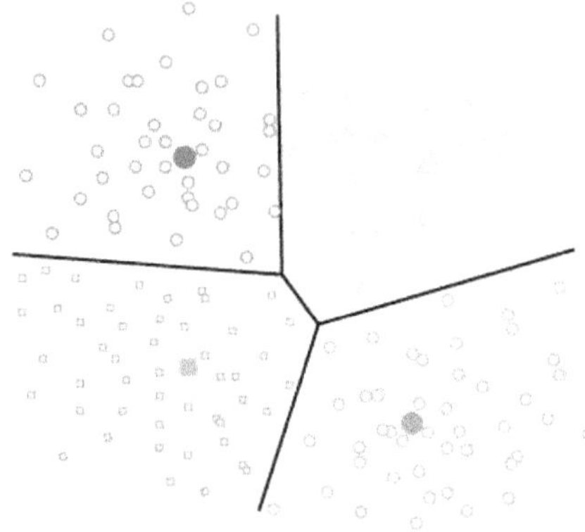

Figure 13.2 Partition-based clustering.

13.3.2 Density-Based Clustering

Density-based clustering methods aim to identify clusters of data points in dense regions of the data space while distinguishing them from low-density areas. This approach links highly dense areas into clusters and can accommodate arbitrarily shaped distributions as long as the dense regions are connected. The algorithm identifies various clusters in the dataset and groups areas of high density into separate clusters. Sparse areas of the data space separate dense regions from one another. These clustering techniques are known for their high accuracy and ability to merge two clusters together.

Its examples include

1. DBSCAN (density-based spatial clustering of applications with noise)
2. OPTICS (ordering points to identify clustering structure)

These clustering methods utilize distance metrics to group objects together. However, they tend to produce spherical-shaped clusters, making it challenging to identify arbitrary-shaped clusters. Additionally, clusters are formed in all directions, as long as the density of objects within a neighborhood exceeds a specific threshold.

Density-based clustering methods are designed to identify the underlying structure of datasets by analyzing the entire density of data points, thereby recognizing functions and features that can impact individual data points. Some algorithms, such as OPTICS and DenStream, are equipped with outlier filtration techniques and can create clusters of arbitrary shapes. However, these methods may face difficulties in clustering high-dimensional datasets with varying densities.

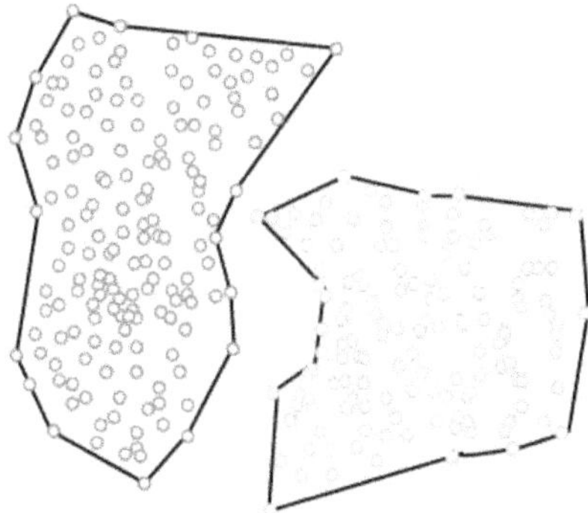

Figure 13.3 Density-based clustering.

Overall, density-based clustering (Figure 13.3) is an effective approach for identifying clusters in complex datasets, particularly those with varying densities and non-spherical clusters.

13.3.3 Distribution Model-Based Clustering

Distribution model-based clustering is a clustering technique that involves using a predefined mathematical model to fit and optimize the data based on the assumption that the data is hybrid in the form of probability distributions. The number of clusters is then computed based on standard statistics, taking noise and outliers into account to ensure robust clustering. This method falls into two categories: statistical and neural network approach methods.

Examples of distribution model-based clustering algorithms are as follows:

1. MCLUST (model-based clustering)
2. GMM (Gaussian mixture models)

The statistical model-based clustering algorithms follow probability measures to determine clusters based on predefined mathematical models, while neural-network approach algorithms utilize unit carrying weights to associate input and output.

The distribution model-based clustering method divides data based on the probability of how a dataset belongs to a particular distribution. This grouping is achieved by assuming some distributions, commonly Gaussian distribution, as a basis for clustering.

One instance of a distribution model-based clustering algorithm is the expectation-maximization clustering algorithm that utilizes Gaussian mixture models (GMM), see Figure 13.4.

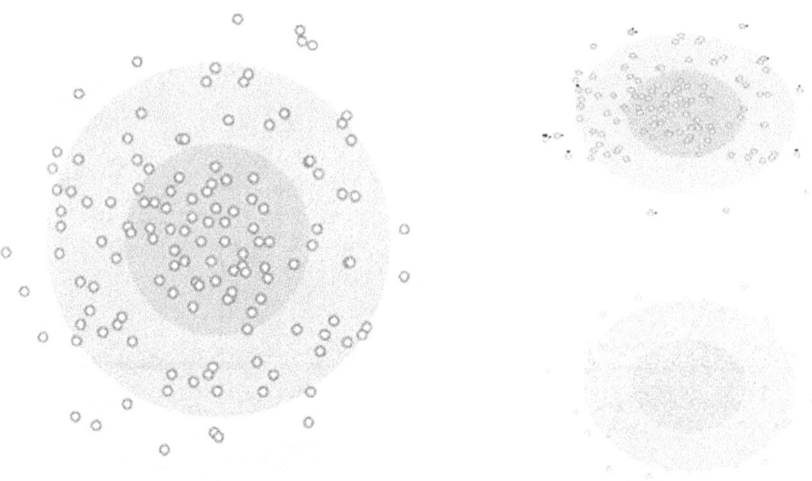

Figure 13.4 GMM model-based clustering.

13.3.4 Hierarchical-Based Clustering

Hierarchical clustering is a clustering technique that creates a hierarchy of clusters using a tree-like structure called a dendrogram, without the need to specify the number of clusters in advance. This can be done using either agglomerative or divisive methods.

Agglomerative hierarchical clustering is the more commonly used approach, where each data point starts as its own cluster, and then pairs of clusters are merged together based on a similarity or distance measure until only one cluster remains. This merging process continues until a dendrogram is formed, which displays the hierarchy of clusters at different levels of similarity.

Once the dendrogram is created, the appropriate number of clusters can be selected by cutting the tree at the desired level. Divisive hierarchical clustering, on the other hand, starts with all data points in a single cluster, and then recursively divides the clusters until each data point is in its own cluster.

Some of the examples of hierarchical clustering are as follows:

1. BIRCH (balanced iterative reducing clustering and using hierarchies)
2. CURE (clustering using representatives)

Overall, hierarchical clustering (Figure 13.5) is a flexible clustering technique that can be useful when the number of clusters is not known in advance or when the structure of the data suggests a hierarchical organization. However, it can be computationally expensive for large datasets, and the choice of similarity or distance measure can impact the resulting clusters.

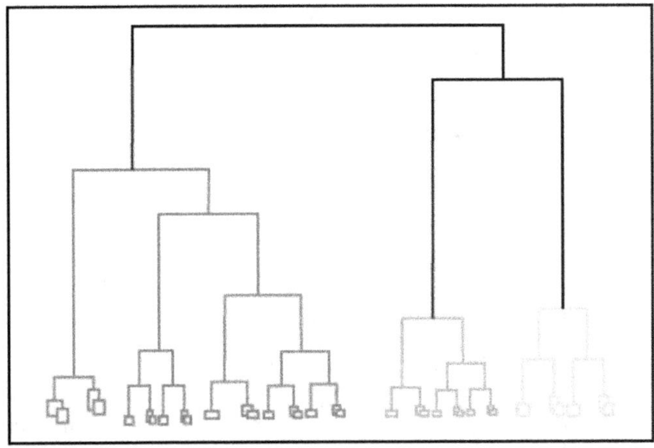

Figure 13.5 Hierarchical-based clustering.

13.3.5 Grid-Based Clustering

Grid-based clustering is a method of partitioning data into a grid-like structure composed of a finite number of cells. In this approach, data objects are assigned to the nearest cell based on their proximity to the cell's center point. This process is known as quantization and creates a discrete representation of the continuous data space.

The grid can then be used to identify clusters of data objects that are grouped closely together. This is achieved by applying various clustering operations to the grid. Grid-based clustering can be computationally efficient, as the number of data objects does not affect the computational complexity of the algorithm.

One example of a grid-based clustering algorithm is the self-organizing map (SOM), which creates a low-dimensional grid of nodes to represent the data distribution. SOM adjusts the position of each node in the grid to better match the data, creating clusters of similar data points in the grid.

Examples of grid-based clustering algorithms are as follows:

1. STING (statistical information grid)
2. Wave cluster
3. CLIQUE (clustering in quest)

By computing statistical measurements for the grids, the speed of the method can be significantly increased. Additionally, grid-based clustering algorithms are highly scalable and require less storage space than the original data stream, making them an efficient choice for large datasets.

However, the quality of the clusters produced by grid-based clustering can be sensitive to the size and shape of the grid, as well as the distance metric used to measure proximity between data objects and grid cells. Careful consideration of these factors is necessary to obtain meaningful clustering results.

13.3.6 Fuzzy Clustering

Fuzzy clustering is a soft clustering technique where a data point can belong to more than one cluster with varying degrees of membership. Unlike traditional hard clustering methods, which assign data points to a single cluster, fuzzy clustering allows for a more nuanced representation of data points and their relationships. In fuzzy clustering, each data point is assigned a set of membership coefficients that reflect the degree of membership in each cluster.

One example of fuzzy clustering is the fuzzy C-means (FCM) algorithm, which is a generalization of the traditional *k*-means algorithm. FCM assigns each data point a membership grade for each cluster and iteratively updates the cluster centers and membership grades until convergence. FCM is widely used in various applications such as image segmentation, pattern recognition, and data mining.

13.4 Clustering Algorithms

Clustering algorithms can be classified based on the underlying models used to cluster the data. Although many clustering algorithms have been proposed, only a few are widely used. The choice of clustering algorithm depends on the type of data being analyzed. For example, some algorithms require an estimate of the number of clusters in the data, while others aim to find the minimum distance between data observations. Each type of clustering algorithm has its own strengths and weaknesses, and may be more suitable for certain types of data or applications.

Overall, selecting an appropriate clustering algorithm for a given dataset is a crucial step in the clustering process, as it can significantly affect the quality of the clustering results.

Here we are discussing mainly popular clustering algorithms that are widely used in machine learning. Facing the increasing diversity of applications, it needs different clustering algorithms for changing scenarios occasionally, consequently, it is essential to propose general acceleration methods for different clustering algorithms, such as *k*-means, PAM, SLINK, and DBSCAN, etc.

13.4.1 K-Means Clustering Algorithm

The *k*-means clustering algorithm is a popular unsupervised learning algorithm used for clustering data. It partitions a set of observations into *k* clusters based on their similarities. This clustering algorithm that works by partitioning data into *k* clusters, where *k* is a user-defined parameter. The algorithm works by iteratively assigning data points to the nearest centroid (center point of the cluster), and updating the centroids based on the mean of the assigned data points.

The objective of this clustering algorithm is to minimize the total sum of squared distances between each data point and its assigned cluster center. This means that the algorithm aims to find cluster centers that are as close as possible to the data points assigned to them, in order to minimize the overall distance between the data points and their respective cluster centers.

The algorithm works as follows:

1. Initialization:
   - Choose the number of clusters k that has to be created.
   - Initialize k cluster centroids randomly by selecting k data points from the dataset. Alternatively, you can use a different initialization method, such as k-means++.
2. Assignment:
   - Assign each data point to the closest centroid based on the Euclidean distance between the data point and the centroid.
   - This creates k clusters of data points that are closest to each centroid.
3. Update:
   - Recalculate the centroid of each cluster based on the mean of all the data points assigned to that cluster.
   - This moves each centroid to the center of its cluster.
4. Repeat:
   - Repeat steps 2 and 3 until either:
   - The centroids no longer move significantly. This means that the algorithm has converged to a stable solution.
   - A maximum number of iterations has been reached. In practice, this is often set to a fixed number, such as 100 or 1000 iterations.
5. Output:
   - The output of the k-means algorithm is the set of k cluster centroids and the set of data points assigned to each cluster.

There are some variations of the k-means algorithm that can improve its performance or overcome some of its limitations. For example, the k-means++ initialization method can lead to better results than random initialization. Another variation is the mini-batch k-means algorithm, which is faster and more scalable for large datasets but can lead to slightly suboptimal results compared to the standard k-means algorithm.

One of the main limitations of the k-means algorithm is that it requires the user to specify the number of clusters k beforehand, which can be difficult in some cases. Also, the algorithm can converge to a local minimum instead of the global minimum, leading to suboptimal results. To overcome this limitation, one can run the algorithm multiple times with different initializations and choose the best result.

13.4.2 PAM Clustering Algorithm

PAM (partitioning around medoids) is a clustering algorithm that is similar to *k*-means, but instead of using centroids to represent clusters, PAM uses representative objects known as medoids. A medoid is a data point within a cluster that minimizes the average distance between all the points in the cluster.

PAM is a more computationally intensive algorithm than *k*-means, as it requires calculating the pairwise distances between all data points and medoids at each iteration. However, PAM is more robust to outliers and can handle non-convex clusters better than *k*-means. PAM is commonly used in applications such as image segmentation, gene expression analysis, and marketing segmentation.

The PAM algorithm works as follows:

1. Initialization:
   - Choose the number of clusters *k* that has to be created.
   - Randomly select *k* data points from the dataset to be the initial medoids.
2. Assignment:
   - Assign each data point to the closest medoid based on the distance between the data point and each medoid.
   - This creates *k* clusters of data points that are closest to each medoid.
3. Update:
   - For each cluster, calculate the total distance between all data points in the cluster and each candidate medoid (which can be any data point in the cluster).
   - Select the candidate medoid with the lowest total distance as the new medoid for that cluster.
   - Repeat steps 2 and 3 until the medoids no longer change or a maximum number of iterations is reached.
4. Output:
   - The output of the PAM algorithm is the set of *k* medoids and the set of data points assigned to each medoid.

The PAM algorithm has some advantages over *k*-means. One of the main advantages is that it is less sensitive to outliers because it uses a representative point that is an actual data point, rather than a mean or centroid that can be influenced by outliers. In other words, if there are a few data points that are very far from the rest of the data, *k*-means may assign them to their own cluster or to an existing cluster that is very far from the other clusters, while PAM will always assign them to the cluster whose medoid is closest to them.

Another advantage of PAM is that it can handle non-convex clusters better than *k*-means. *k*-means assumes that the clusters are convex, which means that they have a round or ellipsoidal shape. However, many real-world datasets have clusters that are not convex, such as clusters that are long and thin, or clusters that have multiple disconnected parts. PAM does not make any assumptions about the shape of the clusters, so it can handle these types of clusters better than *k*-means.

However, PAM also has some disadvantages compared to *k*-means. One of the main disadvantages is that it can be slower and less scalable than *k*-means because it requires a distance calculation between each data point and each medoid at each iteration. In practice, PAM is often used when the number of clusters is small and the dataset is not too large, or when the clusters are expected to have a non-convex shape that *k*-means may not be able to capture well.

13.4.3 SLINK Clustering Algorithm

SLINK (single-linkage) is a hierarchical clustering algorithm that works by starting with each data point as its own cluster and then successively merging the two closest clusters together until all the data points are in a single cluster. SLINK is based on the idea that the similarity between two clusters is the minimum similarity between any two points in the clusters.

The algorithm begins by calculating the pairwise distances between all pairs of data points. It then creates a separate cluster for each data point. At each iteration, the two clusters with the smallest pairwise distance are merged together, and the pairwise distances between the new cluster and all the other clusters are updated. This process is repeated until all data points are in a single cluster.

The SLINK algorithm works as follows:

1. Initialization:
   - Assign each data point to its own cluster.
2. Compute the proximity matrix:
   - Compute the distance between each pair of clusters based on the distance between their closest points (i.e., single linkage).
   - Store the distances in a proximity matrix.
3. Merge the closest clusters:
   - Find the pair of clusters with the smallest distance in the proximity matrix.
   - Merge these two clusters into a single cluster.
   - Update the proximity matrix by removing the rows and columns corresponding to the merged clusters and adding a new row and column for the merged cluster.
4. Repeat step 3 until a stopping criterion is met:
   - The stopping criterion can be a fixed number of clusters, a threshold distance, or a hierarchical tree structure that represents all possible clustering.
5. Output:
   - The output of the SLINK algorithm is a hierarchical tree structure (also called a dendrogram) that represents all possible clustering of the data points.

One of the main advantages of SLINK is that it is computationally efficient for large datasets because it only needs to compute the distance between each pair of clusters once. However, SLINK is sensitive to noise and outliers, as they can cause the algorithm to merge clusters prematurely. Additionally, because SLINK is based on the single-linkage criterion, it can produce long, chain-like clusters in high-dimensional datasets. Another advantage is that it can handle non-convex clusters and clusters of different shapes.

Despite its limitations, SLINK is a widely used algorithm for hierarchical clustering, especially in applications where a large number of data points are involved. It is a simple and intuitive approach that also makes it a good starting point for more complex clustering methods.

These are just a few examples of the many clustering algorithms available in machine learning. The choice of algorithm will depend on the specific requirements of the application and the characteristics of the data being clustered.

13.4.4 DBSCAN Clustering Algorithm

DBSCAN (density-based spatial clustering of applications with noise) is a clustering algorithm that groups together data points that are close to each other in terms of a specified distance metric and have high density, while marking data points that lie alone in low-density regions as noise.

The DBSCAN algorithm works as follows:

1. Parameter Selection:
   - Choose a distance metric (e.g., Euclidean distance) and a radius ε.
   - Choose a minimum number of points m that must be within the radius ε of a data point for it to be considered a core point. These parameters can be set manually, or they can be chosen automatically using a grid search or other optimization technique.
2. Core Point Identification:
   - For each data point, compute the distance to all other data points.
   - Identify all data points that are within the radius ε of each data point.
   - If a data point has at least m other data points within the radius ε, it is considered a core point.
   - A core point is a point that has at least m other points within the radius ε, and it is located in the interior of a cluster.
3. Cluster Assignment:
   - For each core point, create a new cluster and add all data points that are within the radius ε to the cluster.
   - For each non-core point, assign it to the cluster of its nearest core point.

- If a data point is not within the radius ε of any core point, it is marked as noise.

4. Output:
    - The output of the DBSCAN algorithm is a set of clusters and a set of noise points.

DBSCAN has several advantages and disadvantages. One advantage is that it does not require the specification of the number of clusters beforehand, which makes it suitable for datasets with an unknown number of clusters. Another advantage is that it can handle clusters of different shapes and sizes and can identify noise points. Additionally, DBSCAN is efficient for large datasets.

However, one disadvantage of DBSCAN is that it is sensitive to the choice of distance metric and the selection of the parameters ε and m. The values of ε and m should be chosen carefully based on the characteristics of the dataset. Additionally, DBSCAN may not perform well on datasets with varying densities or clusters with significantly different densities. Finally, DBSCAN may not work well on high-dimensional data because the distance metric may become less meaningful in high-dimensional spaces.

To address some of the limitations of DBSCAN, various extensions and modifications have been proposed, such as OPTICS, HDBSCAN, and DBSCAN++, which can handle datasets with varying densities and produce more accurate clustering results.

13.5 Top Clustering Applications

Clustering is a powerful technique with a wide range of applications in many different fields. Here are some examples of top clustering applications:

13.5.1 Image and Video Segmentation

Clustering algorithms are commonly used to group pixels in an image or video into segments based on their color or intensity. This is useful for object detection and recognition, image retrieval, and video surveillance.

13.5.2 Natural Language Processing

Clustering is used to group similar documents, sentences or words, which is useful for topic modeling, document classification, and information retrieval. For example, clustering can be used to identify topics in a large corpus of text, or to group together documents that are similar in content.

13.5.3 Customer Segmentation

Clustering is used to group customers based on their behavior, demographics, or preferences. This is useful for targeted marketing, customer retention, and product development. For example, clustering can be used to identify groups of customers who are likely to respond to a particular marketing campaign or who share similar purchasing patterns.

13.5.4 Anomaly Detection

Clustering is used to find patterns that differ from typical behavior in order to uncover anomalies. This is useful for fraud detection, network intrusion detection, and predictive maintenance. For example, clustering can be used to identify network traffic that is unusual or suspicious, or to identify manufacturing equipment that is likely to fail.

13.5.5 Bioinformatics

Clustering is used to group genes, proteins, or sequences based on their similarity. This is useful for gene expression analysis, protein structure prediction, and drug discovery. For example, clustering can be used to identify groups of genes that are co-regulated or that share a common function.

13.5.6 Social Network Analysis

Clustering is used to group individuals or communities based on their social connections. This is useful for community detection, opinion mining, and recommendation systems. For example, clustering can be used to identify groups of individuals who are closely connected on social media, or to identify communities of users who share similar interests.

13.5.7 Climate Science

Clustering is used to group weather patterns or climate models based on their similarity. This is useful for weather forecasting, climate modeling, and climate change analysis. For example, clustering can be used to identify regions that are likely to experience similar weather patterns, or to group together climate models that share similar features.

13.5.8 Finance

Clustering is used to group stocks, bonds, or other financial instruments based on their performance or risk. This is useful for portfolio optimization and risk management. For example, clustering can be used to identify groups of stocks that share similar performance characteristics or that are likely to be affected by similar market conditions.

Overall, clustering is a versatile technique that can be applied to a wide range of applications, and its popularity and usefulness are likely to continue to grow as more data becomes available and as computational power and algorithms improve.

13.6 Accelerators for Clustering Algorithms

In general, there are three main ways to speed up these algorithms: GPU accelerators, hardware accelerators based on field programmable gate arrays (FPGA) or application-specific integrated circuit (ASIC), and distributed cloud computing systems like Hadoop or Spark. The cloud computing approach, out of these options, is built on a number of computers or data centers; as a result, the overall performance is severely constrained by the network size's scalability, and the cost of power is an important factor to take into account. In comparison, GPU offers simple programming options like CUDA, but the energy-efficiency is still a matter of debate because it makes use of a significant amount of computational power. For machine learning and data mining algorithms, there have recently been a variety of hardware accelerators as DianNao [3] to obtain higher energy-efficiency leverage with affordable hardware costs.

Modern hardware accelerators that are focused primarily on accelerating one particular algorithm must be modified when the algorithm changes, which adds time to market (TTM) and requires a large amount of work. The majority of the time, algorithms in the same research area may have identical or at least similar key functions. As a result, it is possible to examine the key codes of related algorithms and create a widely used hardware accelerator for the key codes. After that, many algorithms can share a hardware accelerator without having to modify the hardware circuitry for every algorithm.

13.7 Types of Accelerators

There are several types of accelerators that are commonly used for cluster applications in machine learning. Each type of accelerator has its own set of strengths and weaknesses, and organizations need to carefully consider their specific machine learning workloads when selecting an accelerator. Factors to consider include the type of computations required, the size and complexity of the dataset, the level of customization required, and the cost and availability of the accelerator.

13.7.1 Graphics Processing Units (GPUs)

GPUs and other accelerators are becoming increasingly important in cluster applications due to their ability to perform complex calculations and computations in parallel. GPUs (graphics processing units) are specialized hardware devices that are designed to handle massive amounts of data in parallel and can perform operations much faster than traditional CPUs (central processing units) on certain types of calculations.

TensorFlow is a widely used framework for running deep learning algorithms on GPUs. It is an open-source software library that enables dataflow and differentiable programming for a variety of tasks. TensorFlow has built-in support for distributed computing, allowing users to scale up their computations across multiple GPUs or even multiple machines in a cluster.

Another popular framework for GPU-accelerated computing is NVIDIA CUDA, which is a parallel computing platform and programming model developed by NVIDIA. CUDA allows developers to write high-performance GPU-accelerated applications in C, C++, or Python using a simple set of API functions.

In addition to GPUs, other types of accelerators are also being used in cluster applications, such as FPGAs (field-programmable gate arrays) and ASICs (application-specific integrated circuits). These devices are designed to perform specific tasks much faster than general-purpose CPUs and can be used for a wide range of applications, from cryptography to machine learning.

Overall, the use of GPUs and other accelerators in cluster applications is becoming increasingly important as more and more data-intensive applications are being developed. These devices can help researchers and scientists to run larger and more complex simulations and computations, leading to new insights and discoveries in a wide range of fields.

13.7.2 Tensor Processing Units (TPUs)

TPUs are becoming increasingly popular in cluster applications as they can significantly speed up the training of machine learning models. In a cluster environment, TPUs can be used to distribute the workload across multiple machines, allowing for faster training times and more complex models. The use ofTPUs in a cluster can also reduce the cost of training models, as it allows for the more efficient use of resources.

Google offers TPUs as a cloud-based service called Google Cloud TPU, which provides access to TPUs through the Google Cloud Platform. Google Cloud TPU allows users to run machine learning workloads at scale and provides tools to manage the infrastructure and optimize performance.

Another option for using TPUs in a cluster is to build a custom cluster with TPU-enabled machines. Google provides instructions and tools for building a TPU-enabled cluster using a combination of standard servers and TPU accelerators. This approach allows for even greater control over the cluster environment and can be customized to meet specific performance requirements.

Overall, the use of TPUs in cluster applications is becoming increasingly important as machine learning becomes more widespread. TPUs can significantly accelerate machine learning workloads, allowing researchers and scientists to train more complex models and analyze larger datasets. As a result, TPUs are likely to become an increasingly important tool in the development of new machine learning applications and technologies.

13.7.3 Field Programmable Gate Arrays (FPGAs)

FPGAs (field programmable gate arrays) are a type of accelerator that are becoming increasingly popular in clustering applications. FPGAs are programmable logic devices that can be reconfigured to perform specific tasks. They are highly parallel and can perform many calculations simultaneously, making them ideal for applications that require a lot of computational power.

FPGAs can be used in clustering applications to accelerate a wide range of tasks, including data analytics, machine learning, and scientific simulations. In these applications, FPGAs can be used to offload certain tasks from the CPU or GPU, freeing up these resources for other tasks. FPGAs can also be used to perform certain tasks that are not well-suited for CPUs or GPUs, such as low-level signal processing or encryption.

One advantage of FPGAs is that they can be reprogrammed as needed, allowing them to be customized for specific applications or workloads. This flexibility makes FPGAs ideal for clustering applications, as different workloads can be optimized for different FPGAs. FPGAs can also be programmed to communicate with other FPGAs or CPUs, allowing them to be used in distributed systems.

In clustering applications, FPGAs can be used in a variety of configurations. They can be used as standalone devices, as part of a GPU-accelerated system, or as part of a hybrid CPU-FPGA system. Hybrid systems, in particular, have become increasingly popular, as they offer the benefits of both CPUs and FPGAs.

There are several programming languages and development tools available for programming FPGAs, including VHDL, Verilog, and OpenCL. OpenCL is particularly useful for clustering applications, as it provides a standard API for programming both CPUs and FPGAs.

Overall, FPGAs are becoming an increasingly important tool in clustering applications. They offer high performance, flexibility, and the ability to customize hardware for specific workloads. As a result, FPGAs are likely to play an important role in the development of new clustering applications and technologies.

13.7.4 Application-Specific Integrated Circuits (ASICs)

In clustering applications, ASICs can be used to accelerate a wide range of tasks, including machine learning, data analytics, and scientific simulations. ASICs are particularly useful for tasks that require a lot of computational power and that can be parallelized. They are also well-suited for tasks that require low-latency communication between nodes in a cluster.

ASICs are typically designed using hardware description languages such as VHDL or Verilog. The design process can be complex and time-consuming, as the ASIC must be optimized for a specific task or set of tasks. However, once the ASIC is designed and manufactured, it can provide significant performance gains over traditional CPUs or GPUs.

One advantage of ASICs is that they can be customized for specific workloads, making them ideal for clustering applications. For example, an ASIC could be designed to perform a specific machine learning algorithm with high efficiency, or to perform certain types of data analytics tasks with low power consumption.

ASICs can be used in a variety of configurations in clustering applications. They can be used as standalone devices, as part of a GPU-accelerated system, or as part of a hybrid CPU-ASIC system. Hybrid systems, in particular, have become increasingly popular, as they offer the benefits of both CPUs and ASICs.

Overall, ASICs are an important tool in clustering applications. They offer high performance, low power consumption, and the ability to customize hardware for specific workloads. As a result, ASICs are likely to play an increasingly important role in the development of new clustering applications and technologies. However, the high cost of designing and manufacturing ASICs means that they may only be practical for certain high-performance applications.

13.7.5 Digital Signal Processors (DSPs)

In clustering applications, DSP accelerators can be used to accelerate a wide range of tasks, including audio and video processing, image recognition, and scientific simulations. DSPs can be used to offload certain tasks from the CPU or GPU, freeing up these resources for other tasks. They can also be used to perform certain tasks that are not well-suited for CPUs or GPUs, such as low-level signal processing or encryption.

One advantage of DSP accelerators is that they can be reprogrammed as needed, allowing them to be customized for specific applications or workloads. This flexibility makes DSPs ideal for clustering applications, as different workloads can be optimized for different DSPs.

DSP accelerators can be used in a variety of configurations in clustering applications. They can be used as standalone devices, as part of a GPU-accelerated system, or as part of a hybrid CPU-DSP system. Hybrid systems, in particular, have become increasingly popular, as they offer the benefits of both CPUs and DSPs.

There are several programming languages and development tools available for programming DSPs, including C, C++, and Assembly language. Some DSPs also support high-level languages such as MATLAB or Python.

Overall, DSP accelerators are becoming an increasingly important tool in clustering applications. They offer high performance, flexibility, and the ability to customize hardware for specific workloads. As a result, DSP accelerators are likely to play an important role in the development of new clustering applications and technologies, especially those that involve processing digital signals.

13.7.6 Central Processing Units (CPUs)

CPUs are the most common type of processor found in computers and are capable of performing a wide variety of computations. While they are not as specialized as other types of accelerators, they can still be used for certain machine learning workloads that do not require as much parallelism or computational power.

13.7.7 Graphics Processing Clusters (GPCs)

GPCs are clusters of GPUs that are designed to work together to accelerate machine learning workloads. They are commonly used in high-performance computing environments and can provide even greater performance gains than individual GPUs.

Using accelerators, can bring significant benefits to cluster applications in machine learning, such as faster training and inference times, higher accuracy, and better scalability. However, there are also risks and challenges associated with using accelerators, such as increased complexity in programming and hardware management, higher energy consumption, and increased costs. Organizations need to carefully weigh these factors when deciding whether to use accelerators and which type of accelerator to use for their specific machine learning workloads.

In addition to these accelerators, there are also software-based techniques for accelerating machine learning workloads, such as distributed training and quantization.

13.8 Clustering Algorithm Analysis

In this section, we will give a brief introduction about the four selective clustering algorithms, and then analyze the hot spot of the algorithms for accelerator design guidance. When performing a cluster analysis, the goal is to organize a set of objects into clusters that are more related to one another than to objects in other clusters. Of the state-of-the art clustering algorithms, *k*-means, partitioning around medoids (PAM), SLINK, and DBSCAN algorithms are the most commonly used clustering algorithms in recent decades. In order to explore the common features of these algorithms, Figure 13.6 illustrates the execution time of the functions in each algorithm.

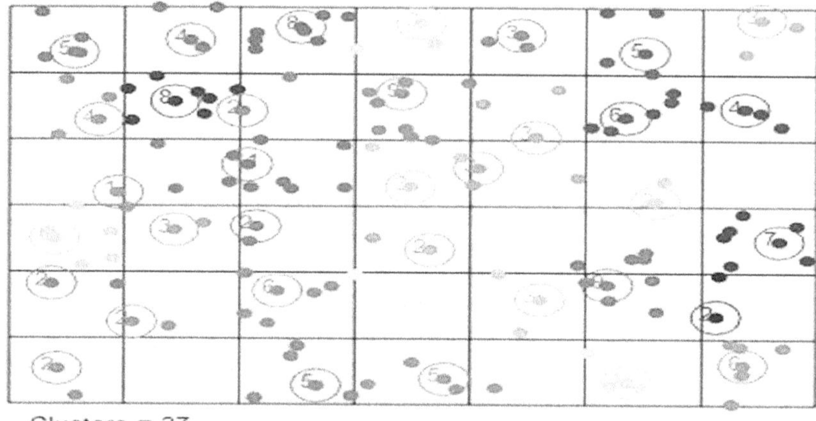

Figure 13.6 Grid-based clustering.

The results report that all the four algorithms contain the same functions including distance calculation and update cluster operations. In particular, the distance calculation is the dominant operation, which takes 87.9%, 70.5%, 83.7%, and 70.1% of the overall execution time in the DBSCAN, SLINK, PAM, and *k*-means algorithms, respectively. Besides the distance calculation, the update cluster operation, takes about 22.5% for SLINK, 16.3% for PAM, 5.3% for DBSCAN, and 20.3% for *k*-means algorithms. Besides these two operations, the overall execution time also includes a small portion contributed by other calculation functions, for example, the find nearest operation takes 9.5% of the *k*-means algorithm. Finally, other operations only take no more than 1.5%, as denoted in the dark yellow legend.

From the profiling exploration of the algorithms, it can be concluded that a ubiquitous accelerator, which includes a hardware module for shared functions like distance calculation, may bring significant speedup to the different algorithms. Regarding the individual operations employed in each algorithm, we can use FPGA to design specific custom modules (denoted as custom instructions) to accommodate different algorithms. Researchers have been working on ASIC and FPGA-based accelerators for a variety of applications over the past few years. Specialized accelerators have been used for machine learning [6], neural networks [7, 8], genome sequencing [9], and graph processing [10, 11]. In particular, Diannao [6, 23] presented the first machine learning accelerator, which was latterly taped out with the specific instruction set architecture [8]. Inspired by the designs of ASIC accelerator, researchers have also been devoted to using FPGAs as the acceleration platform in similar application scenarios [12–14]. Figure 13.7 shows profiling for clustering algorithms. Regarding the clustering algorithms in machine learning, a number of studies have been conducted focused on the FPGA-based accelerators.

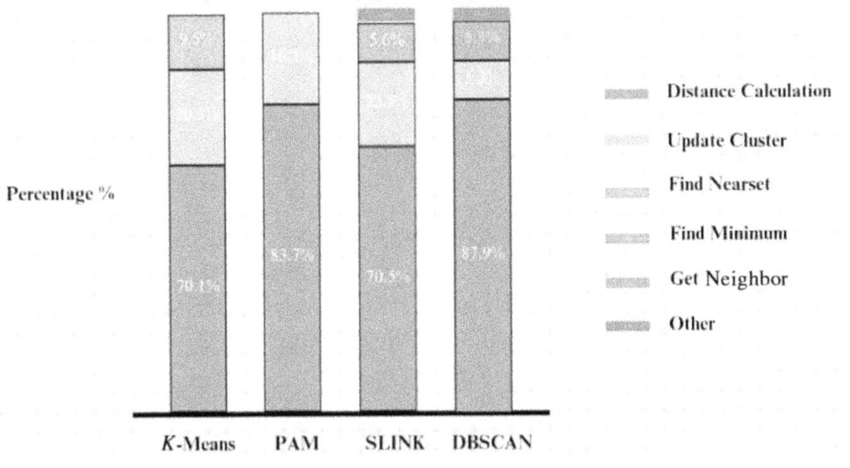

Figure 13.7 Profiling for cluster algorithms.

Winterstein *et al.* present a hardware accelerator for the *k*-means algorithm, which uses a tree-based data structure to store the means and objects for fast mapping [15]. Lin *et al.* designed a *k*-means accelerator for high-dimensional data using triangle inequality [16]. Canilho *et al.* propose a *k*-means accelerator on Zynq-7000 FPGA platform that exploits the parallelism between the ARM processor and the programmable logic and achieves ten performance improvement against ARM-only solution [17]. Similar to this, Abdelrahman *et al.* construct the *k*-means algorithm on the strongly coupled CPU-FPGA platform Intel QuickAssist, where the hardware accelerator collaborates with CPU threads to speed up the process by 3.8 compared to the software-only implementation [18]. Raghavan *et al.* suggested a parameterized and scalable *k*-means accelerator from the standpoint of architecture design, which can be adjusted to accommodate various data sizes, vector numbers, and clustering numbers [19]. Penha *et al.* present a code generator that can automatically generate hardware codes for *k*-means algorithm implemented on GPUs and FPGAs [20]. Bis-KM introduces the idea of data quantification into *k*-means algorithm and deploys corresponding accelerator architecture to make a trade-off between the clustering accuracy and memory bandwidth utilization [21].

This work mainly distinguishes itself from other approaches in the following ways. First, although there have been a number of previous works using FPGA to improve the performance of the clustering algorithms, most literature only focuses on the accelerators for one dedicated algorithm, especially for the *k*-means algorithm. Therefore, it is still an open question to design a ubiquitous accelerator that is able to accommodate multiple algorithms. Another weakness is that in most hardware implementations when the number of the centroids grows, it does not scale regularly due to limited memory blocks. Some studies though claim that they can adapt to unlimited data sizes, the detailed solution is not comprehensively presented to address the frequent off-chip memory access problem.

13.9 Advantages of Accelerators

Accelerators offer several advantages for cluster applications in machine learning. Overall, accelerators are essential tools for organizations that are running cluster applications in machine learning. They offer significant performance benefits, energy efficiency, scalability, and customization, making them a valuable investment for any organization that is looking to improve their machine learning capabilities.

13.9.1 Increased Performance

One of the main advantages of using accelerators for cluster applications in machine learning is the increased performance they offer. Compared to general-purpose CPUs, accelerators are made to be more efficient at doing a given sort of computation. For example, GPUs are designed to handle large matrices and perform parallel computations, which makes them ideal for deep learning applications. By using accelerators, organizations can significantly reduce the time it takes to train machine learning models, which can lead to faster time-to-market and improved productivity.

13.9.2 Better Energy Efficiency

Another advantage of using accelerators is their energy efficiency. Accelerators are designed to perform computations using less energy than general-purpose CPUs. This means that they can significantly reduce the energy consumption of machine learning workloads, which is important for reducing costs and minimizing the environmental impact of computing. For example, Google's Tensor Processing Units (TPUs) are designed to deliver up to 15× higher energy efficiency compared to traditional CPUs.

13.9.3 Scalability

Accelerators can be easily scaled to handle large-scale machine learning workloads. They can be added to a cluster as needed to handle additional workloads, allowing organizations to quickly scale up their computing resources as needed. This is particularly important for organizations that are dealing with large amounts of data or complex machine learning workloads that require significant computing resources [22].

13.9.4 Customization

Many accelerators can be customized to suit specific machine learning workloads. This means that organizations can optimize their hardware resources to achieve maximum performance and efficiency for their specific use cases. For example, NVIDIA's Tensor Cores can be customized to accelerate specific operations in deep learning workloads, such as matrix multiplication.

13.9.5 Support for Complex Workloads

Accelerators are designed to support complex machine learning workloads, including deep learning and neural network applications. They can handle large matrices and perform parallel computations, making them ideal for these types of workloads. This means that organizations can use accelerators to train more complex models and achieve higher levels of accuracy.

13.10 Risks and Challenges

While accelerators offer several advantages for cluster applications in machine learning, there are also some risks and challenges that organizations should be aware of. Here are some of the key risks and challenges:

13.10.1 Complexity

Accelerators can be complex to implement and manage. They often require specialized knowledge and expertise to properly configure and optimize, which can be a challenge for organizations that do not have a dedicated team of experts.

13.10.2 Compatibility

Not all accelerators are compatible with all machine learning frameworks and libraries. This means that organizations may need to modify their existing code or adopt new frameworks to take advantage of accelerators. This can be a significant undertaking, particularly for organizations that have already invested heavily in their existing infrastructure.

13.10.3 Cost

Accelerators can be expensive, particularly for organizations that need to purchase large numbers of them to handle their machine learning workloads. In addition, accelerators often require specialized hardware infrastructure, such as high-performance computing clusters, which can further increase costs.

13.10.4 Vendor Lock-In

Some accelerators are proprietary and only work with specific vendors or software platforms. This can create a vendor lock-in situation where organizations are tied to a specific vendor or software platform for their machine learning workloads.

13.10.5 Integration

Integrating accelerators into existing machine learning workflows can be a challenge. Organizations need to ensure that their existing workflows are compatible with the accelerators and that data can be efficiently transferred between the two.

13.10.6 Maintenance

Accelerators require ongoing maintenance and support. Organizations need to ensure that they have the resources to properly maintain and support their accelerators, including performing software updates and addressing hardware failures.

Although hardware accelerators can offer significant benefits for cluster applications in machine learning, organizations must also be mindful of the potential risks and challenges involved. These include issues related to complexity, compatibility, cost, vendor lock-in, integration, and maintenance. To make informed decisions about whether or not to invest in accelerators for their machine learning workloads, organizations should carefully evaluate these factors and consider strategies for mitigating these challenges. By doing so, they can maximize the benefits of using accelerators while minimizing the potential downsides.

13.11 Conclusions

In this chapter, the importance of machine learning is emphasized, and various clustering techniques that are commonly used in the field are defined. Additionally, the chapter covers different clustering applications, such as image segmentation, speech recognition, and anomaly detection. The chapter proposes a ubiquitous accelerator based on FPGA with custom instructions that is designed to accelerate four representative clustering methods: *k*-means, PAM, SLINK, and DBSCAN algorithms. The researchers analyzed the common functions in these four algorithms and implemented a hardware accelerator that can support various algorithms through customized instructions. The use of accelerators offers several key advantages for cluster applications in machine learning, such as increased performance, energy efficiency, scalability, customization, and support for complex workloads. By leveraging accelerators, organizations can improve their machine learning capabilities and gain a competitive edge in their respective industries. However, there are also risks and challenges associated with using different accelerators. For instance, hardware accelerators like FPGAs and ASICs require specialized programming skills and hardware knowledge, which may limit their adoption in some organizations. Moreover, accelerators may introduce new security risks, and their upfront costs may be higher compared to traditional CPUs. The chapter also discusses the potential of using hybrid computing systems, which combine traditional CPUs with accelerators, to achieve even higher performance and energy efficiency. Hybrid systems can leverage the strengths of both types of computing resources and mitigate their weaknesses. Overall, the paper highlights the potential benefits of using hardware accelerators for clustering applications in machine learning and the challenges that need to be addressed to fully realize their potential. It also presents different approaches that can be used to optimize the use of accelerators and hybrid systems for different applications.

References

[1] Y. LeCun, Y. Bengio, and G. E. Hinton, "Deep learning," *Nature*, 521, pp. 436–444, 2015.

[2] J. O. Awoyemi, A. O. Adetunmbi, and S. A. Oluwadare, "Credit card fraud detection using machine learning techniques: A comparative analysis," in *2017 International Conference on Computing Networking and Informatics (ICCNI)*, pp. 1–9, 2017.

[3] J. Hirschberg, and C. D. Manning, "Advances in natural language processing," *Science*, 349, pp. 261–266, 2015.

[4] M. Eisenstein, "Big data: The power of petabytes," *Nature*, vol. 527, pp. S:2–S:4, Nov. 2015.

[5] H. J. Jay Shendure, "Next-generation dna sequencing," *Nature Biotechnology*, vol. 26, pp. 1135–1145, Nov. 2015.

[6] T. Chen, "Diannao: A small-footprint high-throughput accelerator for ubiquitous machine-learning," *ASPLOS* 'vol. 14, pp. 269–284, 2014.

[7] Z. Du, and et al., "Shidiannao: Shifting vision processing closer to the sensor," in *Proceedings of the 42nd Annual International Symposium on Computer Architecture*, pp. 92–104, 2015.

[8] S. Liu, and et al., "Cambricon: An instruction set architecture for neural networks," in *2016 ACM/IEEE 43rd Annual International Symposium on Computer Architecture (ISCA)*, pp. 393–405, IEEE, 2016.

[9] D. Fujiki, and et al., "Genax: A genome sequencing accelerator," in *2018 ACM/IEEE 45th Annual International Symposium on Computer Architecture (ISCA)*, pp. 69–82, IEEE, 2018.

[10] T. J. Ham, and et al., "Graphicionado: A high-performance and energy-efficient accelerator for graph analytics," in *2016 49th Annual IEEE/ACM International Symposium on Microarchitecture (MICRO)*, pp. 1–13, IEEE, 2016.

[11] M. Yan, and et al., "Alleviating irregularity in graph analytics acceleration: A hardware/software co-design approach," in *Proceedings of the 52nd Annual IEEE/ACM International Symposium on Microarchitecture*, pp. 615–628, 2019.

[12] E. Nurvitadhi, and et al., "Graphgen: An fpga framework for vertexcentric graph computation," in *FCCM*, pp. 25–28, IEEE, 2014.
[13] C. Zhang, and et al., "Caffeine: Toward uniformed representation and acceleration for deep convolutional neural networks," *IEEE Transactions on Computer-Aided Design of Integrated Circuits and Systems*, vol. 38, no. 11, pp. 2072–2085, 2018.
[14] S. Zhou, and et al., "Fastcf: Fpga-based accelerator for stochasticgradient-descent-based collaborative filtering," in *Proceedings of the 2018 ACM/SIGDA International Symposium on Field-Programmable Gate Arrays*, pp. 259–268, 2018.
[15] F. Winterstein, S. Bayliss, and G. A. Constantinides, "Fpga-based kmeans clustering using tree-based data structures," in *2013 23rd International Conference on Field programmable Logic and Applications*, pp. 1–6, IEEE, 2013.
[16] Z. Lin, C. Lo, and P. Chow, "K-means implementation on fpga for high-dimensional data using triangle inequality," in *22nd International Conference on Field Programmable Logic and Applications (FPL)*, pp. 437–442, IEEE, 2012.
[17] J. Canilho, and et al., "Multi-core for k-means clustering on fpga," in *2016 26th International Conference on Field Programmable Logic and Applications (FPL)*, pp. 1–4, IEEE, 2016.
[18] T. S. Abdelrahman, "Accelerating k-means clustering on a tightlycoupled processor-fpga heterogeneous system," in *ASAP*, pp. 176–181, IEEE, 2016.
[19] R. Raghavan, and D. G. Perera, "A fast and scalable fpga-based parallel processing architecture for k-means clustering for big data analysis," in 2017 *IEEE Pacific Rim Conference on Communications, Computers and Signal Processing (PACRIM)*, pp. 1–8, IEEE, 2017.
[20] J. C. Penha, and et al., "A gpu/fpga-based k-means clustering using a parameterized code generator," in *WSCAD*, pp. 61–69, IEEE, 2018.
[21] Z. He, and et al., "Bis-km: Enabling any precision K-mean on fpgas", in *FPGA*, pp.233–243, 2020.
[22] Daniel, A., Partheeban, N., and Sriramulu, S., (2021, March). "Enhanced Ant colony optimization algorithm for optimizing load balancing in cloud computing platform", *International Conference on Computational Intelligence in Data Science*, pp. 64–70, Springer, Cham. https://link.springer.com/chapter/10.1007/978-3-030-92600-7_6
[23] S. Premkumar, and An. Sigappi, "IoT-enabled edge computing model for smart irrigation system," *Journal of Intelligent Systems*, vol. 31, no. 1, pp. 632–650, May 2022. doi: 10.1515/jisys-2022-0046

Chapter 14

Design of a Smart Healthcare Environment with Digital Twinning and Machine Learning

S. Geetha, J. Madhusudanan, and V. Prasanna Venkatesan

14.1 Introduction

The constant growth of digital twin technology in learning new skills and understanding capabilities have made it continue to generate the insights needed to make products better and process them more efficiently. Digital twin has achieved its forefront in the Industry 4.0 revolution with the help of powerful data analytics and Internet of Things (IoT) connectivity. The connection between the physical and virtual twin in digital twin addresses the difficulties that are faced during the integration of the IoT and data analytics process. Since decisions made are reliable in digital twin, analysis of applications that use digital twin technology and making real-time decisions has become a better one. It has provided the healthcare sector with substantial benefits, enabling the development of intelligent healthcare settings that can offer real-time patient monitoring and treatment. A smart healthcare environment uses various approaches to address the critical problems faced by modern healthcare. The solutions for healthcare problems demand has increased more due to the increase in the volume of patient data, more technological support required for these systems. The developers of smart healthcare try to include the suggestions

DOI: 10.1201/9781003469612-14

of specialists of digital twin technology, mobile medicine and IoT technologies to address the various challenges that are faced during patient monitoring and treatment. These IoT connectivity and data analytics-based smart healthcare environments enable the convergence of the real and virtual worlds through the deployment of digital twins. The digital twin is a virtual representation of a real-world thing or process that receives regular updates from sensors and other sources of real-time data. The performance of the physical object or process can then be determined by analyzing this data, which enables decision-making, optimization, and predictive maintenance.

Digital twin is considered to be a technology that can be used for various online monitoring services because of its flexibility and decision-making abilities. Many researchers have identified that digital twin interacts with the virtual environment through the physical objects and integrates their communication to decide a better solution. The addition of machine learning support for the digital twin will increase the performance of technology rapidly. Digital twin with the integration with AI and advanced analytics techniques has the ability to predict how the objects will perform. The accuracy of the digital twin in prediction and decision making is also seen to be increasing with support of these technologies. The use of digital twin technology allows for the development of a virtual counterpart of a physical object or process that can be updated in real-time with data from sensors and other sources. The digital twin can learn from the data collected and generate accurate predictions about the performance of the physical object or process by incorporating machine learning algorithms into this system. This can aid in optimizing decision-making, improving communication between various sensors and devices, and improving quality of service (QoS).

Digital twin technology has the potential to transform many industries, including healthcare. Digital twin technology can be used to create virtual models of patients, medical devices, and healthcare facilities in smart healthcare systems. These digital twins can be continuously updated with data from various sensors and sources, allowing healthcare providers to monitor and optimize the performance of healthcare systems, devices, and processes. The ability to predict how patients will respond to treatment is one of the significant benefits of using digital twin technology in healthcare. Digital twins can learn from collected data and generate accurate predictions about the patient's condition using machine learning and advanced analytics techniques, allowing for early intervention and personalized treatment plans. Furthermore, digital twins can improve communication between different sensors and devices, allowing for more efficient and effective healthcare delivery. Another area where digital twin technology could be used in healthcare is in the design and optimization of medical devices and healthcare facilities. Digital twins can be used to simulate the performance of medical devices and facilities, allowing for design testing and validation prior to physical implementation. This can lower development costs, improve product quality, and shorten product time to market.

Overall, the application of digital twin technology in healthcare has the potential to improve care quality, lower costs, and improve patient outcomes. However, in order to ensure widespread adoption of this technology in healthcare systems, concerns about data privacy, security, and interoperability must be addressed.

Large volumes of sensor data may be used to train machine learning algorithms to find patterns and anomalies. When sensors are expected to fail or when a patient is having a health condition, these algorithms may then be utilized to make predictions. Data from various sensors and devices may be evaluated and processed in real-time by incorporating machine learning algorithms into the smart healthcare environment, enabling quicker response times and more accurate decision-making. Machine learning algorithms can help to enhance the QoS offered by smart healthcare settings in addition to enhancing communication and interoperability. It can also enable real-time patient monitoring, which can increase diagnosis and treatment accuracy. This is especially significant in critical care situations, where making rapid and precise decisions is critical to saving lives. The ability of a system to provide a constant level of service to its users is referred to as QoS. Response speed, data accuracy, and service availability are a few examples of QoS elements that might be included in smart healthcare systems. In order to guarantee that patients receive the best treatment possible, QoS may be enhanced by utilizing machine learning algorithms to examine data from sensors and devices. Overall, the incorporation of machine learning algorithms into smart healthcare environments has the potential to revolutionize the way healthcare is delivered. Machine learning algorithms can help to ensure that patients receive timely and accurate care by improving communication and interoperability and optimizing QoS.

A smart healthcare environment dynamically accesses the patient's information through the sensors connected in the environment and transmits it to the monitoring devices to manage the issue and respond in an intelligent manner. Healthcare data is generated by multiple sources and in various formats, which can make it difficult to integrate and analyse. Interoperability challenges arise when healthcare data systems are not designed to communicate with each other, resulting in data silos. The collection and sharing of sensitive patient data in a smart healthcare system raise privacy and security concerns. Health data is attractive to hackers, and any data breaches can compromise patient confidentiality and safety. Smart healthcare systems raise ethical questions about the use of patient data and the potential for bias in decision-making. There is also the potential for over-reliance on technology, which could result in the dehumanization of healthcare. The implementation of a smart healthcare system requires a robust and reliable infrastructure to support data collection, storage, and analysis. The lack of appropriate infrastructure can limit the functionality of a smart healthcare system and reduce its effectiveness. Smart healthcare systems require significant changes in healthcare delivery and workflows. There may be resistance to change, particularly from healthcare providers who may be skeptical of the benefits of these systems. Implementing a smart healthcare system can be expensive, especially for small healthcare providers. The costs of technology, infrastructure, and training can be significant barriers to adoption. One of the major challenges faced by a smart health environment is the communication between the various sensors available in the environment. If sensors are from different manufacturers they cannot communicate effectively with each other. The interoperability between these devices requires additional functionalities for their communication. Inaccurate data collected from sensors will lead to risk of the patient's life. To address the issues that smart healthcare environments encounter, a comprehensive solution that enhances communication, interoperability, and data accuracy is necessary. One of the most promising alternatives is the utilization of digital twins with machine learning algorithms, which can assist in improving the healthcare system's performance and dependability. This work is such an attempt to create a virtual resource (logical resource) to connect with the physical resources and address the connectivity, data transfer, and QoS services with the help of machine learning algorithms in decision-making.

14.2 Literature Review

Digital twin technology plays a vital role in reinvention of digitalization for various industries. As the use of cognitive powers is increasing more of digital twin their future is said to be limitless. The digital twinning concept is improved with the ability to constantly learn new skills and capabilities, which makes them more efficient for better future outcomes. According to Parate [1], the market of digital twins will be increasing by around 30% from 2020 to 2025 every year. The combination of IoT and cloud platforms along with digital twins will increase the demand more in future. In the healthcare industry, digital twins are used to create models that are developed based on the data collected from various wearable devices, patient records, etc., Maddahi [2] states that more training on how the values are read and how they interact with digital twins is required. If this is performed the healthcare experts can easily interact with the virtual object of the digital twin instead of the physical object. Digital twins help in creating twins for the various monitoring devices and sensors used in the healthcare industry. This twinning of devices in healthcare will improve the diagnosis of diseases. They also increase the performance of monitoring patients at a lower cost. Thus personalized healthcare services can be achieved using this twinning technology. The ever changing and digitally expanding world requires these digital twinning services for providing advanced healthcare services. They also assist in achieving an efficient healthcare management with the same available resources. In the future, twinning of organs and cells can also be performed. Virtual surgery facilities can also be developed with the help of AI and machine learning techniques.

Kamel Boulos [3] discusses the digital twin as a simulation model or digital shadow that maps the physical entity with the digital domain. Digital twins in healthcare provides the facility to collect the real-time data about the patients through the sensors in the environment and facilitates better communication between the devices and humans. It also provides research opportunities in developing precise public healthcare services also. As per Volkov [4] the researches related to digital twins in healthcare industry should determine the evaluation of predicted data that produces better outcomes from the collected data and should also analyze the cost-effective methods to apply this approach to healthcare sectors. If digital twin is implemented in healthcare services it will increase the performance of patient monitoring services from virtual spaces and also will improve the treatment options used for different patients effectively. Mobile-based healthcare services with the help of digital twinning technology can be developed as everyone is using mobile applications widely.

As per Hassani [5], the gap between the rich and poor could widen as digitalization develops more quickly. It is crucial to consider if today's quick-moving digital developments are lagging more people behind or advancing them. In addition to worries about the digital divide, there are worries about accessibility, privacy, ethics, security, and suitability for a range of needs, especially given how quickly AI technology is developing across industries. Voigt claims that [6], the improvement of clinical decision-making for specific patients, patient communication, shared decision-making, and consequently quality of treatment, is feasible with the establishment of a DTMS (digital twin for multiple sclerosis). DTs need to be validated by studies, experts, and field trials to demonstrate the efficacy and safety of their approaches before they can be implemented in patient treatment.

Zeadally states that [7], currently, information and communication technologies are advancing quickly. It is a well-known fact that the adoption and deployment of these technologies in the healthcare industry result in major advantages for all parties involved in the industry (affordable healthcare, cost-effective healthcare services, and many others). Islam claims that [8], the system offered "smart healthcare" to monitor patients' vital signs like heart rate and body temperature as well as some indicators of the health of hospital rooms like humidity levels and gas concentrations of CO and CO_2. For all instances of the existing healthcare system, there is an approximately 95% success rate between observed data and real data. Although the tests are carried out outside of the hospital, genuine medical staff can still access and monitor the data in real-time. In times of emergencies or epidemics, the technology can also help nurses and medical professionals because it can quickly examine raw medical data. The created prototype is very easy to use and design. In this scenario, the system is really helpful. Reference [13] says that technology of many new kinds is being created and used more and more to combat diabetes and its complications. By detecting glucose and other biomarkers of glycemic control and connecting glucose levels with insulin delivery, new technologies will enhance the lives of individuals with diabetes.

Croatti shares his idea on integration of agents and states that [16] technology of many new kinds is being created and used more and more to combat diabetes and its complications. By measuring, new technology will enhance the lives of people with diabetes. According to the conceptual design model, a very early prototype of the system has been created. It has a service-oriented architecture (SOA). Every digital twin is created as a micro-service that exposes an ad hoc RESTful API for data and information access. Each micro-service was created using the Vert x library and the Java programming language. The hospital's local area network, where the entire system is installed, acts as a conduit for security issues and access control.

Montagna discussed his view [34] that accurate documentation is essential in trauma resuscitation in order to raise the standard of trauma care. Hospital emergency rooms frequently use faulty handwritten paper records and flow sheets for data collection. They discuss TraumaTracker, a computer-based method for tracking and documenting trauma, in this article. Findings show that adopting TraumaTracker considerably increased the completeness and quality of trauma documentation by enabling the addition of data and information that were not documented in paper documentation, particularly specific times and locations of episodes. Armeni *et al.* observe [21] there is a growing trend of applications in healthcare, with a primary focus on precision medicine, for digital twins (DTs), which are utilized in a variety of different industries (for example, manufacturing, construction, automotive, and aerospace). If DTs reach their full potential, they will enable linked care's as-yet-unrealized promise and change how chronic illness, lifestyle, and health will be treated in the future. Nevertheless, there is currently no agreement on the extent to which DTs in healthcare can provide breakthrough applications in the coming ten years due to technological, regulatory, and ethical barriers. Finally, we review the results to date, prospects and challenges, and offer suggestions to support the further advancement of DT use in healthcare.

Taylor [24] states that the domains have similar ideas about the requirements that digital twins must meet. Several frameworks that provide more in-depth assistance for the conceptual creation and use of digital twins have been developed in the manufacturing area. Yet, the maritime industry has generated useful design patterns for solutions including digital twins. It is intriguing that significant industry players are creating and promoting open platforms for the deployment of digital twins in both domains, which offer the essential infrastructure and support for user communities. Future work will focus on creating and implementing digital twins in the marine and industrial sectors, and eventually, assessing how well they perform in terms of real-world applications. Dmytro states that [27] a variety of advantages of using digital twins make their integration into the digital environments of industrial businesses justifiable. Asset models provide consistent documentation across the course of the plant's life. This gives simulations, optimizations, expansions, or replanning a better place to start. Preventive maintenance is made feasible by combining historical data, planning, real-time data, and their modeling. Also, the maintenance staff has to have access to all the information they require about the test and maintenance plans in one location, step by step. The expenses of building and maintaining the digital twin are currently outweighed by the cost reductions.

Although the links between the components and data cannot be examined, building a system-based digital twin is very expensive for complicated systems and for systems that are already in operation. There are two methods that may be used to create a digital twin. The data-based digital twin is the first method. The alternative strategy uses a system-based digital twin. The best functionality and value are obtained by combining the two strategies. There are several software solutions available from different vendors for both methods. Depending on the precise uses of the digital twin are to be made, a solution must be selected. Access to sensor data or access to the system is a vital requirement as well. The infrastructure, software, and processing power employed restrict the potential and applications of the digital twin as well. High-resolution models are not feasible due to inadequate infrastructure and computer resources. It is reasonable to anticipate that the digital twin design process will become even more standardized. Zweber states that [37] the DSM, DT, and DTw are important components of the DoD DE concept, which aims to leverage digital, model-based methods to enhance the generation, management, and application of data, information, and knowledge throughout the lifespan of a military programme. Sahal [23] investigated the idea of a personal digital twin (PDT), an improved iteration of the DT with the capacity for actionable insight. PDT may benefit patients in particular by facilitating more accurate decision-making, better therapy selection, and optimization.

Reference [45] suggested architecture seeks to deliver a clinical information integration model of the behavior of lung cancer in patients undergoing treatment in the form of a digital twin. As a result, there is study in the medical area, primarily based on clinical data, on the suitability of therapies for patients. The integration of clinical data and diagnostic test results, ultimately with additional data from the patient's outpatient setting and gathered through intelligent sensors, appears ideal in order to produce advances in both the diagnosis and the therapies, though.

Angulo [45] suggested architecture seeks to deliver a clinical information integration model of the behavior of lung cancer in patients undergoing treatment in the form of a digital twin. As a result, studies, primarily based on clinical data, on the suitability of therapies for patients exist. The integration of clinical data and diagnostic test results, ultimately with additional data from the patient's outpatient setting and gathered through intelligent sensors, appears ideal in order to produce advances in both the diagnosis and the therapies, though. The development of digital twins will enable the use of data in a way that is specifically tailored to the requirements of each researcher or scientist, and it will also integrate ICT tools for data analysis, study, validation, visualization, etc. These virtual patients will develop into a potent ICT tool that enables the suggestion of novel disease biomarkers made up of data aggregation, as well as the display and analysis of the findings by physicians and researchers. Last but not least, the ability to create an endless number of samples using a GAN is also desired. This quality could make it possible to continuously simulate some patient-specific variables for research reasons. By increasing the amount of samples, which are typically insufficient in private datasets, this might also aid in the training of other types of models. Due to a lack of examples, it may also be practical in educational settings to minimize subject misunderstandings.

In order to better understand how the healthcare industry is changing into a new era known as Healthcare 4.0 and the potential issues that may arise as a result of this transformation, Hasselgren [44] outlines a theoretical basis for the need for a blockchain enhanced trust model in a virtual healthcare setting. As a result of this investigation, we created and put into use the ground-breaking VerifyMed technology, which could be trusted in a virtualized healthcare setting. Bruynseels [47] gives the idea that "digital twins" offers a sound intellectual and ethical framework for debating future medical advancements and human augmentation. It does this by contrasting augmentation with very detailed, customized information about each person's biological make-up, physiology, way of life, and nutritional preferences. In addition, the topic of digital twins in medicine is still in its infancy but has the potential to develop into a testing ground for therapies and enhancements. Comparing digital twins across whole populations enables a far clearer understanding of health vs sickness, which in turn clarifies the argument between therapy and augmentation. Digital twins might be a valuable resource for discovering cutting-edge engineering approaches for both therapy and augmentation. Digital twins can therefore be used to determine desired physical well-being metrics. Due to the fact that the patterns in the data may be given meaning, digital twins also have the ability to affect an individual's identity.

According to [53], digital twins can boost industrial operations' productivity and efficiency by enabling predictive maintenance, lowering downtime, and enhancing equipment performance. A system that enables the construction and maintenance of digital twins across several industries, such as aviation, power production, and healthcare, is presented as part of the Predix platform from GE. The document underlines the significance of data security and privacy while describing the essential elements of a digital twin, including data intake, data processing, analytics, and visualization. The possible drawbacks of using digital twin technology are also discussed in the article, including the requirement for precise data and data analytic competence. Reference [48] offers case examples of various anonymization methods used to health data, such as eliminating identifying, generalizing or masking the data, and introducing noise. Also, the authors explore the trade-offs between anonymity and data value and emphasize the necessity of striking a balance between these two elements when choosing an anonymization technique. The paper presents a step-by-step approach to implementing an anonymization process, including data inventory, risk assessment, and choosing the appropriate anonymization technique. The writers also go into the ethical and legal ramifications of managing health data, including HIPAA compliance.

Hofmann [56] states that the traditional sources of limitation for human enhancement, such as nature, therapy, and sickness, may have drawbacks, but they may not be required. Inherent in the idea of human improvement itself, the specification-of-betterment dilemma offers ways to control its unjustified spread. We merely need to demand specific, measurable enhancement objectives that are grounded on data and not grandiose guesses, hypes, analogies, or flimsy associations. Human improvements that outline what would improve and offer sufficient justification are desirable and ought to be pursued. It is inappropriate to accept others.

14.3 Drawbacks of Existing IoT-Based Healthcare Applications

In the present scenario [4] the data collected from various wearable devices and sensors are not sufficient enough to interpret the health conditions of the patients. There are no direct mechanisms currently to merge the data seamlessly and there is no centralized data management by any vendor. The increase in the number of wearable devices has also increased the amount of data generated but there is no appropriate network architecture that transmits the data received to the other parties in a stable and timely manner. Though there are mobile applications developed specifically for healthcare that are available in the market, for example Apple Health it is isolated with the Apple Ecosystem. This app cannot be accessed with an open API which makes it narrow the range of users who use it. Similarly other healthcare applications like Google Fit, Microsoft Healthvault, Health Box, Open mHealth all have some limitations, which makes them unsuccessful due to their lack of data integration services needed for healthcare services. One of the major drawbacks of these healthcare applications is their price, which is huge and these healthcare applications are not very compatible with most of the available modern electronic goods.

14.4 Challenges to Digital Twinning

Currently digital twinning in health care faces the challenge of clear data visualization, privacy issues, accessibility of data, flaws in data provided and integration with clinical information. Security of data shared among different platforms and reproducibility issues are some of the challenges for digital twin. Digital gap in providing treatments for individuals and groups is a major challenge. Aggravating the use of digital twins to people who are not familiar with using digital technologies at a specific situation would be a very challenging task for the researchers. Still there are no standard development methods or norms that have been defined for digital twinning. Healthcare professionals cannot blindly accept the decision made by digital twinning in a patient's condition, so creating a trusted digital twinning application is a challenging task. If digital twinning is supported with AI or machine learning algorithms then the knowledge about how the algorithm evaluates the data and does the prediction is to be given for the healthcare professionals. This requires a challenge to digital twinning in educating the professionals about its accuracy and prediction methods.

14.5 Proposed System

The discussion done in the literature review clearly shows that a better and trusted system with digital twins in the healthcare industry is present requirement. The digital twinning layer plays a crucial role in monitoring the patient's physical surroundings and their medical condition in the proposed healthcare system architecture. Digital twinning is a strong technology for constructing a virtual replica of the real environment and the patient that may be used to monitor, analyze, and optimize healthcare services.

Digital twinning can aid in the prediction of possible difficulties and risks in the healthcare setting, which can then be avoided. The digital twinning layer can detect patterns and trends in patient data, assisting in the prediction of prospective health issues. The usage of digital twin technology can also aid in the optimization of medical procedures and treatment plans based on data from patients. Another benefit of employing digital twin technology in healthcare is that it can deliver real-time data analysis and insights. This can aid in the provision of proactive healthcare services to individuals by allowing the system to recognize future health risks and take preventive steps ahead of time. Furthermore, digital twinning can aid in enhancing the efficiency and effectiveness of healthcare services. The system can optimize medical procedures and treatment plans to give better and faster healthcare services to patients by monitoring the patient's physical surroundings and medical state.

To make the digital twinning concept be a trusted one and its results to be accepted with a high accuracy, a better decision-making mechanism according to the context of the environment and patient condition is to be developed. To improve the accuracy and reliability of decision-making mechanisms, the proposed methodology for healthcare is creating a system that makes decisions based on the context identified with the help of digital twinning. Overall, the application of digital twin technology in healthcare has the potential to dramatically improve the quality of treatment offered to patients. The suggested healthcare system design, which includes digital twinning, can aid in the creation of a seamless and proactive healthcare service capable of preventing potential health issues and providing better healthcare to patients. The architecture is designed with various layers where each layer is used for different purposes. The layers of the proposed architectures are human interface layer, device layer, digital twinning layer, context layer, decision-making layer, and application layer. The purpose and the activities performed by each layer is discussed one by one in detail.

14.5.1 Human Interface Layer

The bottom layer is the human interface layer and is responsible for providing an interface between the patients and healthcare professionals. This layer is connected with the next layer, that is, the device layer where the various devices used for monitoring the patients are available. This layer is an interface between human activities and physical devices.

14.5.2 Device Layer

The device layer is essential in the smart healthcare setting because it offers the hardware required for patient monitoring and data collection. Wearable devices such as fitness trackers, smartwatches, and medical sensors are examples of devices in this tier. These devices collect and communicate real-time data on the patient's health to the system for analysis and decision-making. The device layer is in charge of gathering data from sensors and other devices utilized in the healthcare environment. Sensors and devices used to monitor patients, such as blood pressure monitors, pulse oximeters, and EKG machines, are included in this layer. The changes in the behavior and health conditions of patients like blood pressure, heart rate, sugar level, etc., will be monitored and sent to the next layer for analysis of their criticality levels. Any changes in the regular activities or threshold values are also noted and will be informed to the higher layers to make necessary and preventive actions on time. These devices' data is transferred to the digital twinning layer for examination and processing.

14.5.3 Digital Twinning Layer

The digital twinning layer is the main part of the system, as it is in charge of constructing a virtual reproduction of the actual environment and the patient. The physical devices present in the device layer are made as twinning objects in the twinning layer. The twinning layer will monitor the patients from remote places. Figure 14.1 shows proposed architecture for digital twinning.

If any physical device is not working in the device layer [71]. The twin objects available in this layer will provide the service that the failed device has to perform. The twinning layer also performs various functions like monitoring the activities of patients, analyzes the data received from the devices, validates the result according to the situation and optimizes the information and sends it to the next layer called the context layer. The result validation function will validate the information before it is sent to the next layer so the quality of data will be accurate and correct. This layer analyses data obtained from the device layer using powerful algorithms and machine learning techniques to produce a digital twin that properly depicts the physical surroundings and patient status. This will identify what device is facing the issue in sensing the environment and the clone of that device in the twinning layer will be

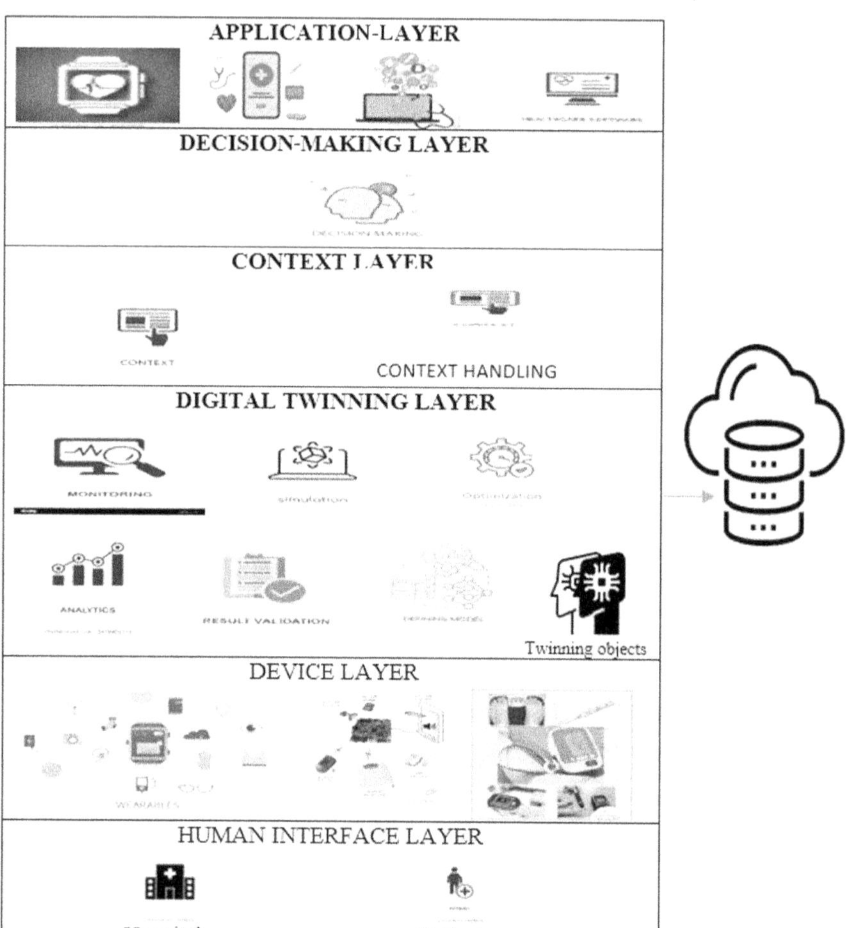

Figure 14.1 Proposed architecture with digital twinning.

activated to do the task without any failure in service [72]. The details of the patients collected and analyzed are stored in a database server to retrieve it whenever needed.

14.5.4 Context Layer

The context layer is the system's fourth layer, and it is in charge of determining the context of the surroundings and the patient's condition. This layer uses data from the digital twinning layer to find patterns and trends in the patient's data, such as blood pressure or heart rate fluctuations. The system can identify the context of the patient's condition and make smart judgements based on this knowledge.

14.5.5 Decision Making Layer

The fifth layer in the system is the decision-making layer will decide what action is to be taken based on the context of the situation prevailing. This layer incorporates a decision-making algorithm that considers the patient's medical history, present state, and other pertinent information while making treatment decisions. The decision made is sent to the application layer, which can be accessed by the users. As the decision on what preventive measure or treatment is to be provided for the patient is done according to the context, seamless service can be achieved in the smart healthcare environment. The decision-making layer will assist in making decisions based on the context and it will consider the previous history of the situation and will give suggestions accordingly. Therefore, a better and proactive service can be provided to the patients.

14.5.6 Application Layer

The application layer is the topmost layer, which will be used by the users. The various activities carried out in the below layers will be visualized in the application layer. It is in charge of communicating the findings of the decision-making process to healthcare professionals and patients. This layer comprises various devices that display real-time information about the patient's state, as well as alerts and notifications that notify healthcare providers of any changes in the patient's condition. This allows the users to access and retrieve the information from the remote location. The interface is also user-friendly and intuitive, allowing users to easily navigate and retrieve the information they require. The application layer includes various kinds of applications, which could be a mobile app, smart watches, laptop application, and smart healthcare software. Emergency alert applications will also be available in this layer to inform the critical situations to the nearby hospitals and the members of the house who are outside. This will help to carry out the smart healthcare service in a smart way during critical conditions. As immediate reaction and measures are available the patient's life can be protected from risk.

14.6 Discussion and Future Research Direction

The widespread use of digital twin in various fields has improved its success ratio in recent years. Healthcare can also achieve this advantage by using digital twinning. Patients can be operated remotely without physical availability of the doctors. They can also access, view, and share their health information from anywhere so immediate service can be provided by health professionals from their location. Emergency conditions can also be handled in a better manner. If digital twinning is allowed to make decisions based on the context of the situation accurate and proper preventive measures can be done. The proposed system provides such a mechanism of

Table 14.1 Comparison of Features of Digital Twinning with Existing Models in Healthcare Industry

Features Compared	*Smart Healthcare Applications*	*Digital Twin for pwMS*	*Smart Healthcare System with IoT*	*Proposed System*
Simulation	No	Yes	Yes	Yes
Integration	Yes	No	No	Yes
Testing	No	Yes	Yes	Yes
Monitoring	Yes	Yes	Yes	Yes
Maintenance	No	No	No	Yes

making decisions based on the context of the data received. This will improve the performance of the smart healthcare environment by providing continuous service to patients and the quality of the service can also be increased. As the service is provided without any interruptions the risk of patients' life-facing problems due to unavailable service at the right time can be reduced. In future AI-based applications can also be included in result validation and model defining. This helps in improving the performance of the model to be more accurate and be readily available from anywhere and at any time. A comparison of the proposed system with the existing healthcare architectures using digital twinning concept is discussed below to understand whether the existing models support the required features that are to be used to perform digital twinning. Features such as simulation, integration of results, testing, monitoring, and maintenance were taken to compare with other models. In future some more example models can be taken for comparison when the model is tested for its application purpose. Table 14.1 shows a comparison of features of digital twinning with existing models in the healthcare industry.

14.7 Conclusion

The use of digital twins in health services have increased more in recent years. In order to realize the Industry 4.0 goal, enabling technologies will be combined with the help of digital twins. Digital twins will be used in several industries, similar to Industry 4.0. The smart healthcare environment should be monitored continuously in order to provide continuous and readily available service to the patients. The capacity to enhance patient care and research is one of the advantages of building a digital twin in healthcare. Some major pros of using digital twins include expanding the availability of healthcare to historically underserved communities, reduced costs for patients' and providers' healthcare, improved patient results in terms of health thanks to individualized treatment programs, and increased operational effectiveness in hospitals and other medical facilities. Any failure in the physical devices present in the environment will create a risk to the patient's life. Therefore, a system with better quality of service and continuous data transfer capability is the need for a healthcare environment. New inclusive laws and regulations will need to be drafted in order to guarantee that the benefits of the new technology reach every level of society, that humans remain important, and that ethics, privacy, and security are not compromised. A significant effort should be made to democratize technology. The General Data Protection Regulation (GDPR), which was approved by the European Parliament and Council in April 2016, is one positive step in the right direction for data protection and privacy regulations. The various governmental entities have to start feasibility studies for using digital twins in their respective industries.

This research is one such attempt to provide a good quality service to the patients without any failure in their service. Data validation will help in providing accurate data to the model. If the data is correct and accurate the decision made for the situation based on context will improve the quality of service provided by the system. This will ultimately improve the diagnosis mechanism, patient well-being in a better way, proactive treatment services and also support for economic cost of healthcare services.

Bibliography

1. Sakshi, P., Bhivgade, A., Manisha Bharati, Digital Twin Technology Transforming Modern Healthcare Industry. *International Journal for Research in Applied Science & Engineering Technology (IJRASET)*, May 2022, 10(V). ISSN: 2321-9653
2. Maddahi, Y., Chen, S. Applications of Digital Twins in the Healthcare Industry: Case Review of an IoT-Enabled Remote Technology in Dentistry. *Virtual Worlds* 2022, 1, 20–41. https://doi.org/10.3390/ virtualworlds1010003
3. Kamel Boulos, M.N., Zhang, P. Digital Twins: From Personalised Medicine to Precision Public Health. *J. Pers. Med.* 2021, 11, 745. https://doi.org/10.3390/ jpm11080745
4. Volkova, I., Radchenkoa, G., Tchernykh, A., Digital Twins, Internet of Things and Mobile Medicine: A Review of Current Platforms to Support Smart Healthcare. *Program. Comput. Softw.* 2021, 47(8), 578–590. ISSN 0361–7688. DOI: 10.1134/ S0361768821080284
5. Hassani, H., Huang, X., MacFeely, S. Impactful Digital Twin in the Healthcare Revolution. *Big Data Cogn. Comput.* 2022, 6, 83. https://doi.org/10.3390/ bdcc6030083
6. Voigt, I, Inojosa, H, Dillenseger, A, Haase, R, Akgün, K, Ziemssen, T. Digital Twins for Multiple Sclerosis. *Front. Immunol.* 2021, 12, 669811. doi: 10.3389/ fimmu.2021.669811
7. Sherali, Z., et al., Smart Healthcare Challenges and Potential Solutions Using Internet of Things (IoT) and Big Data Analytics. *PSU Res. Rev.* 2020, 4(2), 93–109. Emerald Publishing Limited 2399–1747 DOI 10.1108/PRR-08-2019-0027

8. Islam, Md. M., Rahaman, A., Islam, Md. R. Development of Smart Healthcare Monitoring System in IoT Environment. *SN Comput. Sci.* 2020, 1, 185. https://doi.org/10.1007/s42979-020-00195-y
9. Khandekar, G., Padole, Dr. VB. Development of Smart Healthcare Monitoring System In IoT Environment. *International Research Journal of Engineering and Technology (IRJET)*, April 2021, 08(04). ISSN: 2395-0056
10. Case study: Type 1 diabetes – Open mHealth. www.openmhealth.org/features/case-studies/case-study-type-1-diabetes/ Cited 28.04.2021
11. ELPP 2016: Big Data for Healthcare. http://scet.berkeley.edu/wp-content/uploads/Big-Data-for-Healthcare-Report-ELPP-2016.pdf Cited 14.10.2020
12. Case study: Post-Traumatic Stress (PTSD) – Open mHealth. www.openmhealth.org/features/case-studies/case-study-post-traumatic-stress-ptsd/Cited 28.04.2021
13. Wearable device for blood glucose level automatic diagnostics and correction in diabetic blood was created. https://habr.com/ru/company/icover/blog/392085/ Cited 14.10.2020
14. Digital Twin: Enabling Technologies. *Challenges and Open Research – Aidan Fuller, Zhong Fan*, Charles Day, Chris Barlow, November 2020.
15. Digital Twin approach to clinical DSS Explainable AI- Persistent System Ltd. – Dattaraj Jagdish Rao, Shraddha Mane, September 2020.
16. Croatti, A., Gabellini, M., Montagna, S., Ricci, A. On the Integration of Agents and Digital Twins in Healthcare. *J. Med Syst.* 2020, 44, 161.
17. Subramanian, K. Digital Twin for Drug Discovery and Development—the Virtual Liver. *J. Indian Inst. Sci.* 2020, 100, 653–662.
18. Braun, M., Represent Me: Please! Towards an Ethics of Digital Twins in Medicine. *J. Med. Ethics* 2021, 394–400.
19. Rahaman, A., Islam, M., Islam, M., Sadi, M., Nooruddin, S. Developing IoT Based Smart Health Monitoring Systems: A Review. *Rev Intell Artif.* 2019, https://doi.org/10.18280/ria.330605
20. Patil, A.A., Suralkar, Dr. S.R. "Review on IOT Based Smart Healthcare System". *Int. J. of Adv. Research in Engineering and Technology (IJARET)* May – June 2017, 8(3), 37–42.
21. Armeni, P., Polat, I., De Rossi, L.M., Diaferia, L., Meregalli, S., Gatti, A. Digital Twins in Healthcare: Is It the Beginning of a New Era of Evidence-Based Medicine? A Critical Review. *J. Pers. Med.* 2022, 12, 1255. https://doi.org/10.3390/jpm12081255
22. Rasheed, A., San O., Kvamsdal T. Digital Twin: Values, Challenges and Enablers From a Modeling Perspective. *IEEE Access* 2020, 8. doi:10.1109/ACCESS.2020.2970143
23. Sahal, R., Alsamhi, S.H., Brown, K.N. Personal Digital Twin: A Close Look into the Present and a Step towards the Future of Personalised Healthcare Industry. *Sensors* 2022, 22, 5918. https://doi.org/10.3390/s22145918
24. Elkefi, S, Asan, O. Digital Twins for Managing Health Care Systems: Rapid Literature Review. *J Med Internet Res.* Aug 16 2022, 24(8), e37641. doi: 10.2196/37641. PMID: 35972776; PMCID: PMC9428772

25. Taylor, N., Human, C., Kruger, K., Bekker, A., Basson, A. Comparison of Digital Twin Development in Manufacturing and Maritime Domains. In *International Workshop on Service Orientation in Holonic and Multi-Agent Manufacturing*. pp. 148–170. Springer, Cham, 2019.
26. Fuller, A. et al., Digital Twin: Enabling Technologies, Challenges and Open Research. *IEEE Access, Multidisciplinary, Rapid Review, Open Access Journal.* DOI 10.1109/ACCESS.2020.2998358
27. Adamenco, D. et al., Review and Comparison of the Methods of Designing the Digital Twin. *Procedia CIRP* 2020, 91, 27–32. Science Direct.
28. Bönsch, J., Elstermann, M., Kimmig, A., Ovtcharova, J. A Subject-Oriented Reference Model for Digital Twins. *Computers & Industrial Engineering* 2022, 172, 108556. Part A. ISSN 0360-8352, https://doi.org/10.1016/j.cie.2022.108556
29. Boyes, H., Watson, T. Digital Twins: An Analysis Framework and Open Issues. *Computers in Industry* 2022, 143, 103763. ISSN 0166-3614, https://doi.org/10.1016/j.compind.2022.103763
30. Jones, D., Snider, C., Nassehi, A., Yon, J., Hicks, B. Characterising the Digital Twin: A Systematic Literature Review. *CIRP Journal of Manufacturing Science and Technology* 2020, 29, 36–52, Part A. ISSN 1755-5817, https://doi.org/10.1016/j.cirpj.2020.02.002
31. Riazul Islam, S.M, Daehan, K., Humaun Kabir, M, Hossain, M, Kyung-Sup, K. The Internet of Things for Health Care: A Comprehensive Survey. *IEEE Access* 2014, 3, 678–708. https://doi. org/10.1109/ACCESS.2015.2437951
32. Lin, T., Rivano, H., Le Mouel, F. A Survey of Smart Parking Solutions. *IEEE Trans Intell Transp Syst*. 2017, 18, 3229–53. https:// doi.org/10.1109/TITS.2017.2685143
33. Van Houten, H. *The Rise of the Digital Twin: How Healthcare Can Benefit.* 2018. www.philips.com/a-w/about/news/archive/blogs/innovation-matters/20180830-the-rise-of-the-digital-twin-how-health care-can-benefit.html
34. Montagna, S., Croatti, A., Ricci, A., Agnoletti, V., Albarello, V., Gamberini, E. Real-Time Tracking and Documentation in Trauma Management. *Health Informatics Journal*, 2019.
35. Ricci, A., Tummolini, L., Piunti, M., Boissier, O., Castelfranchi, C. Mirror Worlds as Agent Societies Situated in Mixed Reality Environments. In: *Revised Selected Papers of the International Workshops On Coordination, Organizations, Institutions, and Norms in Agent Systems X*, Vol. 9372, pp. 197–212. New York: Springer-Verlag New York, Inc., 2015.
36. IBM. *IBM Knowledge Center – Getting Started with Data Management by usind REST APIs. IBM*. 2019.
37. Zweber, J.V., Kolonay, R.M., Kobryn, P., Tuegel, E.J. Digital Thread and Twin for Systems Engineering: Requirements to Design. *55th AIAA Aerospace Sciences Meeting*, 2017.
38. Kuhn, D.T. *Informatics Dictionary – Digital Twin*. Kaiserslautern: Fraunhofer IESE, 2017.
39. Vogt, O. *Selection of a Suitable Platform for Building up a Digital Twin*. Unpublished master thesis. Duisburg, 2019.
40. Matinez, V., Ouyang, A., Neely, A., Burstall, C., Bisessar, D. *Service Business Model Innovation: The Digital Twin*. Cambridge Service Alliance, 2018.

41. Vogt, O. *Selection of a Suitable Platform for Building up a Digital Twin*. Unpublished master thesis. Duisburg, 2019.
42. Fuller A., Fan Z., Day C., Barlow C. Digital Twin: Enabling Technologies, Challenges and Open Research. *IEEE Access* 2020, 8, 108952–108971. doi: 10.1109/ACCESS.2020.2998358.
43. Akash, S.S., Ferdous, M.S. A Blockchain Based System for Healthcare Digital Twin. *IEEE Access* 2022, 10, 50523–50547. doi: 10.1109/ACCESS.2022.3173617.
44. Hasselgren, A., Rensaa, J.A.H., Kralevska, K., Gligoroski, D., Faxvaag, A. Blockchain for Increased Trust in Virtual Health Care: Proof-of-Concept Study. *J. Med Internet Res*. 2021, 23, e28496. doi: 10.2196/28496. - DOI - PMC - PubMed
45. Angulo, C., Gonzalez-Abril, L., Raya, C., Ortega, J.A. A Proposal to Evolving towards Digital Twins in Healthcare. *Proceedings of the International Work-Conference on Bioinformatics and Biomedical Engineering*, Granada, Spain. 6–8 May 2020; pp. 418–426.
46. Soria, L.M., Gonzalez-Abril, L., Ortega-Ramirez, J.A., Alvarez, M.A. Low Energy Physical Activity Recognition System on Smartphones. *Sensors* 2015, 15, 5163–5196.
47. Bruynseels, K., Santoni de Sio, F., van den Hoven, J. Digital Twins in Health Care: Ethical Implications of an Emerging Engineering Paradigm. *Front. Genet.* 2018, 9, 31.
48. El Emam, K., Arbuckle, L. *Anonymizing Health Data: Case Studies and Methods to Get You Started*. Cambridge: O'Reilly Media, Inc., 2013.
49. Feutry, C., Piantanida, P., Bengio, Y., Duhamel, P. Learning Anonymized Representations with Adversarial Neural Networks. 2018. https://arxiv.org/pdf/1802.09386.pdf
50. Ford, R.A., Price II, W.N. Privacy and Accountability in Black-Box Medicine. *Mich. Telecommun. Technol. Law Rev.* 2017, 23(1).
51. Feutry, C., Piantanida, P., Bengio, Y., Duhamel, P. *Learning Anonymized Representations with Adversarial Neural Networks*. https://arxiv.org/abs/1802.09386. Accessed 12 Dec 2019
52. Kahane, G., Savulescu, J. Normal Human Variation: Refocusing the Enhancement Debate. *Bioethics* 2015, 29, 133–143. doi: 10.1111/bioe.12045
53. General Electric. *Predix Technology Brief – Digital Twin*. Boston, MA: General Electric; 2017.
54. PMI Working Group. *The Precision Medicine Initiative Cohort Program – Building A Research Foundation for the 21st Century Medicine*. 2015. Available at: www.nih.gov/sites/default/files/research-training/initiatives/pmi/pmiworking-group-report-20150917-2.pdf
55. Husain, M., Mehta, M. Cognitive Enhancement by Drugs in Health and Disease. *Cell* 2011, 15, 28–36. doi: 10.1016/j.tics.2010.11.002
56. Hofmann, B. Limits to Human Enhancement: Nature, Disease, Therapy or Betterment? *BMC Med. Ethics* 2017, 18, 56. doi: 10.1186/s12910-017-0215-8
57. Parens, E. The Goodness of Fragility: On the Prospect of Genetic Technologies Aimed at the Enhancement of Human Capacities. *Kennedy Inst Ethics J.* 1995;5(2), 141–53.
58. Agar, N. *Truly Human Enhancement: A Philosophical Defense of Limits*. Cambridge: MIT Press; 2013.

59. Cohen, E. Conservative Bioethics & the Search for Wisdom. *Hastings Cent Rep.* 2006, 36(1), 44–56.
60. Karakra, A., Fontanili, F., Lamine, E., Lamothe, J. HospiT'Win: A Predictive Simulation-Based Digital Twin for Patients' Pathways in Hospital. In *Proceedings of the 2019 IEEE EMBS International Conference on Biomedical Health Informatics (BHI)*, Chicago, IL, USA, 19–22 May 2019; IEEE: Piscataway, NJ, USA, 2019; pp. 1–4.
61. Boje, C., Guerriero, A., Kubicki, S., Rezgui, Y. Towards a Semantic Construction Digital Twin: Directions for Future Research. *Autom. Constr.* 2020, 114, 103179.
62. Kuwabara, A., Su, S., Krauss, J. Utilizing Digital Health Technologies for Patient Education in Lifestyle Medicine. *Am. J. Lifestyle Med.* 2020, 14, 137–142.
63. Van Dijk, M.R., Oostingh, E.C., Koster, M.P.H., Willemsen, S.P., Laven, J.S.E., Steegers-Theunissen, R.P.M. The Use of the mHealth Program Smarter Pregnancy in Preconception Care: Rationale, Study Design and Data Collection of a Randomized Controlled Trial. *BMC Pregnancy Childbirth* 2017, 17, 1–7.
64. Di Guardo, F., Palumbo, M. Immersive Virtual Reality as a Tool to Reduce Anxiety During Embryo Transfer. *J. Obstet. Gynaecol.* 2022, 4, 1–7.
65. Chakshu, N.K., Sazonov, I., Nithiarasu, P. Towards Enabling a Cardiovascular Digital Twin for Human Systemic Circulation Using Inverse Analysis. *Biomech. Modeling Mechanobiol.* 2021, 20, 449–465.
66. Chavez, O.L., Rodríguez, L.F., Gutierrez-Garcia, J.O. A Comparative Case Study of 2D, 3D and Immersive-Virtual-Reality Applications for Healthcare Education. *Int. J. Med. Inform.* 2020, 141, 104226.
67. Rodrigues, T.K., Liu, J., Kato, N. Application of Cybertwin for Offloading in Mobile Multiaccess Edge Computing for 6G Networks. *IEEE Internet Things J.* 2021, 8, 16231–16242.
68. Sepasgozar, S.M. Differentiating Digital Twin from Digital Shadow: Elucidating a Paradigm Shift to Expedite a Smart, Sustainable Built Environment. *Buildings* 2021, 11, 151.
69. Silva, E.S., Hassani, H., Madsen, D.Ø., Gee, L. Googling Fashion: Forecasting Fashion Consumer Behaviour Using Google Trends. *Soc. Sci.* 2019, 8, 111.
70. Van Deursen, A.J., Van Dijk, J.A. The First-Level Digital Divide Shifts from Inequalities in Physical Access to Inequalities in Material Access. *New Media Soc.* 2019, 21, 354–375.
71. Rajavel, R., Ravichandran, S.K.,Harimoorthy, K.,Nagappan, P.,Gobichettipalayam, K.R.IoT-Based Smart Healthcare Video Surveillance System Using Edge Computing. *Journal of Ambient Intelligence and Humanized Computing*, 2022, 13(6), 3195–3207.
72. Ullah, S., Partheeban, N., Sriramulu, S., Soni, R.K., Gupta, S., Daniel, A. Identifying and Analyze the Face Mask Detection for The Person During Covid. *3rd International Conference on Advances in Computing, Communication Control and Networking (ICAC3N)*, 2021, pp. 2036–2040. https://doi.org/10.1109/ICAC3N53548.2021.9725678

Chapter 15

FloodWatch

Suggesting an IoT-Driven Flood Monitoring and Early Warning System for the Flood-Prone Cuddalore District in the Indian State of Tamilnadu

R. Indrakumari, Srinivasan Sriramulu, N. Partheeban, and Rajkumar Rajavel

15.1 Introduction

15.1.1 Floods

Around the world, floods are considered deadly disasters, as they causes severe loss and damage, including human lives. When water soaks or overflows dry land, a flood will happen. Floods will not occur suddenly; hence, residents have time to evacuate or to prepare alternative measures [1]. In rare cases, a flood develops quickly with little warning and causes an enormous number of losses that are hard to reverse. Floods can be developed in many ways, in which the overflow of streams or the river bank is the most common reason. This type of flood is called a riverine flood. Rapid ice melt, heavy rain, or a broken dam can overwhelm the rivers to cause a flood. Tsunami causes coastal flooding and it rushes the seawater to land. The destruction made by the flood is countless and unimaginable. When flood enters the inland with vigorous force, many structures are inept at overcoming the power of the water stream. The flood picks up houses, cars, trees, and bridges and carries them away. Soils erode from

DOI: 10.1201/9781003469612-15

the basement of the buildings and make them tumble and sometimes crack [2]. In this chapter, an IoT-driven flood monitoring and early warning system for the flood-prone Cuddalore district in the Indian state of Tamilnadu is discussed.

15.1.2 Cuddalore District

The Cuddalore district is situated in the eastern coastal region in the state of Tamilnadu, India in the latitude of 150 5' /110 11' and 120 35' N, longitude: 780 38' to 800 00'. Cuddalore is bordered by the districts of Perambalur on the west, Ariyalur, Tanjavur, and Nagapattinam on the south, Villupuram on the north, and the Bay of Bengal on the East [3]. The Cuddalore town is located at the estuary of the rivers Pennaiyar and Gedilam in the coastal line of the Bay of Bengal. The district's total area is 3678 sq. km with an average altitude of 4.6 MSL. Cuddalore district consists of 10 taluks, Cuddalore, Kurinjipadi, Panruti, Chidambaram, Srimushnam, Bhuvanagiri, Virudhachakam, Kattumanarkoil, Veppur, and Tittagudi. Out of the 10 taluks, Cuddalore, Bhuvanagiri, Kurinjipadi, and Chidambaram are coastal taluks, often prone to heavy wind and cyclones, while the other six taluks are considered flood-prone zones [4]. The population of the district is 26,05,914, as per the 2011 Census, in that the urban population is 8,05,781 and the rural population is 18,00,133.

The Cuddalore district has a 2000-year history in trade with the Roman Empire. The Dutch first captured Cuddalore in Southern India followed by Portugal, France, and later Britain.

From 1746 A.D. Cuddalore was adopted as the headquarters of the British in South India. The Britishers constructed St. David's Fort, near Devanampattinam and ruled southern India from this port. The Britishers extensively used the Cuddalore port for their trade. In 1717, the Britishers started St. David School followed by St. Joseph's Higher Secondary School in 1868, and St. Mary's School in 1914.

The Cuddalore district is a combination of natural beauty, cultural heritage, and history. The popular tourist places in Cuddalore district are, Silver Beach, Mangrove Forest in Pitchavaram, Padaleeshwarat Temple, Thiruvananthapuram Temple, Natarajar Temple in Chidambaram, Bhuvaraha Swami Temple, Neyveli Lignite Corporation Limited (NLC), and E.I.D Parry Sugar mill in Nellikuppam established in 1842.

The Cuddalore district is classified as one of the multi-hazard-prone districts of Tamilnadu. The Cuddalore district has a coastal line of 68 km, hence it is vulnerable to cyclones, and heavy rainfall, in turn causing floods. This district witnessed several cyclones and floods in the past. In 2008, the Nisha cyclone, and in 2011, the Thane cyclone created havoc in the district. On top of it, in 2004, this district experienced a Tsunami that caused billions of dollars' worth of devastation and loss of human lives.

The Cuddalore district experiences a dry and wet climate, and during North East Monsoon season the district witnessed heavy rainfall from October till December, and the southwest monsoon sets in June and continues till September. Table 15.1 and Figure 15.1 show the annual rainfall for the past 12 years.

S.No	Year	Annual Rainfall
1	2009	1,219.52
2	2010	1,461.84
3	2011	1,397.93
4	2012	793.09
5	2013	988.18
6	2014	1,186.72
7	2015	1,748.14
8	2016	588.32
9	2017	1,449.68
10	2018	936.62
11	2019	1,242.18
12	2020	1,494.00

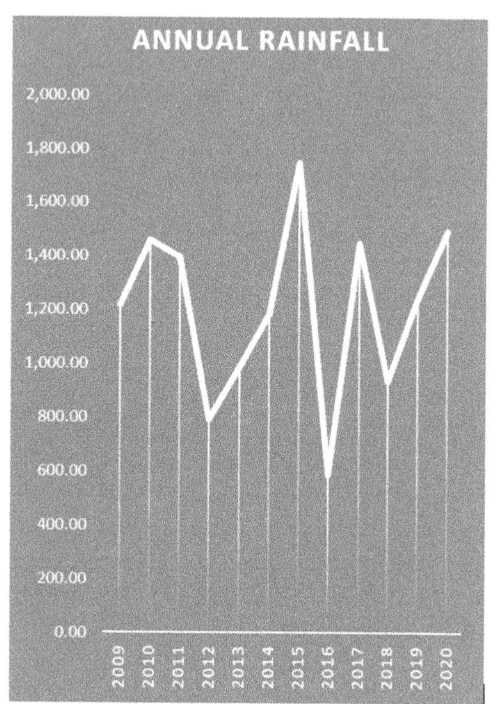

Figure 15.1 Annual rainfall in the Cuddalore district for the past 12 years.

Table 15.1 Annual Rainfall in Cuddalore District for the Past 12 Years

S.No	*Year*	*Annual Rainfall*
1	2009	1,219.52
2	2010	1,461.84
3	2011	1,397.93
4	2012	793.09
5	2013	988.18
6	2014	1,186.72
7	2015	1,748.14
8	2015	588.32
9	2017	1,449.68
10	2018	936.62
11	2019	1,242.18
12	2020	1,494.00

Source: [5].

Figure 15.2 Flood in Gadilam River during the 2015 flood.

The Pennaiyar is the principal river in the Cuddalore district which forms the southern boundary of the Villupuram district and the Northern boundary of the Cuddalore district, and empties itself in the Bay of Bengal. Next is the Gadilam river, which originates in the Sankarapuram taluk and runs through Thirukoilur taluk of the Villupuram district, and travels through Panruti and Cuddalore Taluk [6]. The Gadilam and the Pennaiyar are connected by the river Malattar, which carries surplus water from the Pennaiyar to the Gadilam. The old name of Cuddalore is Kudalur, as the three rivers unite here and drain in the Bay of Bengal. From Virudhachalam taluk, the river Paravanar originates flows through Kurinjipadi, and falls into the Bay of Bengal. The Coleroon Rives which bifurcates from the Cauvery River in Tiruchirapalli, flows on the southern boundary of Kattumannarkoil and Chidambaram taluk and ends in the Bay of Bengal at Parangipettai. Vellar River is formed by the junctions of the rivers Swetanadi and Vasishtanadi in Salem District. Through the Perambalur district and Kalrayan hills, the Vasishtanadi enters the Cuddalorre district for 15 miles and unites with the river Swetanadi. These united rivers flow for 29 miles, gather the water from the Mayura River and Manimukta Gomathi River, and join the Bay of Bengal, near Parangipettai, Chidambaram taluk [7].

These five major rivers, namely, Pennaiar, Gadilam, Paravanar, Vellar, and Coleroon flow through the plain of the Cuddalore district and ends in the Bay of Bengal at closer interval. The areas in which the river originates are upstream of the Cuddalore district. The terrain is just 1.50 m above the mean sea level (MSL), and flat, hence all the overflow water accumulates in the land and is not drained into the sea. Another reason for the flood is that the river level lies only minus one meter from the sea, which causes flood. Figure 15.2 shows the flood in the Gadilam River during the 2015 flood. Figure 15.3 shows the India and Tamilnadu map.

The flood-prone areas of the Cuddalore district are shown in Figure 15.4.

Figure 15.3 India and Tamilnadu map.

15.1.3 IoT in Flood Monitoring System

The Internet of Things (IoT) is an inevitable technology in flood monitoring and early warning systems (FMEWS). The FMEWS deploys devices, communication networks, and IoT-based networked sensors to collect data on meteorological conditions, water levels, and other relevant topographies. The real-time data collected from multiple sources is pre-processed, analyzed, and transferred to provide the authorities and societies with useful information and early flood warnings.

A network of sensors plays a vital role in the IoT-based flood monitoring system that is implanted in flood-prone areas, such as lakes, rivers, and urban drainage systems. These sensors measure the temperature, humidity, rainfall, and other related

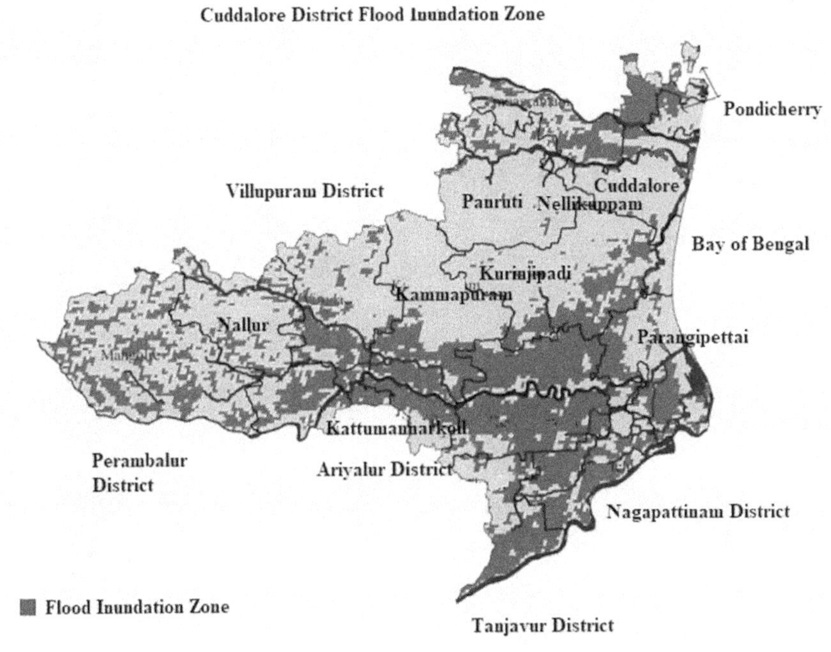

Figure 15.4 Flood-prone zones in the Cuddalore district.

Source: [8].

environmental elements [9]. The collected data are continuously transmitted to a cloud platform or a central server for analysis. Collecting and analyzing real-time data is considered the primary feature of the IoT in flood monitoring [10] as the traditional flood monitoring system may not process the real-time information which delays the warning update [11].

LoRaWAN (long-range wide area network), Sigfox, satellite communication, and cellular networks as shown in Figure 15.5, respectively, are examples of real-time communication technologies. These communication lines ensure data connectivity even in remote places [12]. The collected data are analyzed and processed using advanced data analytics techniques to find hidden patterns, anomalies, trends, and the probability of flood occurrence. These outputs are displayed in a human-understandable format like graphs, charts, and maps using data visualization technologies [13]. When crucial water levels or weather criteria are exceeded, automated alerts are issued to local authorities and inhabitants, allowing them to take essential precautions and evacuate if necessary. By gathering real-time data, the IoT-based pumps and floodgates intimate the community before the water overflows thus preventing further floods.

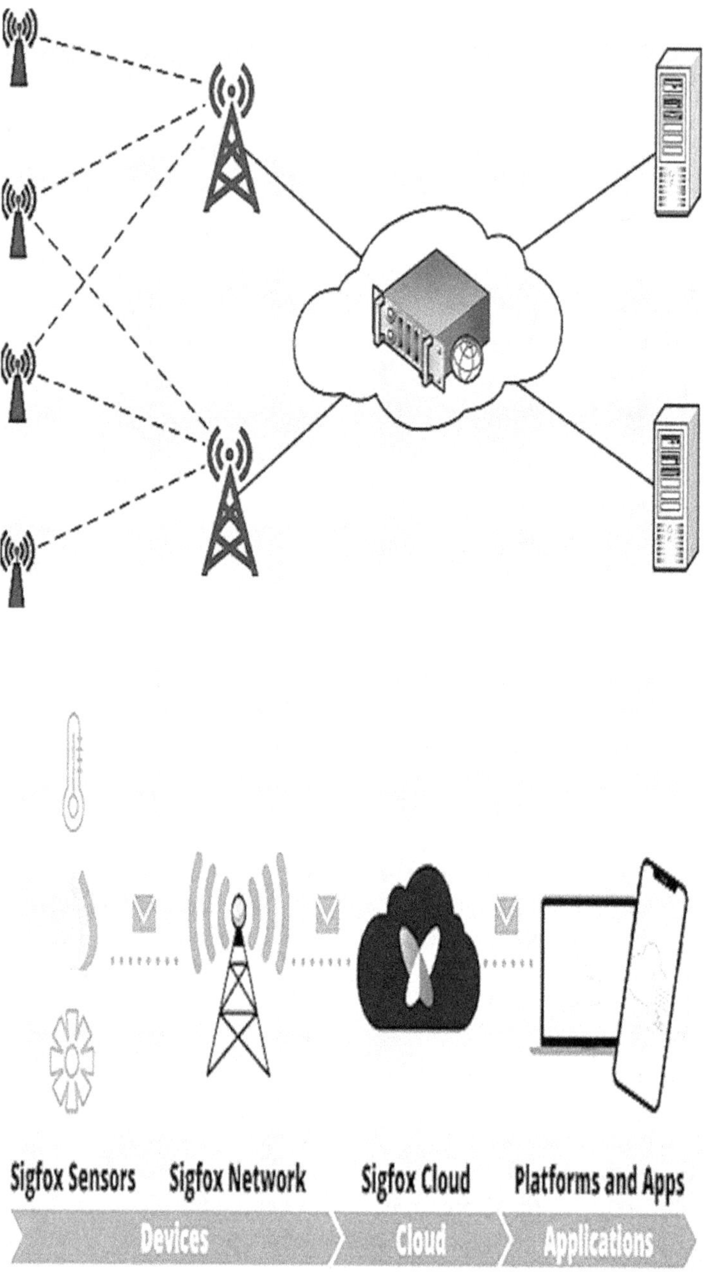

Figure 15.5 (a) LoRaWAN; (b) Sigfox network; (c) satellite communication; (d) cellular network.

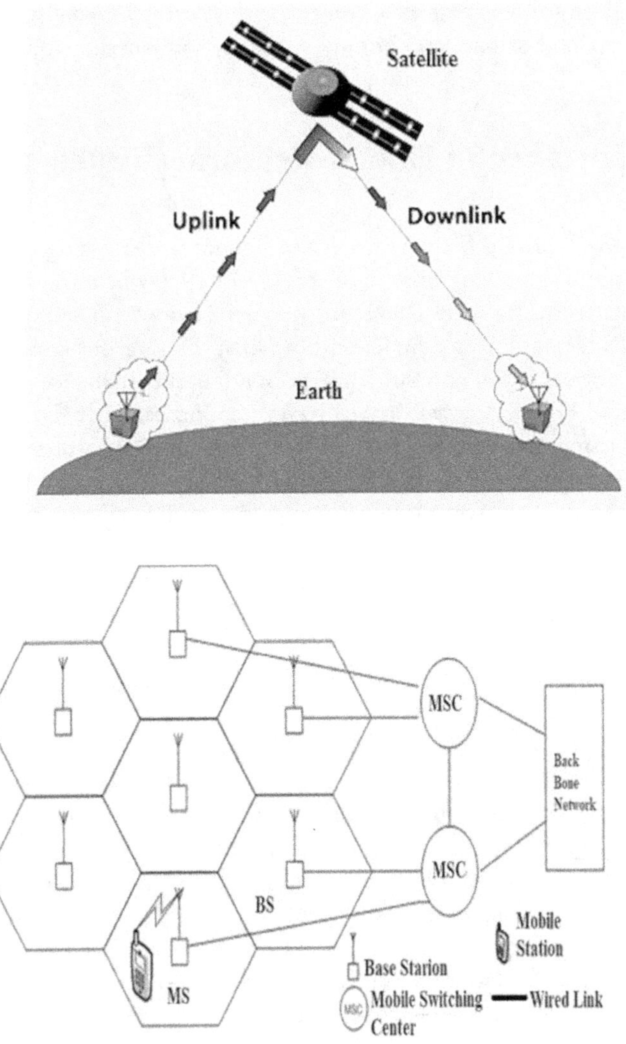

Figure 15.5 (Continued)

Flood monitoring using IoT can be linked to larger smart city programs to enhance traffic management, urban planning, and disaster response methods [14]. The IoT-based flood monitoring system is comparatively cheaper than traditional approaches without allowing much manual intervention for installation and maintainance. IoT-based flood monitoring and early warning systems are scalable and hence minimal extensions are required to cover larger geographic areas [15]. Overall, IoT in flood monitoring has considerable benefits in terms of real-time data collecting, early warnings, cost-effectiveness, and scalability, making it a powerful tool in limiting flood damage and improving disaster response capabilities [16].

15.2 Importance of Flood Monitoring Systems

Background

The Flood Early Warning System was evolved about 2 decades ago. The importance of early warning systems arose in the middle of 1970 when draughts and flood encounters in Africa. Later in 2005, the United Nations organized the Second World Conference on Disaster Reduction in Japan. During this conference, a bill "Hoyogo Framework for Action 2005–2015: Building the Resilience of Nation and Communities to Disasters" was adopted by 158 countries. Early days, the disaster management system was used for post-disaster preparedness measures.

Flood monitoring systems are critical for a variety of reasons, particularly in flood-prone areas. These systems use cutting-edge technology to monitor water levels, weather conditions, and other pertinent parameters [17].

Floods are one of the most damaging natural disasters, causing damage to the environment, infrastructure, and infrastructures. Hence early detection and effective flood management are necessary.

Here are some of the main reasons why flood monitoring systems are so important:

- Ability to detect the flood in advance well before it becomes a disastrous event.
- The weather forecast, satellite imagery, sensor networks, and river gauges are connected to the system to give real-time data on water levels, precipitation levels, and other relevant parameters that alert the authorities to provide early warning to their communities.
- This proactive approach significantly reduces the loss of life and property that often results from sudden and unanticipated flooding.
- It avoids the environmental contamination of degradation like water bodies, soil erosion, and disruption of ecosystems by giving authorities an idea about ecological impacts.
- Flood dynamics and trends can be understood with the help of historical data on floods. It helps the disaster management team, urban planners, and policymakers to incorporate climate-resilient practices into infrastructure development and urban planning.

The IoT-based flood monitoring and early warning system predicts the flood in advance and alerts the authorities by constantly monitoring the rainfall, weather, and water level [18]. This early warning helps the communities to prepare and evacuate the flood-prone zones. The flood monitoring system can also predict the severity and amount of flood. Authorities can identify the areas affected, the depth of flooding, and the possible impact on infrastructure and populations by observing water levels and flood advancement [19]. IoT uses big data to analyze historical weather-related data to predict weather patterns. Data collected from the sensors provides the potential threats, trends, flood patterns, and flood response to the authorities to make decisions. It continuously monitors the level of water in the rivers and the dams, and its flow rate to ensure effective operations [20]. When a flood occurs, the consideration of the ecosystem is a must, and hence the environmental monitoring components are added to the flood monitoring system to measure the impact of floods on ecosystems, and wild habitats. Flood monitoring system teaches the public about flood risks and the preparatory measures in advance.

15.3 Flood Monitoring System Architecture

A flood monitoring system's design consists of different components and infrastructure elements that work together to monitor and manage flood-related data. A flood monitoring system's major components and infrastructure are water level sensors, rainfall sensors, weather sensors, ultrasound sensors, a Raspberry Pi controller, an LED, a Buzzer, a Database, an online API, and a GSM Module comprise the flood monitoring architecture [21].

Weather sensors collect meteorological data such as humidity, temperature, atmospheric pressure, and wind speed to comprehend the pattern of the weather that causes floods [22]. Rainfall sensors measure the amount of rainfall in crucial flood-prone areas [23]. The water level sensor is set at all four corners of the river bed, streams, and flood-prone areas at the height relevant to the depth and width. When the water level reaches the defined height of scale, the light emitting diode (LED) emits and triggers the alarm to buzz to alert the communities and the authorities. The sensors may be pressure sensors, radar-based sensors, or ultrasonic sensors [24]. The flood-related data collected from various sources is processed according to the end application, stored, and analyzed by the server in the dedicated or cloud server. The Cloud server accommodates data and provides scalability and accessibility [25]. Edge devices collect data from various sensors, process the data, and transform it into a central system or cloud platform [26]. The edge system uses filtering and compression operations to limit the amount of data transferred to the cloud server. The real-time data are analyzed using powerful algorithms. These algorithms do real-time analysis of the data, and find the anomalies and trends that could predict the flood occurrence [27]. Based on the geographical location and the range, the communication method varies. Some of the communication methods are NB-IoT, satellite communication, cellular networks, and LoRaWAN [28]. Through web-based dashboards, desktop interfaces, or mobile applications,

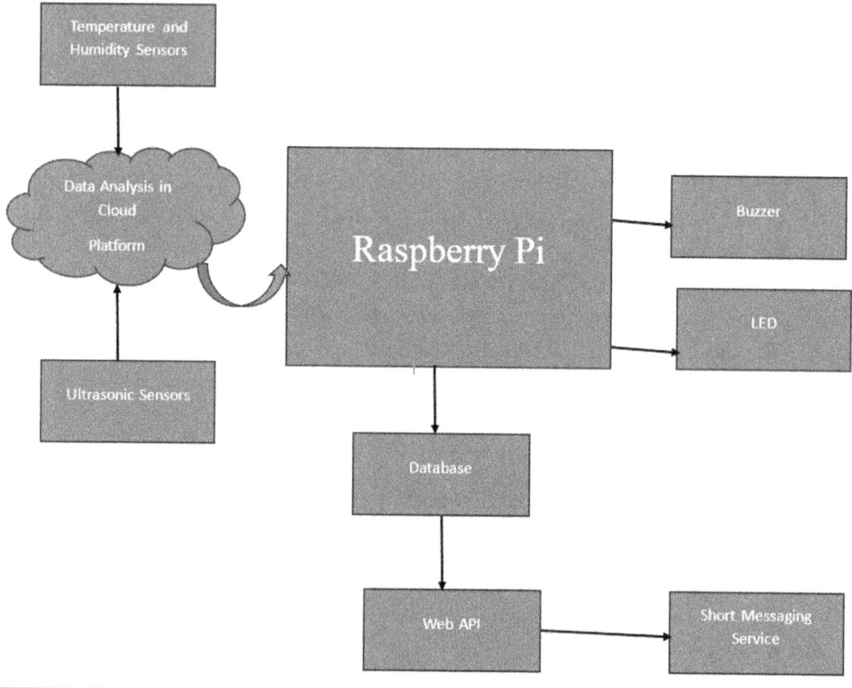

Figure 15.6 Blueprint for flood monitoring and alert system architecture.

authorized persons can monitor flood-related data, checks the historical data, and receives warning and alerts [29]. Geographical Information System (GIS) integration enables flood-related data to be visualized on maps, allowing for better decision-making and resource allocation during flood occurrences [30]. Figure 15.6 shows the blue print for flood monitoring and alert system architecture.

15.3.1 Managing Data in Flood Monitoring System

Data plays a crucial role in the prediction of flood occurrence. Collecting data from various sensors and transmitting it to the server is a critical process [31]. The sensors collect the real-time data or at regular intervals, based on the sensor specification and the end application. For example, weather sensors gather the metrological data continuously whereas water level sensors collect the data at regular intervals, a few minutes. In advanced data collection methods, the edge devices are located near the sensors to perform the pre-processing operation initially. The pre-processing steps include data filtration, data compression, and calibration to ease the transmission of data over the communication network. Edge computing devices reduce the amount of data sent to the cloud while increasing performance. The transmitted data is stored in the centralized cloud servers as it is well suited to accommodate the IoT data due to its flexibility, scalability, and accessibility [32].

Machine learning algorithms are used to analyze the data to find the correlation and the hidden patterns in the flood monitoring data. These algorithms are trained to find anomalies, forecast flood events, and improve flood management measures. Machine learning does predictive maintenance by forecasting the health and maintenance of flood monitoring equipment and infrastructure for consistent operation. In flood monitoring systems, cloud platforms provide a centralized and cost-effective option for data storage, real-time data processing, and data analysis. The flood monitoring system makes use of cloud infrastructure from cloud service providers such as Amazon Web Services (AWS), Microsoft Azure, Google Cloud Platform, and others. To ensure redundancy and high availability, the cloud architecture comprises a network of servers, data centers, and storage resources deployed across multiple geographic regions. Cloud service providers provide data security features like compliance certifications, encryption, and access control.

15.4 Conclusions

The FloodWatch system is a significant innovation in the flood monitoring and early warning system using IoT technology. This chapter suggests a flood monitoring and early warning system for the Cuddalore district, in the Indian state, of Tamilnadu. FloodWatch is a complete solution that portrays the requirements for an accurate flood warning system with the help of sensors, cloud servers, machine learning algorithms, data analytics, and communication channels. This chapter emphasizes FloodWatch's usefulness and applicability in minimizing flood damage.

FloodWatch provides society with the tools it needs to build resilience and adapt to the increasing difficulties posed by floods and other climate-related catastrophes by combining technology innovation with community engagement. As we face a changing climate and a shifting environment of natural dangers, FloodWatch serves as a beacon of progress, improving our collective ability to protect people, infrastructure, and ecosystems.

References

1. Denchak M. Flooding and Clinate Change: Everything you need to know. *Natural Resourses Defence Council.* 2019 Apr.
2. Das SK, Gupta RK, Varma HK. Flood and drought management through water resources development in India. *Bulletin of the World Metrological Organization.* 2007 Jul;56(3):179–88.
3. Sakthivel R. *Deciding Flood Premium Rates based on Flood Risk Zones using Remote Sensing and GIS Techniques: A Case Study of Cuddalore District,* Tamilnadu, India.

4. Nithya SE, Priyanka S. Vulnerable area assessind due to Flood in Cuddalore District by morphometric analysis method. *International Journal of Engineering Research & Technology*. 2019;7(2):1–6.
5. About district [Internet].[cited 2023 May 28]. Available from: https://cuddalore.nic.in/about-district/
6. [Internet]. [cited 2023 May 28]. Available from: https://tnpcb.gov.in/pdf/ph/ExeSumEngVadarangam271020.
7. Shankar K, Aravindan S, Rajendran S. Hydrogeochemistry of the paravabar river Sub-basin, Cuddalore district, Tamilnadu, India. *E-Journal of Chemistry*. 2011;8(2):835–45.
8. Ravikumar P, Baskaran G. Delimiting the flood risk zones in Cuddalore district, Tamilnadu, India. *International Journal for Research in Applied Science and Engineering Technology*. 2018;6(5):569–73.
9. Esposito M, Palma L, Belli A, Sabbatini L, Pierleoni P. Recent advances in Internet of Things solutions for early warning systems: A review. *Sensors*. 2022 Mar 9;22(6):2124.
10. Moreno C, Aquino R, Ibarreche J, Pérez I, Castellanos E, Álvarez E, Rentería R, Anguiano L, Edwards A, Lepper P, Edwards RM. RiverCore: IoT device for river water level monitoring over cellular communications. *Sensors*. 2019 Jan 2;19(1):127.
11. Khedo KK. "Real-time flood monitoring using wireless sensor networks". *Journal of the Institution of Engineers Mauritius-IEM Journal*. 2013 Sep;1:59–69.
12. Azid SI, Sharma BN, Raghuwaiya K, Chand A, Prasad S, Jacquier A. "SMS-based flood monitoring and early warning system". *ARPN Journal of Engineering and Applied Sciences*. 2015 Aug 15;10(15):6387–91.
13. Zambrano AM, Calderón X, Jaramillo S, Zambrano OM, Esteve, M, Palau. "C3 community early warning systems. In *Wireless Public Safety Networks 3*", Câmara, D., Nikaein, N., Eds.; Elsevier: Amsterdam, The Netherlands, 2017; pp. 39–66.
14. Sinha RS, Wei Y, Hwang SH. A survey on LPWA technology: LoRa and NB-IoT. *ICT Express*. 2017 Mar 1;3(1):14–21.
15. Srivastava D. An introduction to data visualization tools and techniques in various domains. *International Journal of Computer Trends and Technology*. 2023;71:125–130. DOI: 10.14445/22312803/IJCTT-V71I4P115
16. Indrakumari R, Poongodi T, Sumathi D, Suganthi S, Suresh P. IoT-Based biomedical sensors for pervasive and personalized healthcare. *Internet of Things, Artificial Intelligence, and Blockchain Technology*. 2021;1:111–30.
17. Patil S, Pisal J, Patil A, Ingavale S, Ayarekar P, Mulla PS. A real-time solution to flood monitoring systems using IoT and wireless sensor networks. *International Research Journal of Engineering and Technology*. 2019 Feb;6(02):1807–11.
18. Indrakumari R, Poongodi T, Suresh P, Balamurugan B. The growing role of the Internet of Things in healthcare wearables. In *Emergence of Pharmaceutical Industry Growth with Industrial IoT Approach* 2020 Jan 1 (pp. 153–194). Academic Press.
19. Kuller M, Schoenholzer K, Lienert J. Creating effective flood warnings: A framework from a critical review. *Journal of Hydrology*. 2021, 602:126708.
20. Mourato S, Fernandez P, Marques F, Rocha A, Pereira L. An interactive Web-GIS fluvial flood forecast and alert system in operation in Portugal. *International Journal of Disaster Risk Reduction*. 2021 May 1;58:102201.

21. Tehrany MS, Pradhan B, Jebur MN. Flood susceptibility analysis and its verification using a novel ensemble support vector machine and frequency ratio method. *Stochastic Environmental Research and Risk Assessment.* 2015, 29:1149–1155.
22. Loudyi D, Hasnaoui MD, Fekri A. Flood Risk Management Practices in Morocco: Facts and Challenges. In: Sumi, T., Kantoush, S.A., Saber, M. (eds) *Wadi Flash Floods. Natural Disaster Science and Mitigation Engineering: DPRI reports* 2022. Springer, Singapore.
23. Orozco MM, Caballero JM. Smart disaster prediction application using flood risk analytics towards sustainable climate action. In *MATEC Web of Conferences* 2018 (Vol. 189, p. 10006). EDP Sciences.
24. Kar AK, Lohani AK, Goel NK, Roy GP. Rain gauge network design for flood forecasting using multi-criteria decision analysis and clustering techniques in the lower Mahanadi river basin, India. *Journal of Hydrology: Regional Studies.* 2015 Sep 1;4:313–32.
25. Fowdur TP, Beeharry Y, Hurbungs V, Bassoo V, Ramnarain-Seetohul V, Lun EC. Performance analysis and implementation of an adaptive real-time weather forecasting system. *Internet of Things.* 2018 Oct 1;3:12–33.
26. Ugwuanyi S, Paul G, Irvine J. Survey of IoT for developing countries: Performance analysis of LoRaWAN and cellular nb-IoT networks. *Electronics.* 2021 Sep 10;10(18):2224.
27. Haxhibeqiri J, De Poorter E, Moerman I, Hoebeke J. A survey of LoRaWAN for IoT: From technology to application. *Sensors.* 2018 Nov 15;18(11):3995.
28. Alam T. Cloud-Based IoT applications and their roles in smart cities. *Smart Cities.* 2021; 4(3):1196–1219. https://doi.org/10.3390/smartcities4030064
29. Habeeb RA, Nasaruddin F, Gani A, Hashem IA, Ahmed E, Imran M. Real-time big data processing for anomaly detection: A survey. *International Journal of Information Management.* 2019 Apr 1;45:289–307.
30. Damaševičius R, Bacanin N, Misra S. From sensors to safety: Internet of Emergency Services (IoES) for emergency response and disaster management. *Journal of Sensor and Actuator Networks.* 2023;12(3):41. https://doi.org/10.3390/jsan12030041
31. Lammers R, Li A, Nag S, Ravindra V. Prediction models for urban flood evolution for satellite remote sensing. *Journal of Hydrology.* 2021 Dec 1;603:127175.
32. Meissen U, Voisard A. Increasing the effectiveness of early warning via context-aware alerting. In *Proceedings of the 5th International Conference, on Information Systems for Crisis Response and Management (ISCRAM)* 2008 May (pp. 431–440).
33. Daniel A, Partheeban N. "An improved and efficient and dynamic load balancing approach in cloud computing environment", *Advance Innovation and Technology with Sustainability Engineering, International Journal of Social Ecology and Sustainable Development (IJSESD)* 2022. IGI Global. www.igi-global.com/article/an-improved-and-efficient-and-dynamic-load-balancing-approach-in-cloud-computing-environments/302469 DOI: 10.4018/IJSESD.302469

Index

For Product Safety Concerns and Information please contact our EU representative GPSR@taylorandfrancis.com
Taylor & Francis Verlag GmbH, Kaufingerstraße 24, 80331 München, Germany

www.ingramcontent.com/pod-product-compliance
Lightning Source LLC
LaVergne TN
LVHW020601110826
845149LV00002B/343

* 9 7 8 1 0 3 2 7 4 5 1 7 6 *